# Spatial Histories of Radical Geography

# Antipode Book Series

Series Editors: Vinay Gidwani, University of Minnesota, USA and Sharad Chari, University of California, Berkeley, USA

Like its parent journal, the Antipode Book Series reflects distinctive new developments in radical geography. It publishes books in a variety of formats – from reference books to works of broad explication to titles that develop and extend the scholarly research base – but the commitment is always the same: to contribute to the praxis of a new and more just society.

## Published

Spatial Histories of Radical Geography: North America and Beyond
*Edited by Trevor J. Barnes and Eric Sheppard*

The Metacolonial State: Pakistan, Critical Ontology, and the Biopolitical Horizons of Political Islam
*Najeeb A. Jan*

Frontier Assemblages: The Emergent Politics of Resource Frontiers in Asia
*Edited by Jason Cons and Michael Eilenberg*

Other Geographies: The Influences of Michael Watts
*Edited by Sharad Chari, Susanne Freidberg, Vinay Gidwani, Jesse Ribot, and Wendy Wolford*

Money and Finance After the Crisis: Critical Thinking for Uncertain Times
*Edited by Brett Christophers, Andrew Leyshon, and Geoff Mann*

Frontier Road: Power, History, and the Everyday State in the Colombian Amazon
*Simón Uribe*

Enterprising Nature: Economics, Markets and Finance in Global Biodiversity Politics
*Jessica Dempsey*

Global Displacements: The Making of Uneven Development in the Caribbean
*Marion Werner*

Banking Across Boundaries: Placing Finance in Capitalism
*Brett Christophers*

The Down-deep Delight of Democracy
*Mark Purcell*

Gramsci: Space, Nature, Politics
*Edited by Michael Ekers, Gillian Hart, Stefan Kipfer, and Alex Loftus*

Places of Possibility: Property, Nature and Community Land Ownership
*A. Fiona D. Mackenzie*

The New Carbon Economy: Constitution, Governance and Contestation
*Edited by Peter Newell, Max Boykoff and Emily Boyd*

Capitalism and Conservation
*Edited by Dan Brockington and Rosaleen Duffy*

Spaces of Environmental Justice
*Edited by Ryan Holifield, Michael Porter and Gordon Walker*

The Point is to Change it: Geographies of Hope and Survival in an Age of Crisis
*Edited by Noel Castree, Paul Chatterton, Nik Heynen, Wendy Larner and Melissa W. Wright*

Privatization: Property and the Remaking of Nature-Society
*Edited by Becky Mansfield*

Practising Public Scholarship: Experiences and Possibilities Beyond the Academy
*Edited by Katharyne Mitchell*

Grounding Globalization: Labour in the Age of Insecurity
*Edward Webster, Rob Lambert and Andries Bezuidenhout*

Privatization: Property and the Remaking of Nature-Society Relations
*Edited by Becky Mansfield*

Decolonizing Development: Colonial Power and the Maya
*Joel Wainwright*

Cities of Whiteness
*Wendy S. Shaw*

Neoliberalization: States, Networks, Peoples
*Edited by Kim England and Kevin Ward*

The Dirty Work of Neoliberalism: Cleaners in the Global Economy
*Edited by Luis L. M. Aguiar and Andrew Herod*

David Harvey: A Critical Reader
*Edited by Noel Castree and Derek Gregory*

Working the Spaces of Neoliberalism: Activism, Professionalisation and Incorporation
*Edited by Nina Laurie and Liz Bondi*

Threads of Labour: Garment Industry Supply Chains from the Workers' Perspective
*Edited by Angela Hale and Jane Wills*

Life's Work: Geographies of Social Reproduction
*Edited by Katharyne Mitchell, Sallie A. Marston and Cindi Katz*

Redundant Masculinities? Employment Change and White Working Class Youth
*Linda McDowell*

Spaces of Neoliberalism
*Edited by Neil Brenner and Nik Theodore*

Space, Place and the New Labour Internationalism
*Edited by Peter Waterman and Jane Wills*

# Spatial Histories of Radical Geography

*North America and Beyond*

*Edited by*

Trevor J. Barnes and Eric Sheppard

*Registered Office(s)*
John Wiley & Sons, Inc., 111 River Street, Hoboken, NJ 07030, USA
John Wiley & Sons Ltd, The Atrium, Southern Gate, Chichester, West Sussex, PO19 8SQ, UK

*Editorial Office*
9600 Garsington Road, Oxford, OX4 2DQ, UK

For details of our global editorial offices, customer services, and more information about Wiley products visit us at www.wiley.com.

Wiley also publishes its books in a variety of electronic formats and by print-on-demand. Some content that appears in standard print versions of this book may not be available in other formats.

*Library of Congress Cataloging-in-Publication data has been applied for*

ISBNs
978-1-119-40471-2 (hardback)
978-1-119-40479-8 (paperback)

Cover Design: Wiley
Cover Image: Collage of the first and the most recent *Antipode* journal cover designs: First issue of *Antipode* edited by Ben Wisner; cover design by Jane Cooper Wisner. Current cover of *Antipode*, designed by Ray Zilli.

Set in 10.5/12.5pt Sabon by SPi Global, Pondicherry, India
Printed in Singapore by C.O.S. Printers Pte Ltd

10  9  8  7  6  5  4  3  2  1

*We dedicate this book to the community of scholars, students and activists who raised the flag to, shaped, and carried radical geography forward to its present wonderfully variegated state, and to future radical thinkers Charlotte and Jonah.*

# Contents

# List of Figures

# Notes on Contributors

**Trevor J. Barnes** completed an undergraduate degree at University College, London, and graduate degrees at the University of Minnesota, where Eric Sheppard was his doctoral supervisor. Barnes has been teaching at the Department of Geography, University of British Columbia, since 1983. Much of his research has been in economic geography, and, recently, in the history of human geography since the Second World War. In 2018, with Brett Christophers, he published *Economic Geography: A Critical Introduction*. He is a Fellow of the Royal Society of Canada, and of the British Academy.

**Nicholas Blomley** teaches, studies, and performs ideas and practices relating to the critical and often oppressive relationship between law and space. He is particularly interested in the political geographies of land and property. He has worked at Simon Fraser University's Department of Geography for 30 years, and has long wanted to explore its sidelined radical past.

**Mark J. Bouman** is the Chicago Region Program Director in the Field Museum's Keller Science Action Center, where he leads the Museum's interdisciplinary conservation and cultural heritage work in the Chicago region. Previously, he was Professor and Chairperson of the Department of Geography at Chicago State University. He led the development of the University's Neighborhood Assistance Center, its Calumet Environmental Resource Center, and GIS Laboratory.

**Yann Calbérac** is a Lecturer at the University of Reims (France). His work mainly concerns the history of geography and social sciences. His current research deals with the spatial turn and the use of the category of space in the social sciences.

**Verónica Crossa** is an Associate Professor in Urban Studies at the *Centro de Estudios Demográficos, Urbanos y Ambientales*, at *El Colegio de Mexico*. Her research, at the intersection of urban, cultural geography and critical theory, examines how "order" is (re)produced and performed on the streets of contemporary cities, with a particular focus on changing notions of urban order triggered by revitalization policies in Mexico City's public spaces. Her recent book, *Luchando por un Espacio en la Ciudad de México: Espacio público urbano y el comercio ambulante* (COLMEX, 2018), examines how street vendors in Mexico City negotiate and struggle over changing power structures in their everyday lives.

**Nik Heynen** is a Professor in the Department of Geography at the University of Georgia. His research interests include urban political ecology, social movement theory and politics of social reproduction. His main research foci relate to the analysis of how uneven social power relations, including race, gender, and class are inscribed in the transformation of nature/space, and how in turn these processes contribute to interrelated connections between nature, space and uneven development. He is currently Editor at the *Annals of the American Association of Geographers* and *Environment and Planning E: Nature and Space*.

**Bryan Higgins** is an Emeritus Distinguished Service Professor of Geography at the State University of New York. Drafted into the U.S. Army in the Viet Nam era and stationed in the Chemical Corps at Edgewood Arsenal, MD, providing Army "volunteers" psychoactive agents in chemical warfare research radically changed his worldview. Under the GI Bill, he received a Ph.D. in geography from the University of Minnesota. His early writing focused on the geographical revolutions of American Indians, radical geographies in the People's Republic of Burlington Vermont and revolutionary geographies in Nicaragua. His later work addresses political ecology and ecotourism. As Director of International Education at SUNY Plattsburgh, he established progressive study abroad programs throughout the world.

**Matthew T. Huber** is Associate Professor of Geography at Syracuse University. His research focuses on energy, climate change, and the political economy of capitalism. His first book, *Lifeblood: Oil, Freedom, and the Forces of Capital* (University of Minnesota Press, 2013), examines the role of oil and suburbanization in the neoliberalization of American politics. He is currently writing a book on the intersection of climate change and class politics.

**Cindi Katz** teaches at The Graduate Center of the City University of New York. Her work concerns social reproduction and the production of space, place and nature; managing insecurity in the domestic and public environment; the cultural politics of childhood; the consequences of global economic restructuring for everyday life; and the politics of knowledge. *Antipode* has been part of her life since she helped affix address labels to them back when they were held together with staples.

**Juan-Luis Klein** (Ph.D. Université Laval) is a full Professor at the Geography Department of the University of Québec at Montreal and the former director of the *Centre de recherche sur les innovations sociales* (CRISES). His research projects are on economic geography, social innovation, and local and community development. He is a member of several editorial boards of scientific journals and is the director of the *Géographie Contemporaine* book series (Presses de l'Université du Québec).

**Chris Knudson** is a Postdoctoral Research Associate in the Institute of the Environment at the University of Arizona. He is a political ecologist whose research centers on the history, governance, and practice of managing ecological crises, particularly through the creation and use of financial risk management tools.

**Audrey Kobayashi** is a Professor of Geography and Queen's Research Chair at Queen's University. Her research and publications address a range of human rights issues including racialization, poverty, housing, and immigration, as well as the history of Geography. She recently co-authored *The Equity Myth: Racialization and Indigeneity at Canadian Universities* and co-edited *The International Encyclopedia of Human Geography: People, the Earth, Environment, and Technology*. She is currently the Editor for *The Encyclopedia of Human Geography*, 2nd edition, to be published in 2020.

**Mickey Lauria** is a Professor of City and Regional Planning and Director of the transdisciplinary Ph.D. program in Planning, Design, and the Built Environment at Clemson University. He has served as President of the Association of Collegiate Schools of Planning, has edited the *Journal of Planning Education and Research* and *Town Planning Review*, and serves on the editorial boards of four planning journals. He has published articles on urban schooling, urban redevelopment, and politics and planning. His recent research interests include professional planners' ethical frameworks, neighborhood conditions, and planning issues involving race and class, and conservation easements and affordable housing.

**Brij Maharaj** is an urban political geographer at the University of KwaZulu-Natal, Durban, South Africa. He has received widespread recognition for his research on megaevents and social impacts, segregation, local economic development, xenophobia and human rights, migration, and diasporas. He has published over 150 scholarly papers on these themes in journals such as *Urban Studies, International Journal of Urban and Regional Studies, Political Geography, Urban Geography, Antipode, Polity and Space, Geoforum, Local Economy*, and *GeoJournal*, as well as five co-edited book collections. He is a B-rated NRF researcher.

**Kent Mathewson** is the Fred B. Kniffen Professor in the Department of Geography and Anthropology at Louisiana State University. His research interests include cultural, historical, and Latin American geography and the history of geography. He is author, editor, and coeditor of a number of books including *Re-Reading Cultural Geography, Culture, Form and Place, Concepts in Human Geography, Dangerous Harvest*, and *Carl Sauer on Culture and Landscape*. Topics of current collaborative book projects include photography of the Guatemalan civil war, Elisée Reclus' travels in Latin America, and a culture history of the castor plant (emphasizing the African diaspora).

**Eugene McCann** is University Professor of Geography at Simon Fraser University. An urban political geographer, he researches policy mobilities, urban policy-making, development, public space, and planning. He is co-editor, with Kevin Ward, of *Mobile Urbanism* (Minnesota, 2011) and of *Cities & Social Change*, with Ronan Paddison (Sage, 2014). He is co-author, with Andy Jonas and Mary Thomas, of *Urban Geography: A Critical Introduction* (Wiley, 2015). He is managing editor of the journal *EPC: Politics & Space*.

**Fujio Mizuoka** received his Ph.D. in geography from Clark University in 1986 after teaching at the University of Hong Kong from 1979 to 1981. He then joined the Faculty of Economics, Hitotsubashi University to teach economic geography, retiring in 2016. His research interests focus on the critical theories of economic geography, the history of critical thought in geography, Marxian economics, central-place theory, British colonialism in Hong Kong, the concepts of contemporary capitalism, and human rights issues in Japan's child welfare policies. He has published various articles in English; his recent book *Contrived Laissez-faireism* (Springer Verlag) examines the unique art of colonial rule in post-WWII Hong Kong.

**Joanne Norcup** is a historical and cultural geographer currently based at the Centre for Caribbean Studies, University of Warwick. Informed by feminist and intersectional ideas, her research centers on geographies of education, learning, and knowledge-making, focusing on vernacular, dissenting, and transgressive practices, ideas, and productions. Current research areas include public library geographies, DIY education publications & archives, popular culture (including TV comedy and crime fiction), and lay nature, and environmental knowledges. Jo is the founding director of Geography Workshop, a radio and education resource production company whose recent work has been broadcast by BBC Radio and Resonance FM.

**Linda Peake** is a professor in Urban Studies and director of the City Institute at York University, Toronto. She has written widely in the field of critical human geography with interests in urban theory, feminist methodologies and, more recently, mental health. She is chair of the AAG Task Force on Mental Health, co-editor of the special issue on "An engagement with planetary urbanization" in *Environment and Planning D: Society and Space* (2018) and of *Urbanization in A Global Context* (2017, OUP) (with Alison Bain) and principal investigator of the SSHRC funded GenUrb: Gender, urbanization and the global south.

**Jamie Peck** is the Canada Research Chair in Urban & Regional Political Economy and Professor of Geography at the University of British Columbia, Canada, where he is a Distinguished University Scholar. His recent books include *Doreen Massey: Critical dialogues* (2018, Agenda, co-edited with Marion Werner, Rebecca Lave and Brett Christophers); *Offshore: Exploring the Worlds of Global Outsourcing* (Oxford, 2017); *Fast Policy: Experimental Statecraft at the Thresholds of Neoliberalism* (Minnesota, 2015, with Nik Theodore); and *Constructions of Neoliberal Reason* (2010, Oxford University Press).

**Eric Sheppard** is a Professor of geography and Alexander von Humboldt chair at UCLA. His scholarship embraces geographical political economy, uneven geographies of globalization, urban transformations, and informality in Indonesia, neoliberalism and its urban contestations, social movements and their spatialities, geographic information technologies and society, geographical philosophies and methods, and environmental justice. He has co-authored or co-edited nine books, most recently *Limits to Globalization* (Oxford University Press, 2016). He was active in the Union of Socialist Geographers (1977–1982), and co-edited *Antipode* (1980–1986) before it became a highly profitable enterprise.

**Renee Tapp** is the Pollman Fellow in Real Estate and Urban Development at the Harvard University Graduate School of Design. She received her

Ph.D. in Geography from Clark University in 2018. Her research interests include the geographies of tax and finance, urban redevelopment, urban politics, and modernist architecture. Currently, she is examining the impact shell companies and tax evasion have on rental markets in the United States.

**Gwendolyn C. Warren** is a long-time public sector administrator, who has distinguished herself as a leader in the areas of education, health, social and community services. She has worked in executive level capacities in city and county government in California, Florida, and Georgia, working with diverse populations that add unique and challenging issues to the provision of quality government services, and serving with the goal to maximize human capacity. She became Co-Director of the Detroit Geographical Expedition and Institute at the age of 18, shaping many of its mapping projects, and serving as a key author of its Field Notes and leader of its educational component.

# Series Editors' Preface

The *Antipode Book Series* explores radical geography "antipodally," in opposition, from various margins, limits, or borderlands.

*Antipode* books provide insight "from elsewhere," across boundaries rarely transgressed, with internationalist ambition and located insight; they diagnose grounded critique emerging from particular contradictory social relations in order to sharpen the stakes and broaden public awareness. An *Antipode* book might revise scholarly debates by pushing at disciplinary boundaries, or by showing what happens to a problem as it moves or changes. It might investigate entanglements of power and struggle in particular sites, but with lessons that travel with surprising echoes elsewhere.

*Antipode* books will be theoretically bold and empirically rich, written in lively, accessible prose that does not sacrifice clarity at the altar of sophistication. We seek books from within and beyond the discipline of geography that deploy geographical critique in order to understand and transform our fractured world.

Vinay Gidwani
*University of Minnesota, USA*

Sharad Chari
*University of California, Berkeley, USA*

**Antipode Book Series Editors**

# Preface

The idea for this book, marking the 50th Anniversary of *Antipode: A Journal of Radical Geography*, was hatched when one of us (Sheppard) was completing an essay with Linda Peake, commissioned for a putative edited book by Lawrence Berg.[1] Our essay sought to deconstruct the Clark University-centered history of Anglophone North American radical geography that is commonly narrated. Realizing that some of the early figures of the 1960s generation of radical geography had passed or were rapidly aging, Sheppard approached Barnes to tap his experience in interviewing first generation quantitative geographers and narrating their history. For both of us, it felt exciting to extend this methodology to the history of radical geography that had so profoundly shaped our lives. But the project felt too big. Gradually, we recruited others, notably that included bringing their own ideas to the project. We were able to benefit from a workshop in Vancouver in 2013 with Nik Heynen, Audrey Kobayashi, Linda Peake, Jamie Peck, and Bobby Wilson, funded by the Antipode Foundation. When we finally brought this project to the 2016 San Francisco Association of American Geographers' annual meeting, the enthusiastic response from an engaged audience, including ghosts emerging from radical geography's deep past, encouraged us to bring the project to print. Over the years, we broadened this multi-nodal account to also incorporate voices from beyond Canada and the U.S., recruiting a second round of authors. Notably, virtually no-one who we asked declined to participate; Linda Peake was unable to help with the editing, but contributed her own chapter. Authors went beyond any temptation to craft second-hand accounts, interviewing early radical geographers and digging into gray literatures and departmental archives. Thus this book took some time to come together as a final product. The result, we think, catches the complex spirits of these times before they fade into the past, and their influence on the contemporary discipline is lost to memory.

## Note

1   Our cover image captures this history by melding the very first Antipode cover with the latest.

# Acknowledgments

We would like to acknowledge a $6,000 grant from the Antipode Foundation, making possible an early workshop that brought energy to this collective. We also wish to thank the many interviewees contacted during the research reported here for their willing and informative collaboration, as well as Matt Zebrowski, cartographer in the UCLA Department of Geography, for his help with improving some of the images.

# Introduction

## Trevor J. Barnes and Eric Sheppard

'*Something Better Change,*' *The Stranglers (1977)*[1]

Both of us have lived our entire academic lives under the aegis of radical geography. In 1971, Sheppard, a geography undergraduate at Bristol University, remembers the newly hired junior lecturer, Keith Bassett, having freshly returned from completing an M.A. degree at Penn State University, carrying into the classroom to show students a stack of *Antipodes* he recently brought back from America. Renowned for sardonic humor, even Bassett cracked a hopeful smile, unabashedly enthusiastic, when he showed and talked about *Antipode* and the new movement of radical geography in America and its possibilities.

Certainly, Barnes was enthusiastic when in 1976 as a second-year undergraduate in geography and economics at University College, London (UCL), he held in his hands for the first time a copy of *Antipode*. It felt as if he was doing something illegal, perusing a smuggled underground publication, probably best done under the bed covers, read with a flashlight.[2] The librarians in the Geography Reading Room at UCL treated it as seditious at least. It was kept behind the counter in a sturdy wooden cabinet under lock and key. The journal could be signed out but for just two hours and read at only designated tables under the scrutinizing gaze of the library beadle. Although such constraints permitted only relatively short snatches of reading, *Antipode* captured brilliantly the riven England in which Barnes lived, of strikes, protest marches, and Orwellian grimness. It connected even to punk rock, born during that same mid-to-late 1970s

*Spatial Histories of Radical Geography: North America and Beyond*, First Edition.
Edited by Trevor J. Barnes and Eric Sheppard.
© 2019 John Wiley & Sons Ltd. Published 2019 by John Wiley & Sons Ltd.

dyspeptic period, and the background music of Barnes' undergraduate and, on occasion, academic life (Barnes 2019). *Antipode* looked like a punk publication, a fanzine of radical geography. The early issues were home-made, DIY publishing, its typographical-error-strewn contents bound between one punk discordant garish cover or another: electric yellow, vibrant scarlet, pulsating green, shimmering gold. Bernard Sumner, a member of the band *Joy Division*, after first hearing punk rock said, it was "terrible. I thought [it was] … great. I wanted to get up and be terrible too" (quoted in Marcus 1989: 7). Reading *Antipode* for that first time made Barnes also want to get up and be terrible, but to be great too: to be a radical geographer.

Our edited volume is a history, or rather a set of histories of radical geography. It includes the beginning of *Antipode* and its lurid covers (Huber *et al.*, this volume), but also much, much more. Geographically, the central focus of the book is the United States (US) and Canada. The first nine (long) chapters of the collection – Part I, Histories of Radical Geography in North America – are concerned with the emergence and practices of radical geography at a set of specific U.S. and Canadian sites (six chapters are mostly about the U.S., three mostly about Canada). The last five shorter chapters – Part II, International Perspectives – offer a set of histories, experiences and reflections about radical geography undertaken outside the U.S. and Canada: France, Japan, Mexico, South Africa, and the U.K. Radical geography in the U.S. and Canada had some influence in all those places, but it was not the same in each, and exactly how it influenced was a consequence of specific prior conditions – political, social, cultural, institutional, intellectual – found in each place, as well as often the presence of catalytic individuals. There certainly was no simple process of spatial diffusion. Even if it is granted that the most recent form of radical geography developed first in the U.S. and Canada,[3] it did not steamroll across the world, crushing native intellectual traditions, turning every place into Clark or Johns Hopkins Universities. Rather, its course was contingent and variable, geographically and historically. Radical geography requires sensitive historical and geographical narration, a central purpose of this volume.

Historically, the volume covers the period from the origin of radical geography in the U.S. and Canada sometime during the mid-1950s through to its intellectual consolidation in the early 1980s. We begin in the mid-1950s with the first stirrings of radical activism by U.S. geographers, although hinged not around class but race. Audrey Kobayashi (this volume) recounts the involvement of the geographer Thelma Glass, based at the University of Alabama, in the Montgomery bus boycott of 1955–1956 (best associated with Rosa Parks). Race continues as a key theme during the early-to-mid-1960s albeit within the unlikely formal structure of the Association of American Geographers, involving both geographers of color like Don Deskins and Harold Rose, and white geographers like Jim Blaut, Ron Horvath, and Richard Morrill (Kobayashi; and Peake, both this volume). Also in the early 1960s, issues of race and activism were central to William Bunge's work in Detroit that began in the academy, at Wayne State University, but shifted to his own black inner-city neighborhood of Fitzgerald and to community activists like Gwendolyn Warren (Warren *et al.*, this volume). In 1969, *Antipode* was founded at Clark University (Huber *et al.*, this volume). Initially eclectic in its topics and approaches, by the mid-1970s it became increasingly aligned with a Marxism focused on capital and class, and best associated with David Harvey at Johns Hopkins University (Sheppard and Barnes, this volume). That said, even during this period there were other radical geographical organizations and publications, such as the Socially and Ecologically Responsible Geographers (SERGE) (founded in 1971) and its journal *Transition*, as well as the Union of Socialist Geographers (USG) (established 1974) and its *Newsletter* that typically published on a broader range of topics and approaches than *Antipode* (Peake, this volume). It was also then that radical geography expanded and consolidated in centers outside Clark and Johns Hopkins: Simon Fraser University in Vancouver (Blomley and McCann, this volume), the U.S. Midwest (Lauria *et al.*, this volume), Quebec (Klein, this volume), and the University of California, Berkeley (Peck and Barnes, this volume). By the early 1980s with the publication of David Harvey's 1982 Marxist theoretical compendium, *The Limits to Capital*, radical geography had unquestionably arrived.

It wasn't as if radical geography was then set in stone, however, 1982 was just the end of the beginning. For the form that radical geography took from the early 1970s to the early 1980s especially in *Antipode*, and associated with classical Marxism, began to braid and diverge. Elements of the older Marxist geography were taken apart, critiqued, some thrown out, others joined with new elements, and put together again in novel combinations. This new version, increasingly known as critical geography, more and more became how human geography in the round was done (Castree 2000). The subsequent capaciousness and variegation of critical geography makes telling its story more difficult compared to the earlier radical geography, however. Presenting its history will likely require many volumes, many editors, and many contributors. We very much hope it will be undertaken, but it is not our project. While individual chapters in this book trace how earlier events helped shape critical geography, and our Conclusion will explicitly recount the relation between radical and critical geography, this volume is limited to the early development of radical geography. While we realize this period is only part of a larger story, it is no less necessary to recount, and has some urgency.

There already exist some excellent individual essays about the early history of radical geography. These tend, though, either to give the complete North American story based on secondary literature (often found in textbooks like Cloke *et al.* 1991, ch. 2, or Johnston and Sidaway 2016, ch. 6), or to focus on just one element or episode or individual within it (for example, on William Bunge's contribution found in Merrifield 1995, or Heyman 2007). In contrast, our volume intends to provide not only the larger North American story of radical geography, but also to follow its relationship with selected places outside that core (the purpose of Part II). Further, rather than resting on secondary literature, many of the chapters draw on primary material. In this sense, the book aims to provide both the broader view and specificity, a larger story arc infused by history but also geography.

The use of primary source material in this volume is especially important. While the authors in our volume sometimes draw on traditional material archival sources (for example, in Huber *et al.* and Norcup, both this volume) a lot of information is gleaned from oral histories conducted with leading protagonists (for example, in Kobayashi, and Peck and Barnes, this volume). In part, the reliance on oral history is necessitated by a lack of formal archival sources. Materials relating to histories of radical geography have never been systematically collected, but remain scattered, found in people's garages, or forgotten filing cabinets in university departmental basements. Of course, oral histories have their problems. Memories are fallible – Hemingway said memory is never true – they are only one person's view, they can't capture large-scale historical, political, social and geographical events, and they are unsuitable for relating abstractions, conceptual schema, and dialectical niceties. They must be triangulated with other kinds of information, as our authors do. But given the dearth of other sources, oral histories remain one of the most important bases for telling histories of radical geography. Further, with aging and death – the earliest radical geographers are now in their eighties with Bunge, Deskins, and Rose having all recently passed, and Blaut and the relatively young Neil Smith (at 58) having died some time ago – the ability to gather this type of information is itself diminishing.

There is one other distinctive feature driving the organization of this collection. While we are concerned to provide histories of early radical geography, we want just as much to provide geographies of it too. John Agnew and David Livingstone (2011: 16) contend we must "think geographically about geography, and thereby 'geographizing' geography itself." This book is an attempt to do just that. Strangely, this has often been a missing element in histories of the discipline told by geographers. The geographical setting becomes at best only color and background atmospherics for the history. The essays in this volume make a stronger claim, however: Geography goes all the way down. That imperative explains why we organized the book geographically, by place and nation. Running throughout the collection of essays are three fundamental organizing geographical ideas, making this book not just a history but also a geography of radical geographical knowledge.

- The first is place, by which is meant the internal conditions at a site that enter into and shape the production in this case of radical geographical knowledge. Place might affect knowledge through: a specific geographical relation among participants, for example, between homeplace and workplace, or, as Peck and Barnes (this volume) explore in their chapter on Berkeley, the collapse of the two; or as a particular site of investigation that then structures the development of a conceptual framework, for example, the relation between Baltimore and David Harvey's theoretical agenda (Sheppard and Barnes, this volume); or as a specific mix of pressing social issues found in a given urban neighborhood, along with the presence of galvanizing, energetic individuals eager to take them on, the case in Detroit's inner-city Fitzgerald neighborhood during the late 1960s (Warren, Katz, and Heynen, this volume).
- The second is geographical connectivity. Knowledge does not remain fixed in place but circulates, moving from one site to another. Further, the very process of circulation reshapes the ideas that circulate. This is partly because they interact with other ideas, partly because they are interpreted differently in different locations, and partly because they are put to diverse uses at the various sites among which they travel. This is especially clear in Part II, but it also occurs as radical geographical knowledge travels within the U.S. and Canada, for example, as the idea of industrial change moves from the U.S. East Coast to the U.S. West Coast (Peck and Barnes, this volume); or as the idea of the "geographical expedition" is taken from Detroit to Vancouver and later to Sydney (Blomley and McCann, this volume); or as the idea of the circuit of capital migrates from Baltimore to Quebec City (Klein, this volume).
- The third is geographical scale. We focus especially on cases where radical geographical knowledge is originally articulated at one scale, say, the urban (e.g., Baltimore), or the region (e.g., the San Francisco Bay Area), but then scales up and is applied nationally, or globally. While this bears particularly on theoretical concepts and frameworks, it also applies to the very project of radical geography. It

begins at select urban sites like Detroit, Worcester or Baltimore, but scales up, becoming a global movement, replete with international conferences and journals, drawing readers and participants from around the world.

The remainder of this chapter is divided into four parts. In the first and longest section we lay out what we call the conditions of possibility for the development of radical geography in the U.S. and Canada. We describe the social, cultural, political, and intellectual ferment of the "long 60s"[4] that provided fertile ground for the development of radical geography and set out some of key moments in that unfolding development (further elaborated in subsequent chapters). Second, we describe some of the tensions within the project of radical geography, often there from the beginning, which contorted and disrupted it, making it heterogenous, preparing it for what it was to later to become. Third, we discuss the rationale for the organization of the book, providing capsule descriptions of each chapter. Finally, we provide a short Conclusion.

## "You Say You Want a Revolution:"[5] American Radicalism and Radical Geography During the Long 1960s

In writing about the social sciences since 1945, Roger Backhouse and Philippe Fontaine (2010a: 11) make use of the idea of "the degree of [disciplinary] permeability to social change." They categorize academic disciplines according to their differential social porosity, that is, their internal responsiveness to social events, movements, and interests that lay outside the academy. They argue that social permeability is highly variable by discipline. Some subjects like economics have hermetically sealed themselves from outside social change. While other subjects, including geography, act more like sponges, continually sopping up society's discharge, leaks and spillage, which shape its internal structure and intellectual agenda.

That social porosity can change over time, however, as was the case for geography. Before the Second World, geography was isolated, seemingly immune from social change, doing its own thing. As Neil Smith (1989: 92) argued, geography's strange hybrid form that rolled into one subject natural science, social science and humanities had isolated the subject, given it "a museum-like existence," as if it were some rare entity preserved under glass. From the mid-1950s, however, that glass was smashed. As Smith (1989: 9) puts it, "The museum perimeter" that had been "jealously fenced by a ring of [past] conceptual distinctions, [which] kept geographers in and effectively discouraged would be intruders" was breached. Not only did geographers breakout in large numbers, taking ideas from and working with non-geographers, but the external social and political world came crashing in. Initially, it was a Cold-War-inspired behavioral science and social physics that geographers transformed into spatial science or the "quantitative revolution," with concomitant practices of mathematical modeling, abstract theorization, and statistical verification (Barnes and Farish 2006). During the 1960s, and our concern, it was loud outside social demands for relevance, activism and revolution that produced radical geography. That outside came rushing in at pell-mell pace – Christopher Hitchens (1998: 101) said, "to blink was to miss something" – and was profoundly unsettling – David Horowitz (1970: 185) said America was "shaken to its roots."[6] And so was geography.

## *"The times they are a-changin'"*[7]

If World War II invigorated and shaped American social sciences, the ensuing Cold War Americanized them. Before the Second World War, Germany was the most important home for social science. But after Germany's crushing defeat, the center of gravity for social science moved across the Atlantic, especially to U.S. East Coast centers like Harvard, Yale, and Princeton universities. The Second World War also demonstrated to the U.S. that social science could be effectively mobilized to achieve military and political strategic ends (Barnes and Farish 2006). In this new model, social science was folded into the aims, interests, and bureaucracy of a militarized state. That state brought together different social scientists, setting them to work often in interdisciplinary groups on instrumental projects in the state's interests, fully funding and resourcing them. That same model continued once the hot war of WWII ended, becoming even more entrenched, systematized and formalized within a Cold War in which two superpower states, America and the Soviet Union, played nuclear chicken. Further, the state became a larger assemblage that blurred lines between the military, industry and the academy, forming "a military-industrial-academic complex" in Senator William Fulbright's term (Kay 2000: 10–11).[8] American social sciences, and the sites where they were undertaken, became fully integrated within that complex (Solvey and Cravens 2012).

Social science was thus just another element in the American Cold War boom. Public funding for social sciences increased enormously, and membership in professional societies burgeoned (Crowther-Heyek 2006). American social science became a client of the state, working for it sometimes directly, at other times at arm's length. Further, the state demanded knowledge of a particular form, more scientific than social. Given that the natural sciences had seemingly won the Second World War with the radar, the earliest computer, and of course the atom bomb, social scientists should mimic the same scientific method. There was an attempt even to change the name of social science to behavioral science to indicate a more rigorous, clinical approach, shunning the messy and politically infused term, society.

That was impossible. Messiness and the political kept on reasserting themselves, pushing for attention, on occasion violently screaming for notice, and no more so than during the long decade of the 1960s. While that period highlighted continual experimentation and change, and the overturning of hitherto older imposed forms of constraint – intellectual, cultural, economic and especially political[9] – strangely it hung together as a whole. As Watts (2001: 162) puts it, the long 1960s had

> a homologous coherence across political philosophy, cultural production, economic cycles and political practice. Johnson's decision to bomb North Vietnam, Dylan's decision to go electric at Newport, and the appearance of Pynchon's *Crying of Lot 49* were somehow all of a piece.

Katznelson (1997) argues that this multiform roiling of the long 1960s increasingly seeped into at least the more permeable social sciences and humanities, producing significant intellectual change. Protests on the street entered the university lecture theater. "The sixties," he writes, "were a 'a volatile moment of madness' when ordering rules, civilities, limits and expectations were suspended with important effects inside the academy" (Katznelson 1997: 312). That volatile moment was produced by a perfect storm of social unrest, fueled by the heated atmosphere around four turbulent social movements: civil rights, second-wave feminism, environmentalism, and anti-Vietnam War demonstrations. Each was itself powerful; collectively, working together, they constituted a force of cyclonic magnitude. Where they touched down – and American university campuses were a prime site – they could turn existing relations and ideas topsy-turvy, inside-out. Some social sciences and humanities, including human geography, quickly felt the effect. Things fell apart, the center no longer held. In Katznelson's terms, volatile madness was awash.

- Civil Rights in America were a longstanding issue, going back before even the formal declaration of the U.S. as a nation state. They became increasingly urgent and visible especially from the mid-1950s and associated with escalating acts of non-violent civil disobedience to protest against hateful, violent and sometimes murderous forms of racial prejudice and bigotry held by a predominantly white population against people of color. That disobedience included boycotts, sit-ins, marches, and mass rallies. The larger end was to stop legalized racial segregation and discrimination found especially in the U.S. South, but also in large Northern inner cities where African Americans faced harsh prejudice particularly in employment and housing markets. Those acts included the Montgomery bus boycott by Rosa Parks (1955–1956) (Kobayashi, this volume), the Greenborough sit-in at a Woolworth lunch counter (1960), the March on Washington (1963), Freedom Rides to the U.S. South by civil-rights workers to increase voter registration (1964), and three Selma to Montgomery marches (1965). The assassination of Martin Luther King in April 1968 in Memphis provoked a storm of inner-city riots, and mass arson in major U.S. cities. It was the greatest civil unrest since the American Civil War, with disturbances in over 100 cites, leading to 40 deaths and 20,000 arrests (Sheppard and Barnes, this volume, discuss the Baltimore case). Other inner cities had already gone up in flames, most famously Watts within Los Angeles in 1965 and Detroit in 1967 (Warren *et al.*, this volume). If Martin Luther King's strategy to realize civil rights was non-violent disobedience, other black political movements urged more direct and confrontational tactics. These included the Nation of Islam and its leaders Elijah Muhammed and Malcolm X and, from 1966, the Black Panther Party that spread from Oakland, CA, to other cities, including Baltimore in 1968 (Sheppard and Barnes, this volume). As both Kobayashi and Peake (this volume) document, having voices of people of color heard in geography proved slow and difficult, including in radical geography, with the larger topic of race and civil rights marginalized despite their manifest geographies and radical political purpose.

- The feminist movement was also longstanding, but from the early 1960s it became increasingly active in the form of second-wave feminism, taking to the streets as the "women's liberation movement." Inspired in part by Simone de Beauvoir's (2011 [1949]) *The Second Sex*, where she argued women were "Other" to men in a patriarchal society, and so evident in early post-War, Father-Knows-Best America, the feminist movement demanded legal changes and shifts in social and cultural norms around a series of issues that continued to produce gender inequality: sexuality, the family, the workplace, education, and reproductive rights among others. Betty Friedan's (1963) *Feminist Mystique* further galvanized the movement, and in 1966 she founded and became the first President of the *National Organization of Women* (NOW). NOW saw itself as a necessary organization, like the *National Association for the Advancement of Colored*

*People*, to lobby for changes in civil rights, albeit based on gender rather than race. Through marches and rallies, but also through political lobbying, speeches, writings, cultural performances, and myriad other interventions, feminism contributed to the tumult of the long 1960s. Even during this phase there were internal tensions, as in the civil-rights movement, with second-wave feminism criticized because of its narrow focus on the interests of predominantly white, middle-class and straight women. By the late 1970s and early 1980s, a third-wave feminism began to emerge that took heterogenous identities of women much more seriously both in practice and in theory. Both forms of feminism entered the academy, becoming part of university curriculum, reshaping existing disciplines including geography. Yet women often struggled to participate, including in radical geography, as many of the chapters document, suffering sexual harassment, discrimination, verbal slurs and put downs, a chilly departmental and university environment, a lack of role models, and often unsympathetic male colleagues intent on competitive superiority and noisy mansplaining.

- The U.S. environmental movement had set down strong roots in the nineteenth century through writers like Henry David Thoreau, the geographer George Perkins Marsh, and the institution-builder John Muir (founder of the Sierra Club in 1892). By the 1950s and early 1960s environmental concerns often turned on air and water pollution caused by large-scale industrial and agricultural producers and their use of a bevy of toxic contaminants. Rachel Carson's 1962 *Silent Spring* brought especially widespread attention both to the heavy use of insecticides like DDT in American agriculture and its mortal effect on wildlife. Mixed into the 1960s environmental concerns was also the Malthusian worry that a rapidly burgeoning world population would exhaust the world's resources within two or three generations. In 1968, the Stanford biologist Paul Ehrlich published *The Population Bomb*. Don Meadows' subsequent co-authored 1972 Club of Rome Report, *Limits to Growth*, elaborated on Ehrlich's concerns, using leading-edge computer simulation models to confirm its bleak conclusions, in some cases suggesting that they should be even bleaker (over 30 million copies of the book were sold, still a record for an environmental publication). The resulting environmental protests were less about global-scale environmental Armageddon than about local environmental sites or features: a particular river, a given forested valley, or a specific animal species. On April 4, 1970, there was one mass event for everything environmental. Earthday brought out 22 million people in the U.S., the majority on university and college campuses. Environmental study had long been part of the very definition of academic geography, and so 1960s environmentalism easily entered disciplinary discussions, including in radical geography. It was a theme preoccupying early issues of *Antipode*, with David Harvey,[10] Richard Walker, and later, Neil Smith, all at Johns Hopkins at the time, writing from a Marxist perspective about nature and the environment. In 1971, Wilbur Zelinski (not a Marxist) and Larry Wolfe (who was), founded the journal *Transition* (1971–1986) to provide a critical understanding of the geography of environment (Peake, this volume).

- Opposition to the Vietnam War was likely the most important of the four movements in setting U.S. radical geography in motion. Tariq Ali (2018: 7) said, "the anti-war movement in the United States … has no equivalent in any other imperialist country. It was the highpoint of dissent in U.S. history." Domestic anti-war protests ensued quickly after the first U.S. air bombing of Vietnam on August 5, 1964 – the *de-facto* beginning of a war that was framed as a retaliation for two North Vietnamese Navy boats having allegedly fired torpedoes the previous day at the USS Maddox, a U.S. Navy destroyer in the Gulf of Tonkin.[11] By the end of 1964 the radical Students for a Democratic Society, the most important of the War's student oppositional groups, had approved a proposal to organize anti-War demonstrations often but not exclusively on university campuses.[12] These began the following year. One of their features was "teach-ins," pop-up classes run by professors and graduate students providing political, geographical, and historical information about Vietnam, and America's increasingly belligerent involvement in southeast Asia. The anthropologist Marshall Sahlins led the first teach-in at the University of Michigan in March 1965 (also see Watts' 2012 essay about being at Michigan). As the War expanded during the 1960s, as more and more young American men were drafted (close to 650,000), as more and more U.S. troops were sent to Vietnam (at one point well over half-a-million), and as more and more American soldiers died or were wounded (by the War's end over 58,000 died and over 153,000 wounded – although nothing compared to the millions of casualties in North and South Vietnam), the protests became larger, more violent and more inclusive, involving not just students. Over 400,000 protested in the march on the Pentagon in October 1967;[13] in August 1968 the nation watched on live TV as riot police pounded demonstrators with Billy clubs at the Democratic Party convention in Chicago, and there primarily to protest the continuation of the War; and in May 1970 Americans saw in their newspapers black and white photos of the lifeless bodies of four student protesters at Kent

State University shot dead by the National Guard. Increasingly these protesters were not only students but politicians (like Eugene McCarthy, who ran against Richard Nixon in the 1972 Presidential race), Vietnam War veterans (most famously John Kerry), and Civil Rights leaders including Martin Luther King, until he himself was murdered. North American Geographers did not study the Vietnam War but many mightily protested it – young faculty, graduate students, and undergraduates – coloring their expectations about the purpose and nature of geographical research.[14]

These four social movements, and their increasing entanglement as the 1960s turned into the 1970s, did not directly cause academic geography to swerve to the political left, to revolutionize its methodological and theoretical underpinnings and substantive foci. But it provided conditions of possibility for those changes to occur. They unsettled and loosened existing nostrums, and began to unmoor the discipline from existing intellectual anchors. The wider disciplinary matrix of social sciences and humanities no longer seemed up to the task of understanding contemporary culture and society; indeed, it often seemed an impediment to that end. It was necessary for the turmoil and the passion of the street, as Katznelson (1997: 312) put it, to "burst the bounds of the lecture hall."[15] It did. He (1997: 312–313) continues:

> However, inchoate and unfocused, there was [from the late 1960s] … a powerful revolt against the silences, limits of method, smug confidence, and regime enhancing functions of post-war scholarship in the humanities and social sciences … Civil rights and student (soon anti-war) movements electrified American campuses. In these hot house environments, graduate students and younger faculty … achieved significant scholarly work produced in exasperation of the indifference of their teachers and the inequalities and tumult in society, in anger at the entanglement of the disciplines and the university as institutions with power and privilege, in revolt against particular standards of objectivity that relied too heavily on models inappropriately drawn from the natural and biological sciences, and in the quest of moving class, race, gender, the national security state, and other neglected subjects from the margin to the center of systematic inquiry.

Disciplines changed across the social sciences and humanities. This was not a mechanical process. Because of contingent factors, Backhouse and Fontaine (2010b) suggest, fields like Anthropology, Sociology, Political Science, and English Literature, were more porous and receptive to the outside context of the street than others (like Philosophy, and especially Economics). Yet even Economics couldn't entirely resist the potencies of the late 1960s (Coates 2001). The Union of Radical Political Economy (URPE) formed in 1968, the same year that the radical economist Kenneth Boulding was President of the American Economic Association. URPE aimed to address issues that orthodox Economics woefully neglected, such as "war, race, gender, justice and poverty.... Indeed, many [of URPE's members] saw economics as complicit in the problems of American society" (Backhouse 2010: 57). In the end, URPE never transformed economics as a discipline in the same way radical geography transformed human geography. Over time, at least in the United States, radical economics was reduced at best to a rump. Disciplinary forces had disciplined.

Katznelson (1997) suggests that in those disciplines that allowed in the street change was of three main types. First, there was a "forced revision to existing approaches" (Katznelson 1997: 318). Second, "by exploring new theory and new ways of working, the [new radicalized discipline] directly altered the character of their field" (Katznelson 1997: 318). Finally, the change in a discipline "created space for subsequent insurgencies" (Katznelson 1997: 319). All three of these types of changes played out in North American human geography from the late 1960s to which we now turn.

## "What's going on"?[16]

Academic geography originated during the mid-to-late nineteenth century, serving the interests of various Western European imperial states (Hudson 1977). From almost the beginning, however, there were subversives within the project. The Russian anarchist geographer Kropotkin (1842–1921), who spent over 40 years in Britain as in effect a political refugee, was one, and the French anarchist Élisée Reclus (1830–1905), banished to Switzerland for much of his professional life because of his political activities, was another (Blunt and Wills 2000: 4–5; Ferretti 2018). There was also Frank Horrabin (1884–1962). Not an academic geographer but an English socialist journalist and graphic artist, his "workers' text," *An Outline of Economic Geography* (1923), sold 20,000 copies and was translated into eight languages (Hepple 1999: 81). As important as these individuals were, none of them, as Jim Blaut (1979: 59) writes, "moved the discipline out of its conformist course." For Blaut (1979: 160), geography in the U.S. at least until 1945 was a staid "culturally monotonic" profession.[17] The Cold War, along with the House UnAmerican Activities Committee and McCarthyism, only further stifled dissent. It was not until the long 1960s that radical geography could emerge as a larger movement. Driven by the period's social turmoil, its subsequent institutionalization occurred with "surprizing suddenness" (Blaut 1979: 160).

Radical geography was never a *tabula rasa* creation but in part drew upon elements found within "existing approaches." Following Katznelson (1997: 318), radical geographers wanted to "alter the character of the field" and they did, propelling a "forced revision of existing approaches." The radical geographical alternative constructed was variegated, plural not singular, its different versions containing at least traces of pre-existing forms of geography, albeit knitted together in new ways. Existing human geography came in two main varieties. Its older form was an atheoretical descriptive regional geography, oriented to field work, and frequently concerned with the relationship between humans and their environment. The more recent (and alternative) incarnation, developed from the mid-1950s, was spatial science. Driving it was a concern with formulating the kind of geographical theory found in natural science as well as Cold War social science: abstract, mathematical, logically rigorous, explanatory, predictive, and formally tested against an empirical world. Even as radical geography claimed to separate itself from both traditions, particularly the second, it remained marked by them.

That was clear in William Bunge's work. In the late 1950s, as a graduate student Bunge was an original member of the "space cadets" at the University of Washington that spearheaded geography's quantitative revolution (Barnes 2016; Bergmann and Morrill 2018). His Ph.D. thesis, *Theoretical Geography* (Bunge 1960), was likely the most evangelical of any of the theses written by a cadet in favor of the scientific method. He poured scorn on the earlier regional approach to geography, and those who provided its intellectual justification such as Richard Hartshorne (a life-long nemesis, Barnes 2016). Bunge was already on the political Left, however, even owning a specially bought suit to wear at the many progressive political demonstrations he attended. Yet, his private political radicalism did not make its way into his initial public academic writing.

That changed from the mid-1960s as the external context of civil rights began pushing his work in a different direction. In an autobiographical essay, Bunge (1979: 170) recalls that in 1965 he was writing "the logical extension of *Theoretical Geography, Geography: The Innocent Science* ..." It was jointly written with William Warntz, and bereft of any politics, let alone radical politics. That book was never completed, though, because, as Bunge continues, "The Crime ... [had] started" (Bunge 1979: 170). "The Crime" was the violent abuses of civil rights, which made him "throw himself" into every protest going: "Iwent to Selma. I went to everything. Peace demonstrations in New York, in Washington, Civil Rights demonstrations in Jackson, Mississippi" (Bunge 1979: 170). It also led him to become involved in local community politics, to start undertaking research differently. By the mid-1960s he was teaching at Wayne State University in Detroit. He began sending his students into the city on field trips to map social conditions, becoming involved in community organization especially in his own residential neighborhood, the increasingly black inner-city district of Fitzgerald (Barnes 2018; Bunge 1969; Warren *et al.*, this volume). The July 1967 Detroit rebellion was catalytic, and his own neighborhood was embroiled in it. As he put it, during "the smoke of [that 1967] revolution … I lived in everyone's definition of freedom – No state … [It] had been driven out … I was free to think freely, so I did, I wrote a peace book, *Fitzgerald*" (Bunge 1988: xix).

Bunge's (1971) *Fitzgerald* was the first landmark volume of this emergent version of radical geography. It certainly reflected "the street" in Katznelson's terms, demonstrating Geography's social porosity. It also revealed the continuing influence of repurposed elements of existing disciplinary practices. Geographical theory remained part of the Fitzgerald project. It usually did not take the same form as found in Bunge's earlier spatial science writings (although he managed to work von Thünen's model of land use and rent into the text; Bunge 1971: 132–135). But it was couched in the same language of abstraction, empirical testing, and explanation. Moreover, Bunge recognized that the regional tradition he previously spurned also had its uses. His intensive study of the one square mile of Fitzgerald used maps, field trips, and the deep knowledge of residing in place, leading him to realize his work was in the tradition of his dreaded enemy, Richard Hartshorne. In an interview, Bunge (1976: 2) said that his earlier dismissal of regional geography had been "wrong … which is very painful to admit."

Fitzgerald was rooted in Detroit. Bunge worked with other local activists, including especially the then teenage organizer, Gwendolyn Warren, although it was fraught relationship (Warren *et al.*, this volume; and the videoed conversation between Warren and Katz in 2015 at the City University of New York[18]). Warren found him insensitive, arrogant, on occasion a bully and misogynistic. She resented that Bunge acted as if he had discovered Detroit's black inner city (reflected in his terminology of "geographical expedition").[19] That vocabulary of "expedition" stuck, though, and was taken up and practiced elsewhere: In Toronto, at Simon Fraser University in Vancouver, Canada, and later still in Sydney, Australia (Peake, this volume; Blomley and McCann, this volume). Bunge's work traveled and scaled up.

Another vital place to the dissemination of the radical geography project was Clark University in Worcester, MA (Hubert *et al.*, this volume). Neil Smith said that by the late-1970s Clark became "the center of the radical universe…. The buzz about the School of Geography was palpable."[20]

The three of its faculty most important to radical geography each had intellectual roots in "existing approaches" – Jim Blaut in regional geography, especially its environmental strain; Richard Peet, in geographical theory (his Ph.D. dissertation with Allan Pred was on the von Thünen model); and David Stea through his training in psychology (second only to economics in its formalism as a Cold War social science).

Graduate students also played a crucial role at Clark. Ben Wisner, one of them, says he was radicalized by "student politics …, a reflex against the Vietnam War, racism and environmental pollution."[21] He continues,

> we were groping for root causes of the problems, contradictions and hypocrisies with which we had grown up in the 1940s and 1950s … We had grown up under the nuclear specter of the nuclear Cold War and had been subjected to the flattening out of perspectives described so well by Herbert Marcuse.[22]

The graduate students were for revolution. Their revolutionary act: to publish *Antipode: A Radical Journal of Geography*. *Antipode* was a student-led initiative that came out of Stea's graduate seminar (Mathewson and Stea 2003). In 1969, on the first page of the first issue, Stea (1969: 1) wrote: "Our goal is radical change – replacement of institutions and institutional arrangements in our society that can no longer respond to changing societal needs." Wisner, its first editor, describes the inaugural issue published in 1969 as "a shambolic, amateurish adventure."[23] For volume 2, Dick Peet took over the editorship, although the continued publication of the journal rested fundamentally on the free labor of graduate students, as well as pressures put on those graduate students to contribute under what was in effect a patriarchal system of power (Huber *et al.*, this volume).

The central importance of *Antipode* was in enabling radical geography to travel; to be in Bruno Latour's (1987) lexicon an immutable mobile, taking its message from Worcester, MA, to far-away places such as Bristol or London in the UK as in our opening stories. As Phil O'Keefe, a later editor put it, the importance of "*Antipode* … [and] the work of the Clark graduate students [was in producing] an international impact from a cottage industry."[24] In our language, *Antipode* enabled scaling up.

There were other techniques to widen the audience for early radical geography. Holding special sessions at national meetings was one. The first national AAG meeting at which radical geographers collectively met was at Ann Arbor, August 10–13, 1969.[25] It was there, as Blaut (1979: 161) says, "that most of the local movements – including the Detroit Geographical Expedition and the *Antipode* group at Clark University – suddenly became aware of one another's existence."

Clark Akatiff (2007: 6) called Ann Arbor the "insurrectionist ... meeting." Bunge transported three buses of local participants from his Geographical Expedition project in Fitzgerald, bringing "the presence of the black streets to the walls of Academe." (Akatiff 2007: 7). Oddly, perhaps that led to Bunge being invited to give a plenary session at the following year's annual meeting in San Francisco, August 23–26, 1970. He was given a prized evening slot, 8 pm – 10.30 pm, at the Gold Room, Sheraton-Palace Hotel, on the theme, "Toward survival geography: Reports in human exploration." Even Clark Akatiff, the session organizer, recalls that did not go so well. Presentations, he said, were "confusing," "rambling," and "discursive," some with "long and complicated statistical analyses, illustrated by unreadable and incomprehensible slides" (Akatiff 2007: 9–10). People walked out. When the Toronto geographer Jim Lemon asked Akatiff "'What was that about?'" he said, "I wasn't sure" (Akatiff 2007: 10).

Things went better the next year at the Boston AAG meeting, April 18–21, 1971. It did not start propitiously, with "flurries of snow" the first day. Hugh Prince (1971: 150), one of two on-the-scene correspondents commissioned by the British geographical journal *Area* to report on the meeting for a U.K. audience, observed that the brisk temperature outside the convention hall was mirrored by a "chilly ... atmosphere" inside. It was the "winter of discontent" for status quo geography, Prince (1971: 150) said, the moment when in Katznelson's terms, "existing approaches" were about to be confronted. David Smith (1971: 155), the other correspondent for *Area*, reported that "the radical movement was evident even at the AAG's Business Meeting, ... [where] a strongly worded resolution opposing the Indochina War was passed along with others concerned with the status of women, graduate students and Spanish-speaking minorities." And then there were the papers themselves. Peet organized a series of sessions on poverty, one featuring David Harvey who had left Bristol two years earlier to take up a position at Johns Hopkins (Sheppard and Barnes, this volume). After speaking, Harvey distributed mimeographed copies of what was to become the hinge fourth chapter in his forthcoming book, *Social Justice and the City* (1973): "Revolutionary and counterrevolutionary theory in geography and the problem of ghetto formation." In an interview, Peet (2002) recalled:

> the room was full, like hundreds of people, you know. And it was clear then what was happening. I remember David said he had ten copies of his paper. It was the first mimeographed version of *Social Justice*, and he said he had ten copies, and it was like a dogfight to get them. Maybe 70 people rushed the stage at the end and grabbed these things. I thought, "God, we've arrived."

Harvey was to become a central figure in radical geography, of course. Yet he too began within the existing tradition of geographical theory and the quantitative revolution, his tome *Explanation in Geography* (1969) providing a philosophical rationale for that movement based on logical empiricism. But like other radical geographers, he too was affected by the street, initially participating in political demonstrations against the Vietnam War before he left England. Once he moved to America in 1969, the streets of Baltimore proved catalytic in his transformation from logical empiricist to Marxist, from spatial scientist to radical geographer. "The travails of Baltimore have formed the backdrop to my theorizing" (Harvey 2002: 170). That theorizing has likely traveled more widely than any other radical geographer's. He once likened its geographical circulation to a "viable … globalized commodity" (Harvey 2002: 160). Given its enormous influence and mobility (scaling up), Harvey's radical geographical theory in this early period came closest in Katznelson's (1997: 318) terms to be the "new theory and new way of working" that replaced "existing approaches."

That new theory and new way or working was classical Marxism, which Harvey took directly from Marx's primary texts, typically without any interposing secondary literature. As Peet (1977: 17) wrote: "From 1972 onwards the emphasis…changed from an attempt to engage the discipline in socially significant research to an attempt to construct a radical philosophical and theoretical base…increasingly found in Marxian theory." Harvey's contributions became foundational, providing a larger theoretical blueprint, its working conceptual parts smoothly integrated, joined, and placed. It was an object of beauty, its logic and design breath-taking. It also had tremendous power, which Harvey would demonstrate by running through it some seemingly off-the-cuff real-world fact or event that was then strikingly illuminated and irrefutably explained.

At the same time, that construction sometimes seemed a bit too good to be true. Everything was explained by a single theorist, by someone who lived a good hundred years before, and who was not a geographer. This is not to gainsaid Harvey, but it contrasted with the beginning of *Antipode* that was eclectic and pluralist. The first serious discussion in its pages of a radical theorist was Mahatma Gandhi not Karl Marx (Philo 1998: 4). Chris Philo (1998: 2) also argues that the mandate of intellectual diversity was etched into the very subtitle of the journal: *A Radical Journal of Geography*, and not how it is sometimes remembered, a "Journal of Radical Geography." There was another issue, discussed further in the next section and this book's Conclusion. The Marxism that dominated early radical geography was mostly about large movements of capital, less about social class, even less about race and gender, with the key theorists all white men. It pushed radical geography, according to Peake and Sheppard (2014: 310), toward a "progressively masculine discourse, dominated by confident…, assertive …, imposing … and difficult personalities." Bunge was likely the most extreme, accounting for some of the strain existing between him and Gwendolyn Warren (Warren *et al.*), but tensions existed at other sites of early radical geography as many of the chapters make clear. This is neither to subtract from the enormous excitement, creativity, and energy generated by individuals working at those sites, nor the importance of their work, but they also were not immune from forms of prejudice, unsavory conduct, hurtful words, and poor personal choices that also occurred during the North American long 1960s.

While from the early 1970s the dominant strain of Marxist radical geography emphasized the primacy of capital movement and accumulation, there were other topics, theorists who were not white men, and on occasion non-economic based radical theories. They were rare, though, usually on the margins. Race was one of the topics peripheralized during much of the 1970s and early 1980s. As already discussed, the AAG had recognized it at least as a topic requiring further research and attention during the mid-1960s, convening the Commission on Geography and Afro-America (COMGA) (1964) (Kobayashi, and Peake, both this volume). Bunge (1971) contributed through his study of Fitzgerald, and there was also work by Blaut who linked issues of race to imperialism, and studies by Richard Morrill of urban ghettoization. In fact, the very first issue of Antipode contained Fred Donaldson's (1969) paper on the "invisible … black American." But sustaining an interest in race during this first phase of radical geography was difficult. In 1972, a report in *Antipode* by graduate student organizers of a symposium on race recognized a continued "inconsistency between the *Black Imagination* and the *Geographical Imagination*" (original emphasis, Wilson and Jenkins 1972: 42). That was certainly born out in the publication record of *Antipode*. In its first ten volumes up to 1979, there were well over 250 separate contributions, but only a dozen were explicitly and solely concerned with race, with most of them about one ethnic group, the "Native American Indian." It wasn't until well into the 1980s, even later, that race became part of the intellectual furniture of a radical geography that by then was sliding into critical geography.

Gender was even a less prominent topic. Women graduate students were involved in the physical production of *Antipode* at Clark from the very first issue, from typing copy, editing, stapling, stuffing envelopes, making snacks (Huber *et al.*, this volume). But women rarely made it inside the covers of *Antipode* as authors. The first issue of the journal included one female (joint) author, Ruby Jarrett (Jarrett and Wisner 1969). The issue's remaining 10 authors were men. The first papers explicitly about women were published five years later: a piece by Pat Burnett[26] (1973) about women in the city, with a reply by Irene Breugel (1973). There were two more papers by women, about women, the following year, but nothing more until 1978, when again there were two more papers. So, only six papers directly about women over the journal's first 10 years. It wasn't until the early 1980s that those numbers increased and focused on the debate around domestic reproduction under capitalism.

Non-economic based theory was also scarce. One exception was work at the University of Michigan by Gunnar Olsson and his students, including Bonnie Barton, Jack Eichenbaum, Stephen Gale, Adrian Pollock, and Michael Watts (see Lauria *et al.*, this volume; and Watts' 2012 evocative essay about his Michigan years). Olsson published four papers in *Antipode* during its first 10 years,[27] as well as a long interim summary statement in 1975, *Birds in Egg* (Olsson 1975). While Marx and Hegel were two of his sources, even more so were Aristotle, Wittgenstein, Kierkegaard, Popper, and von Wright, along with Beckett, Joyce and stories from the Old Testament and Homer's Odyssey (Abrahamsson and Gren 2012). What was produced at Michigan was highly eclectic: from fuzzy-set theory to magic theater, from deontic logic to studies of nineteenth century British colonialism in West Africa. It was a profound radicalism, the plumbing, exacting scrutiny and often overturning of all foundations. Everyone went down the rabbit hole. As Watts' (2012: 147, fn. 8) says, for the graduate students "it was a very risky business… [the] anxiety … a constant spectral presence." It is hard to think of anything more radical, although it was rarely couched as political economy.

Despite the lacunae and patchiness, radical geography had established itself as one of the theoretical strands of geography by *Antipode*'s tenth anniversary. Clearly it was not (yet) the establishment, however. It couldn't be. Its origins were with the anti-establishment, with the various social movements that upended the 1960s, with the cohort that said, "don't trust anyone over 30."[28] For that generation existing geography had to change: It was counter-revolutionary, as Harvey (1972c) put it. The disciplinary emergency brake needed to be applied. Yet, at the same time, radical geography was not and could not be completely new, a *de novo* discipline. By holding on to the noun in its title, radical geography necessarily continued to set its inquiries within prior longstanding disciplinary concerns such as the environment, place and region, as well as more recent interests in space and theory. Of course, there were differences between existing and radical geography. The latter was concerned to make the discipline an instrument of radical social and political change, targeting politically and socially relevant topics like regional and urban poverty, U.S. imperialism and colonialism, international and intra-national uneven development, spatial segregation of all kinds, and environmental despoliation. In this sense, radical geography represented both continuity and disjuncture. Early radical geographers drew on their existing geographical training, but added to it, creating new combinations of ideas and practices. None of them knew how it would end, nor the final shape that radical geography would take. Inevitably, there were arguments, contradictions, *aporia*, cleavages, debates, and unresolved issues. As Katznelson (1997: 319) argued, these internal arguments opened critical space, allowing "subsequent insurgencies." In the next section we turn to some of those tensions within early radical geography, making it an uneven intellectual terrain – striated rather than smooth – which propelled further alterations to the project.

## Tensions within the project

Tensions can be both productive and destructive, with no reason to think that they will necessarily be resolved over time into some sort of happy-ever-after story. In radical geography there have been few resolutions, final or otherwise. A continuing tension has been between radical geography as an activist project, participating directly on-the-ground in transforming the world, and as a theoretical project, developing a corpus of abstract geographical theory to represent and explain the world. That tension was evident from the off in the two signal contributions of early radical geography: Bill Bunge's (1971) *Fitzgerald* and David Harvey's (1973) *Social Justice and the City*. While an activist in his time off, wearing his demonstration suit to marches "on the street," when he went to the office in Smith Hall on the University of Washington campus to do serious work Bunge was a theorist. Bunge's (1960) Ph.D. dissertation was called literally *Theoretical Geography*. Moving to Detroit, though, he realized that the serious work was not done in the office but on the street, as an activist. While that activism itself was admittedly fraught and replete with strains (Warren *et al.*, this volume), Bunge was in no doubt that the prime role of radical geography (not that the term was yet available to him) was to participate immediately and unhesitatingly in direct acts in improving the world. *Fitzgerald* showed how one version of an activist radical geography could be done involving the local community, sharing knowledge, and providing resources. In contrast, *Social Justice and the City*, was always a book about theory. Its crucial pivot chapter (ch. 4), in which Harvey comes out as a Marxist radical geographer, is opposed to his old self as a leftish liberal reformer. That transformation occurred, as Harvey made clear, because of his new theoretical allegiance to Marxism.

Theory made the difference. It was a crossing-the-Rubicon move from counter-revolutionary to revolutionary theory. The remainder of *Social Justice* details that revolutionary theory, culminating nine years later as the theoretical tome, *Limits to Capital*. That said, in his off-time, like the early Bunge, Harvey was involved in various forms of activism (Sheppard and Barnes, this volume), and nowhere has there been any sense that he belittled Bunge's activist contributions. Yet Harvey stood at the opposite end of the activism-theory pole to Bunge, a tension that continued within radical and later critical geography.

A second tension has been around a monistic versus a pluralist approach to theory within radical geography. The earliest formulations were pluralist, albeit admittedly not described as such. Early proponents used what came to hand, what seemed to offer radical geographical possibilities. Bunge (1971) re-adapted theory from spatial science, using von Thünen's concentric agricultural land-rent theory to show how the capitalist market mechanism has a propensity to create spatial inequality: a geography of uneven exploitation and surplus extraction. More generally, early *Antipodes* were characterized by theoretical variety. Philo (1998: 1) noted the "sheer diversity of ... geographical radicalism which coursed through the earliest [*Antipode*] issues." They featured not only Mahatma Gandhi, but also Martin Luther King, Gunnar Myrdal, Murray Bookchin, Robert Coles, Frantz Fanon, Eric Hobsbawm, Petyr Kropotkin, and Élisée Reclus. This diversity began to narrow, however, with what Peet (1977: 16) called the "break through to Marxism." For him, the turning point was the 1972 *Antipode* debate between Harvey (1972b) and Brian Berry (1972) around that pivot chapter in *Social Justice*. Only by systematically pursuing Marxism, suggested Peet (1977: 25), could "radical geography begin to construct a theoretical base capable of providing an analysis of the events of late capitalism and proposing revolutionary solutions." While the focus on Marxist theory may have held through the 1970s and very early 1980s (and even that is not clear), it has not endured. Much subsequent debate within radical geography, and later critical geography, has turned precisely on the theoretical tension between adhering to a monistic Marxism and a wider discussion drawing on pluralist sources. The inclination toward a form of theoretical inquiry based on "not only" Marx, "but also" Foucault, Butler, Fanon, Deleuze and Guattari, Agamben, Derrida, Kristeva ... triggered the development of a critical geography that was ecumenical and eclectic

A third tension has been whether radical geographical research must focus on the economic or whether it can also address the non-economic. And if the latter, how should the non-economic be related to the economic? Within classical Marxism, the material and social relations of production constituting the economy are determining, explaining everything else. In this economistic reading, the economy must and should occupy the center of radical geography's attention. The earliest works within radical geography did not hold to this principle, however. The center of attention was race, connected with battles around U.S. civil rights during the long 1960s. Thelma Glass's involvement in the Montgomery bus boycott was not prompted by concerns about the economy, even though black Americans were subjected historically to hideous forms of economic discrimination. Similarly, propelling Bunge's work was less a concern about the economy, although certainly he had a concern, than moral outrage over how the color of one's skin shaped the life one led in the U.S. This strain carried through in the early volumes of *Antipode*. The economy was there in calls for studying rural poverty or imperialism, but it was not necessarily front and center. That changed with the introduction of Marxist theory. Harvey's study of the housing market in Baltimore, with Lata Chatterjee, made use of Marxist rent theory (Harvey and Chatterjee 1974). In a city riven by racial conflict, the focus was on financial institutions and flows of capital, the economy. Race appeared but was secondary: it was just one of several variables that defined the "Inner City" housing sub-market (Harvey and Chatterjee 1974: 26–29). The market came first along with the financial variables that made it. Struggles over the emphasis to accord the economy continued, with some deconstructing the very idea of an economy, others striving to treat it as a hybrid entity joined with the non-economic, and yet others continuing to assert its theoretical sovereignty. Along with those struggles have also gone attempts to redefine the relationship between the base and the superstructure. There has been a movement away from a simple one-way causality seemingly posited by Marx, to more complicated relationships including, for example, hegemony theorized by Antonio Gramsci, or overdetermination theorized by Louis Althusser, or a structure of feeling as theorized by Raymond Williams, all of which have been taken up by radical and critical geographers.

A final, longstanding pair of tensions are around gender and race. Part of the issue is intellectual. How do feminist scholarship and critical race studies fit within the aims and practices of radical geography? Second-wave feminism, and the associated sophisticated body of theorizing and academic empirical study, developed during the same period as radical geography. Yet there was virtually no consideration of gender issues, or utilization of feminist theory, in radical geography's early works, through the first phase of its Marxist turn. That did not systematically appear until the early 1980s, with discussions about household social reproduction and gendered cities. Critical race theory did not emerge until after the first phase of radical geography was established, but there were longstanding works, for example, by W. E. B. DuBois, Franz Fanon, and James Brown, as well as more contemporary writing by, say, Malcolm X, that could have been incorporated conceptually into radical geography, but were not.

The other issue goes to the gender and racial identities of those who participated within early radical geography. The early leading figures were all white men – Jim Blaut, Bill Bunge, David Harvey, Dick Peet, and David Stea – as were their immediate followers, like Ron Horvath, Phil O'Keefe, Ben Wisner, Neil Smith, and Dick Walker. Many had large personae and big reputations, stamping their work as masculine. That clearly effected the women who also participated, as is clear from the chapters on Bunge and his work in Fitzgerald (Warren *et al.*, this volume), the production of *Antipode* at Clark University along with the classes that were run (Hubbert *et al.*, this volume), the Vancouver collective (Blomley and McCann, this volume), and the bilateral creation between the Geography and Urban and Regional Planning Departments of radical industrial geography at Berkeley (Peck and Barnes, this volume). Women at those sites, faculty and students, reflect on the various forms of violence to which they were subjected by male professors and students, including objectification, condescension, discrimination, and sexual harassment. This came in different forms and degrees and certainly not all men participated, but there was often a charged atmosphere, and sometimes a hostile, even vicious one, in the places where female radical geographers worked. This has rarely been acknowledged, nor the male perpetrators called to account.[29] With respect to race, there were very few people of color who worked within early radical geography; Bobby Wilson, was a graduate student at Clark, and there was also Harold Rose, Don Deskins, and later Joe Darden, but none took up the early 1970s version of radical geography as such, but given its emphasis, maybe not surprising.

## Organization of the Book

The collection is divided into two parts. Part I traces the development of radical geography in the U.S. and Canada, from the mid-1950s until around the early 1980s. That narrative is organized geographically, focusing on geographical centers where the work of creating radical geography occurred. Some of the chapters examine several centers together, but most focus on a single place. Part II very selectively examines how radical geography outside the United States and Canada has co-evolved with these North American sites. Often there was some pre-existing tradition of radicalism in those places that then joined and forged novel combinations with the North American variety. This section narrates how radical geography takes on distinctive regional and national forms as it has globalized, producing a variegated landscape of radical geographical knowledge shaped by place, connectivity, and scale.

## Part I

Chapter 1 is by Audrey Kobayashi, "Issues of 'race' and early radical geography: our invisible proponents." Examining the contributions of early African American trained geographers, from Thelma Glass outside the academy to Don Deskins and Harold Rose from within, as well as attempts led by Saul Cohen to connect with geographers in historically black U.S. post-secondary institutions, Kobayashi carefully reconstructs the multi-faceted relation between race and radical geography. She argues that race was neglected as a topic in favor of class, a reflection of radical geography's turn to Marx, albeit with some exceptions: Bunge and Blaut.

Chapter 2 is by Gwendolyn C. Warren, Cindi Katz, and Nik Heynen, "Myths, cults, memories, and revisions in radical geographic history: Revisiting the Detroit Geographical Expedition and Institute (DGEI)". They reinterpret the work of William Bunge and Gwendolyn Warren in Detroit during the late 1960s and early 1970s. Focusing on the establishment of this community-based organization within the inner-city neighborhood of Fitzgerald where both Bunge and Warren lived, they maintain that the usual history told about DGEI is mythic, reflecting a cult of personality that has formed around Bunge and his work. They work to strip away both the myth and the cult.

Chapter 3 is by Matthew T. Huber, Chris Knudson, and Renee Tapp, "Radical paradoxes: The making of *Antipode* at Clark University." Concerned with the relation between place and knowledge, they examine how the Graduate School of Geography (GSG) at Clark University in Worcester, MA – until the 1950s a bastion of political conservatism and the retrograde geographical theory of environmental determinism – surprisingly emerged in the late 1960s as a center for radical geographical inquiry. From 1969 that position was further cemented by publishing *Antipode*, initially produced in the basement of CSG using free graduate labor. Like the previous chapter by Warren *et al.*, Huber *et al.* also seek to unsettle the usual history of radical geography that highlights the celebratory formative role of Clark and *Antipode*. They trouble that narrative by raising difficult questions around the treatment of women and the changing status of the journal.

Chapter 4 is by Nick Blomley and Eugene McCann, "A 'necessary stop on the circuit': Radical geography at Simon Fraser University."

They narrate the emergence of an offshoot of the DGEI, the Vancouver Geographical Expedition, initiated during the early 1970s at Simon Fraser University located in the Vancouver suburb of Burnaby. Simon Fraser itself was a product of the long 1960s, immediately attracting radical geographers, both faculty and graduate students, after it opened. The Expedition did not last – a victim of tensions between activism and theorization as well as gender – but the Simon Fraser group was vital to the creation of the USG, and the publication of its Newsletter. Key also to Simon Fraser's accomplishments was the radical but controversial Michael Eliot Hurst. Chair of the Geography Department during part of this period, he provided comradely encouragement but also necessary material resources.

Chapter 5 is by Linda Peake, "The life and times of the Union of Socialist Geographers". She provides a comprehensive historical geography of the USG, from its creation at a meeting in Toronto in May 1974, to Vancouver, and thence to multiple sites (with their own locals and regional meetings) across North America and abroad, to its final demise in southern California in December 1982. Its key text was the Newsletter, which transformed from a simple means of communication among members to an ever-more ambitious organ of unruly gray literature, before ultimately collapsing from a combination of its own weight and declining membership support. Peake emphasizes that the USG Newsletter from its start was a venue for a far more variegated tradition of radical geographical thought than *Antipode*. The Newsletter functioned as something of a counter-pole both intellectually and geographically, generating a space within which Canadian and U.S. geographers engaged even-handedly with one another.

Chapter 6 is by Eric Sheppard and Trevor Barnes, "Baltimore as truth spot: David Harvey, Johns Hopkins and urban activism". Following the sociologist of science, Tom Gieryn, they argue that even abstract theoretical knowledge is intensely colored by the place in which it is produced. In David Harvey's case it was the travails of Baltimore, the city to which he moved in 1969, which seeped into his Marxist theorizing. That theory was sparked and subsequently shaped by Harvey's initial research project on Baltimore's housing market. Once produced, the theory rapidly circulated through the travels of Harvey's academic papers, including in *Antipode* and his 1973 book *Social Justice and the City*. The theory scaled up, becoming a global geographical Marxism that rendered invisible its local origins and their persistent influences.

Chapter 7 is by Jamie Peck and Trevor Barnes, "Berkeley in-between: Radicalizing economic geography". They examine how during the late 1970s and early 1980s, the internal geography of a collaborative project linking members of the Geography and City and Regional Planning Departments at Berkeley, in conjunction with the changing external geography of the San Francisco Bay Area, conspired to produce a new, radical geographical rendering of industrial geography. That rendering identified a novel powerful economic geographical dynamic producing new industrial spaces, such as Silicon Valley, while at the same time destroying places of old Fordist manufacturing, such as in Oakland and the East Bay. The Berkeley moment set a new intellectual agenda for radical and industrial geography that circulated widely and scaled up.

Chapter 8 is distinct because it is in effect an autobiographical account by Mickey Lauria, Bryan Higgins, Mark Bouman, Kent Mathewson, Trevor Barnes, and Eric Sheppard, "Radical geography in the Midwest". They describe how graduate students at the University of Minnesota, as well as the University of Wisconsin, Madison, and the University of Michigan, abetted by a few Geography faculty, created nodes of radical geography in the US Midwest. Taking over from Simon Fraser the editing, financing and distributing the USG Newsletter, the Minnesota node catalyzed other Midwestern sites such as Madison, Valparaiso, Iowa City, and Chicago, through the organization of a series of USG regional meetings. Lauria *et al.* also highlight the variegated culture of the radical geography propagated especially from the Twin Cities, combining community activism with conventional scholarship.

Chapter 9 is by Juan-Luis Klein, "Radical geography goes Francophone". His concern is the working out of radical geography from the mid-1970s to the mid-1980s within the French-speaking province of Quebec, Canada. Centered at the Université Laval in Quebec City, the radical geographical *Groupe de recherche sur l'espace, la dépendance et les inégalités* (GREDIN) began both theoretically and empirically to apply their radical ideas to their home province. Led by Rodolphe De Koninck and Paul Y. Villeneuve, they drew directly from Marx and dependency theory, as well as on a developing Anglophone radical geography. The group was especially concerned with the core-periphery spatial relationships that had developed within Quebec, that manifest as regional income inequalities, differential service provision by the state, and forms of sub-regional political and economic dependency.

## Part II

Chapter 10 is by Fujio Mizuoka, "Japan: The Yada Faction versus North American radical geography". He describes how radical economic geography was initiated in Japan as early as the 1930s, well before North

America, by geographers connected with the Japanese Communist Party working in the tradition of Marx. By the 1990s, however, the leader of this group abandoned Marxism. In turn, that contributed to shifting the Japanese Association of Economic Geographers to align its work with a neoliberalizing Japanese state. Since then, geographers returning from or influenced by U.S. radical geography have struggled to change the direction of Japanese economic geographers, notwithstanding the presence of a continuing undercurrent of that earlier work.

Chapter 11 is by Brij Maharaj, "The rise and decline of radical geography in South Africa." He discusses how radical geography in South Africa co-evolved with the country's dramatic internal transformations from an Apartheid state, to a revolutionary African National Congress (ANC), to the neoliberal turn taken by the ANC after coming to power. Radical geography emerged to challenge Apartheid and mainstream European-style South African geography, influenced by key figures trained in and returning from North American nodes of radical geography in the 1980s (particularly Vancouver and Baltimore). He describes how South African radical geographers are split between those who are more and those who are less critical of the current political regime, how Black radical geographers have struggled to gain acceptance within South Africa, and how a new, younger, and dissenting radical geography is now emerging.

Chapter 12 is by Verónica Crossa, "The geographies of critical geography: The development of critical geography in Mexico". She begins by analyzing how and why geography as a discipline remained largely immune from any radical shift shaped by the long 1960s, unlike cognate social sciences in Mexico, or geography elsewhere in Latin America (notably Brazil). Yet, during the last two decades the landscape of what she calls critical geography has evolved rapidly, influenced by younger scholars trained in and returning from North America, carrying with them interdisciplinary approaches for studying socio-spatial relationships. Nevertheless, critical geographers face the challenge of securing academic positions within an emergent neoliberal audit culture that gatekeeps both appointments to major universities and opportunities in research institutions.

Chapter 13 is by Joanne Norcup, "'Let's here [sic] it for the Brits, You help us here': North American radical geography and British radical geography education". She examines how radical geographical ideas developed in the U.S., specifically associated with Bunge, traveled to the UK, and by the 1980s complemented the aspiration of some in that country for the implementation of a radical education curriculum. In turn, that aspiration became aligned and was joined with anti-racist, post-colonial, and feminist politics emerging in London and across

Britain. This powerful combination of radical ideas was then taken up in British educational curricula in universities, colleges and high-schools. Bunge had left his mark, his work traveling and scaling up.

Chapter 14 is by Yann Calbérac, "'Can these words, commonly applied to the Anglo-Saxon social sciences, fit the French?' Circulation, translation and reception of radical geography in the French academic context". Calbérac argues it was initially a struggle to transfer radical geographical ideas from North America to France because of the deep embeddedness of the Vidalian regional descriptive tradition within French geography. While there has been a very gradual change, in part because of the radical Quebecois literature gaining purchase, there has not yet been a wholesale radical makeover of French geography. Nonetheless, there is an embryonic hybrid form, *géographie radicale "à la française"*, that combines traditional French geographical themes of the social and the political with Anglophone critical and radical approaches.

## Conclusion

Over the last 45 years in Sheppard's case, and 40 years for Barnes, we continue to remain shaped, absorbed, and stimulated by radical geography and what it has become. Admittedly, when we pick up the latest copy of *Antipode*, or now more likely read it on-line, it doesn't quite provoke the same febrile excitement, or sense of transgression, that holding in our hands the first paper copy did. It's not that the old is better than the new. Far from it. Precisely because the current version of radical geography remains so intellectually alive and present, so crucial to understanding the present, we think it is vital to remember its past. That's not because of antiquarianism, or Left nostalgia (although that might also be at play). Nor is it because we believe that radical geography is defined by its past, or again, that the past provides clues, hints and anticipations that are fulfilled in the present (this is only another bankrupt version of Whig history). Rather, it is because, as Faulkner put it, "the past is never dead. It's not even past." The past and present are inextricably joined. We cannot escape history because the past never fully passes. Instead, we carry the burden of the past into the present, bringing with us what went before. Such a position was grasped perhaps better than anyone else by Michel Foucault (1970: 219) for whom knowledge is profoundly historical: "the unavoidable element in our thought." We never begin from scratch. The best we can do, the only thing we can do, is to provide a "history of the present" (Foucault 1977: 31). That's what the authors do in our collection. They write compelling histories of radical geography's past in order to understand its present.

# Notes

1  From the double A-side single released by The Stranglers in July 1977, and included on their album, *No More Heroes* released in September. 'Something Better Change' went to number 9 in the British top singles chart.

2  On one occasion *Antipode* was viewed by the authorities literally as contraband. In 1977, Dick Peet and Phil O'Keefe drove from Clark University, Worcester, MA, to Toronto, bringing with them a pile of *Antipodes* that they were going to distribute at a meeting. On the Peace Bridge, Port Erie, Canadian Customs confiscated them claiming they were not really geography.

3  As we discuss in the second section, U.S. and Canadian radical geography clearly drew on prior traditions of radicalism developed elsewhere that go back to Europe and the nineteenth century. Fujio Mizuoka (this volume) identifies a radical tradition of geography in Japan that pre-dated the North American version, although it was subsequently abandoned. Each of the other countries also have long traditions of radical thought and practice, which sometimes included geographers.

4  "The long 60s" is inexact and depends on the particular issue at hand, but generally it begins roughly in the mid-50s and continues sometimes to the early-to-mid-1970s. It marks an extended period of political, social and cultural change, particularly in the U.S., but in other places as well. The term is used by Frederick Jameson (1984) in his influential essay, "Periodizing the 60s." He believed the 1960s were produced by the "enlargement of capitalism on a global scale" that in turn induced a significant "superstructural movement.... The 60s were ... an immense and inflationary issuing of superstructural credit" (Jameson 1984: 208–9).

5  The first line of the Beatles' "Revolution," recorded in 1968 for their *White Album*, and on the B-side of their single, "Hey Jude."

6  The quotes from both Hitchens and Horowitz were taken from Michael Watts' (2001: 160–161) brilliant essay about the 1960s.

7  The 1964 song title of Bob Dylan's anthem of change. It was the title track of his album released the same year, and the A-side of a single put out in March 1965.

8  Senator Fulbright's phrase was a reworking of President Dwight Eisenhower's earlier famous term, "the military-industrial complex," from his farewell presidential address on January 17, 1961.

9  Watts (2001) argues that the 1960s were characterized by a set of parallel political changes across different continents and scales, all of them stamped by an "unstoppable predilection for alternatives" (Said 1983: 247; quoted by Watts 2001: 183). From the Prague Spring to the founding of the *New Left Review* to the Cuban Revolution; in each case, to use Walter Benjamin's metaphor, the emergency brake had been forcefully applied to the speeding train of history (Watts 2001: 160).

10 David Harvey (1996: 117) recalls being at Earthday in Baltimore during the first academic year he was at Hopkins. He grasped then that what was

considered an environmental problem depended on the social class position of those who experienced it. Visiting the Left Bank Jazz Club the next day, "a popular spot frequented by African American families in Baltimore," Harvey realised that for them "their main environmental problem was Richard Nixon" (Harvey 1996: 117).

11   There is much dispute over whether any torpedoes at all were launched by the North Vietnamese Navy on August 4. In Earl Morris's 2004 documentary, *The Fog of War*, the U.S. defense secretary at the time, Robert McNamara, suggests they were not: The radar operator on the Maddox saw only noise on their screen, not torpedoes. McNamara was the most important architect of the Vietnam War; by the time he quit in 1967, he believed it was unwinnable, and later still, wrong-headed.

12   Michael Watts (2001: 168–169) provides global maps of student protests for 1968–1969 divided into three types: anti-bureaucratic, anti-authoritarian, and postcolonial. Those maps make the point that the student protests were not only about the Vietnam War, nor were they were confined to the United States. The most famous example is not a student protest in America but one in Paris in May 1968 that came within a hair's breadth of bringing down the French government. Dan Clayton (2018), provides a wonderful geographical account of Paris 1968 that includes such titbits as Henri Lefevre being pelted by well-aimed rotten tomatoes as he lectured students in Nanterre, and the nervous colonial geographer Paul Pélisser carrying a loaded pistol in his pocket when giving his classes.

13   The three radical geography faculty at Clark University, Jim Blaut, Dick Peet, and David Stea were attendees. Blaut flew everyone down to Washington from Worcester but given the flight there Peet declined to fly back. Clark Akatiff (1974) who was also there provides a geographical analysis of the event. He likened the protest to "the clash of two armies. One army represented established order – powerful, disciplined, marshalled by conscription, and representing the status quo … In opposition was an army of rabble – unarmed, undisciplined, marshalled by the mushrooming clouds of alienation, cultural disintegration, and protestation – representing an emergent, revolutionary force …" (Akatiff 1984: 26).

14   The best known geographical research by a geographer about the Vietnam War is by Yves Lacoste (1973), a French Marxist geographer. Drawing on work by the earlier French colonial geographer, Pierre Gourou, Lacoste exposed as a lie the denial by the U.S. Air Force that it had bombed the Red River Delta (Bowd and Clayton 2013).

15   Watts (2001: 158) tells the story of his friend and colleague at Berkeley's Geography Department, Barney Neitschmann, who as a graduate student at the University of Wisconsin, Madison, "walked out on his final in 1967 scribbling on the examination script: 'more important things are happening on the street'."

16   The title of a song written by Al Cleveland, Renaldo Benson, and Marvin Gaye from the A-side of Gaye's single released by Motown Records in

January 1971, becoming the title track of Gaye's album *What's Going On* released later the same year. Very relevant to our concerns, the lyrics were based on an incident of police violence Benson witnessed against anti-Vietnam War demonstrators at People's Park, Berkeley, in May 1969.

17 Blaut (1979) attributes this cultural uniformity in part to the balloting system for membership to the AAG (along with its members' ability to blackball any applicant). In 1945, the last year of elected membership in the AAG, there were 90 members: 89 white men and 1 white woman.

18 Gwendolyn Warren and Cindi Katz in conversation, CUNY 2015; https://vimeo.com/111159306 (last accessed July 15, 2018)

19 Notes taken from the panel session, "Reflections on the Detroit Geographical Expedition and Institute: A Conversation with Gwendolyn Warren, Co-director," organised by Cindi Katz and Amanda Matles, Association of American Geographers, annual meeting, Tampa, Florida, Friday April 11, 2014.

20 Past Editor's reflections: Neil Smith (1979), http://onlinelibrary.wiley.com/journal/10.1111/(ISSN)1467-8330/homepage/editor_s_past_reflections.htm#Smith (last accessed February 11, 2018).

21 Past Editor's Reflections: Ben Wisner (1969–1970). http://onlinelibrary.wiley.com/journal/10.1111/(ISSN)1467-8330/homepage/editor_s_past_reflections.htm#Wisner (last accessed February 11, 2018)

22 Past Editor's Reflections: Wisner (1969–1970).

23 Wisner (1969–1970).

24 Past Editor's Reflections: Phil O'Keefe (1978–1980). http://onlinelibrary.wiley.com.ezproxy.library.ubc.ca/journal/10.1111/(ISSN)1467-8330/homepage/editor_s_past_reflections.htm#Okeefe (last accessed February 11, 2018)

25 This meeting was abruptly shifted to Ann Arbor from Chicago because of the civil unrest there associated with the Democratic National Convention just shy of a year before (see Lauria *et al.*, this volume).

26 Pat Burnett intellectually aligned herself with spatial science, although she used those methods to show how women living in Global Northern cities were disadvantaged compared to men. She taught for a period at one of the key centres of spatial science, Northwestern University, but later sued the university for breach of contract citing "a climate of sexual discrimination as cause" (Burnett 2002).

27 See Olsson's on-line CV available at: http://katalog.uu.se/empinfo/?id=N96-1826 (last accessed August 4, 2018)

28 Coined by Jack Weinberger, the student activist involved in the Free Speech Movement at the University of California at Berkeley, 1963–1964 (see Peck and Barnes, this volume).

29 Don Mitchell's tribute to the radical geographer Neil Smith faces up to his long history of sexual dalliances: Neil Smith, 1954–2012: Radical geography, Marxist geographer, revolutionary geographer, https://progressivegeographies.com/2013/11/11/don-mitchell-on-neil-smith-long-article-available/ (last accessed February 27, 2018).

# References

Abrahamsson, C. and Gren, M. eds (2012). *GO: On the Geographies of Gunnar Olsson*. Farnham: Ashgate.

Agnew, J. A. and Livingstone, D. N. (2011). Introduction. In J. A. Agnew and D. N. Livingstone, eds, *The Sage Handbook of Geographical Knowledge*, pp. 1–18. London: Sage.

Akatiff, C. (1974). The march on the Pentagon. *Annals of the Association of American Geographers* 64 (1): 26–33.

Akatiff, C. (2007). The roots of radical geography: a personal account. Paper given at the *Association of American Geographers annual meeting, San Francisco*, April 21.

Ali, T. (2018). That was the year that was. Tariq Ali talks to Dave Edgar. *London Review of Books*, 40 (10), May 24: 3–10.

Backhouse, R. E. (2010). Economics. In R. E. Backhouse and P. Fontaine, eds, *The History of the Social Sciences Since 1945*, pp. 38–70. Cambridge: Cambridge University Press.

Backhouse, R. E. and Fontaine, P. (2010a). Introduction. In R. E. Backhouse and P. Fontaine, eds, *The History of the Social Sciences Since 1945*, pp. 1–13. Cambridge: Cambridge University Press.

Backhouse, R. E. and Fontaine, P. (2010b). Toward a history of the social sciences. In R. E. Backhouse and P. Fontaine, eds, *The History of the Social Sciences Since 1945*, pp. 184–233. Cambridge: Cambridge University Press.

Barnes, T. J. (2016). The odd couple: Richard Hartshorne and William Bunge. *The Canadian Geographer* 60: 459–465.

Barnes, T. J. (2018). A marginal man and his central contributions: The creative spaces of William ("Wild Bill") Bunge and American geography. *Environment and Planning A*. 50 (8): 1697–1715.

Barnes, T. J. (2019). Her dark past. In B. Christophers, R. Lave, J. Peck, and M. Werner, eds, *Doreen Massey: Critical Dialogues*, pp. 53–64. Newcastle-upon-Tyne: Agenda.

Barnes, T. J. and Farish, M. (2006). Between regions: Science, militarism, and American geography from World War to Cold War. *Annals of the Association of American Geographers* 96: 807–826.

Berry, B. J. L. (1972). Revolutionary and counter-revolutionary theory in urban geography – a ghetto reply. *Antipode* 4 (2): 31–33.

Bergmann, L. and Morrill, R. L. (2018). William Wheeler Bunge: Radical geographer (1928–2013). *Annals of the Association of American Geographers* 108: 291–300.

Blaut, J. (1979). The dissenting tradition. *Annals of the Association of American Geographers* 69: 157–164.

Blunt, A. and Wills, J. (2000). *Dissident Geographies: An Introduction to Radical Ideas and Practices*. Harlow: Prentice Hall.

Bowd, G. P. and Clayton, D. W. (2013). Geographical warfare in the tropics: Yves Lacoste and the Vietnam War. *Annals of the Association of American Geographers* 103: 627–646.

Breugel, I. (1973). Cities, women and social class: a comment. *Antipode* 5 (3): 62–63.

Bunge, W. W. (1960). *Theoretical Geography*, Ph.D. dissertation, Department of Geography, University of Washington, Seattle (available at the Suzzallo & Allen Library, University of Washington).

Bunge, W. (1969). The first years of the Detroit Geographical Expedition Institute: A personal report. In R. J. Horvath and E. J. Vander Velde, eds, *Field Notes, The Detroit Geography Expedition, A Series Dedicated to the Human Exploration of Our Planet, Discussion Paper No. 1, The Detroit Geographical Expedition Institute*, pp. 1–30. Michigan State University, East Lansing, MI.

Bunge, W. W. (1971). *Fitzgerald: Geography of a Revolution*. Cambridge, MA: Schenkman.

Bunge, W. W. (1976). Interview with Don Janelle, University of Western Ontario, London, ON, November 6th. Geographers on Film Transcription, Maynard Western Dow, 2004. http://jupiter.plymouth.edu/~gof/home.html (last accessed February 11th, 2018).

Bunge, W. (1979). Perspective on *Theoretical Geography*. *Annals of the Association of American Geographers* 69: 169–174.

Bunge, W. (1988). *Nuclear War Atlas*. Oxford: Basil Blackwell.

Burnett, P. (1973). Social change, the status of women and modes of city form and development. *Antipode* 5 (3): 57–62.

Burnett, P. (2002). Interview with Trevor Barnes, May 29th, Cambridge, MA.

Carson, R. (1962). *Silent Spring*. Boston: Houghton Mifflin.

Castree, N. (2000). Professionalisation, activism, and the university: Whither "critical geography"? *Environment and Planning A*, 32: 955–970.

Cloke, P., Philo, C. and Sadler, D. (1991). *Approaching Human Geography: An Introduction to Contemporary Theoretical Debates*. London: Paul Chapman.

Clayton, D. (2018). "With a pistol in his pocket and revolution in the air:" Geographies and geographers in and around May '68. Unpublished paper, available from the author: School of Geography and Sustainable Development, St. Andrew's University, Scotland.

Coates, A. (2001). A.E.A. and the radical challenge to social science. In J. E. Biddle, J. B. Davies, and S. G. Medeno, eds, *Economics Broadly Considered*. pp. 144–158. London: Routledge.

Crowther-Heyck, H. (2006). Patrons of the revolution: ideals and institutions in postwar behavioral science. *Isis* 97: 420–444.

De Beauvoir, S. (2009). *The Second Sex*. Originally published in French in 1949. New York: Vintage.

Donaldson, F. (1969). Geography and the black American: The white papers and the invisible man. *Antipode* 1 (1): 17–33.

Ehrlich, P. (1968). *The Population Bomb*. New York: Ballantine.

Ferretti, F. (2018). Teaching anarchist geographies: Élisée Reclus in Brussels and "The art of not being governed." *Annals of the Association of American Geographers* 108: 162–178.

Foucault, M. (1970). *The Order of Things*. London: Tavistock.

Foucault, M. (1977). *Discipline and Punish: The Birth of the Prison*. Translated from the French by A. Sherridan. New York: Vintage.

Friedan, B. (1963). *The Feminist Mystique*. New York: W. W. Norton & Co.

Harvey, D. (1969). *Explanation in Geography*. London: Arnold.

Harvey, D. (1972a). On obfuscation in geography; A comment on Gale's heterodoxy. *Geographical Analysis* 4: 323–330.

Harvey, D, (1972b). A commentary on the comments. *Antipode* 4 (2): 36–41.

Harvey, D. (1972c) Revolutionary and counter revolutionary theory in geography and the problem of ghetto formation. *Antipode* 6: 1–13.

Harvey, D. (1973). *Social Justice and the City*. London: Arnold.

Harvey, D. (1982). *The Limits to Capital*. Chicago: University of Chicago Press.

Harvey, D. (1996). *Justice, Nature and the Geography of Difference*. Oxford: Basil Blackwell.

Harvey, D. (2002). Memories and desires. In P. Gould and F. R. Pitts, eds, *Geographical Voice: Fourteen Autobiographical Essays*, pp. 149–188. Syracuse, NY: Syracuse University Press.

Harvey, D. and Chatterjee, L. (1974). Absolute rent and the structuring of space by governmental and financial institutions. *Antipode* 6: 22–36.

Heyman, R. (2007). "Who's going to man the factories and be the sexual slaves if we all get PhDs?" Democratizing knowledge production, pedagogy, and the Detroit Geographical Expedition and Institute. *Antipode* 39: 99–120.

Hepple, L.W. (1999). Socialist geography in England: J.F. Horrabin and a workers' economic and political geography. *Antipode* 31: 80–109.

Hitchens, C. (1998). The children of '68. *Vanity Fair* June, 92–103.

Horowitz, D. (1970). The fate of Midas. In A. Lothstein, ed., *All We are Saying… the Philosophy of the New Left*. New York: Capricorn.

Hudson, B. (1977). The new geography and the new imperialism: 1870–1918. *Antipode* 9 (2): 12–19.

Jarrett, R. and Wisner, B. (1969). How to build a slum, part one. *Antipode* 1: 37–42.

Jameson, F. (1984). Periodizing the 60s. *Social Text* 9/10: 178–209.

Johnston, R. and Sidaway, J. (2016). *Geography and Geographers: Anglo-American Human Geography Since 1945*. 7th edn. London: Routledge.

Katznelson, I. (1997). From the street to the lecture hall: the 1960s. *Daedalus* 126: 311–322.

Kay, L. E. (2000). *Who Wrote the Book of Life? A History of the Genetic Code*. Stanford: Stanford University Press.

Lacoste, Y. (1973). An illustration of geographical warfare: Bombing the dikes on the Red River, North Vietnam. *Antipode* 5 (1): 1–13.

Latour, B. (1987). *Science in Action: How to Follow Engineers and Scientists Around Society*. Cambridge, MA: Harvard University Press.

Marcus, G. (1989). *Lipstick Traces: A Secret History of the Twentieth Century*. Cambridge, MA: Harvard University Press.

Mathewson, K. and Stea, D. (2003). In memorium: James M. Blaut (1927–2000). *Annals of the Association of American Geographers* 93: 214–222.

Meadows, D., Meadows, D, Randers, J. and Behrens, W. (1972). *The Limits to Growth. A Report for the Club of Rome's Project on the Predicament of Mankind*. New York: Universe Books.

Merrifield, A. (1995). Situated knowledge through exploration: Reflections on Bunge's "Geographical Expeditions." *Antipode* 27: 49–70.

Olsson, G. (1975). *Birds in Egg*. Ann Arbor, MI: Michigan Geographical Publications, No. 15.

Peake, L. and Sheppard, E. (2014). The emergence of radical/critical geography in North America. *ACME: An International E-Journal for Critical Geographies* 13: 305–327.

Peet, R. (1977). The development of radical geography in the United States. In R. Peet, ed., *Radical Geography: Alternative Viewpoints on Contemporary Social Issues*, pp. 6–30. Chicago: Maaroufa Press.

Peet, R. (2002). Interview with Trevor Barnes. Worcester, MA, May.

Philo, C. (1998). Eclectic radical geographies: Revisiting the early *Antipodes*. www.eprints s gla.ac.uk/116527/ (last accessed February 11th, 2018).

Prince, H. (1971). Questions of social relevance. *Area* 3: 150–153.

Said, E. (1983). *The World, the Text and the Critic*. Cambridge, MA: Harvard University Press.

Smith, D. M. (1971). Radical geography: The next revolution? *Area* 3: 153–157.

Smith, N. (1989). Geography as a museum: Private history and conservative idealism in *The Nature of Geography*. *Annals of the Association of American Geographers* 79: 91–120.

Solvey, M. and Cravens, H. eds. (2012). *Cold War Social Science: Knowledge Production, Liberal Democracy, and Human Nature*. New York: Palgrave Macmillan.

Watts, M. (2001). 1968 and all that … *Progress in Human Geography* 25: 157–188.

Watts, M. (2012). Of bats, birds and mice. In C. Abrahamsson and M. Gren, eds, *GO: On the Geographies of Gunnar Olsson*, pp. 143–154. Farnham: Ashgate.

Wilson, B. and Jenkins, H. (1972). Symposium: Black perspectives on geography, Clark University, March 9–11, 1972. *Antipode* 4 (2): 42–43.

# Part I
## Radical Geography within North America

# 1

# Issues of "Race" and Early Radical Geography
## *Our Invisible Proponents*

### Audrey Kobayashi

No topic illustrates more cogently the question of "What is radical about radical geography?" than that of race. Radical geography began with common cause between the Civil Rights movement and the anti-war movement. One of the first overt and public acts was to move the annual meeting of the AAG from Chicago to Ann Arbor after the use of police force at the 1968 Democratic Convention, a geographical act that was worthy of the origins of radical geography. Haynes Johnson (2008) writes:

> The 1968 Chicago convention became a lacerating event, a distillation of a year of heartbreak, assassinations, riots and a breakdown in law and order that made it seem as if the country were coming apart. In its psychic impact, and its long-term political consequences, it eclipsed any other such convention in American history, destroying faith in politicians, in the political system, in the country and in its institutions. No one who was there, or who watched it on television, could escape the memory of what took place before their eyes.

The relocation of the annual meeting and attendant discussions mobilized concerned geographers to address questions of racism in relation to the discipline of geography, but the radical focus soon shifted to class. A few activist geographers of color, such as Thelma Glass in Alabama, were passionate – and radical – activists within the Civil Rights movement, but their work has been little recognized, much less

*Spatial Histories of Radical Geography: North America and Beyond*, First Edition.
Edited by Trevor J. Barnes and Eric Sheppard.
© 2019 John Wiley & Sons Ltd. Published 2019 by John Wiley & Sons Ltd.

theorized, in the literature. Nor, for the most part, were they strongly represented in the political movement that became radical geography. According to my informal interviews with those who were around at the time, many geographers participated in either civil rights or anti-war marches, or even joined the Students for a Democratic Society, but such activities were in the main considered personal (and political) rather than academic. In this chapter I attempt to recognize the work of such activists, including activists of color, who are seldom acknowledged when the term "radical geography" is invoked, and then to untangle the complex relationship between the concepts of radicalism and anti-racism. I show that, in established urban geography, the dominant issue has been the relationship between race and class, which has tended to leave many of the geographers of color out of the picture. Relatedly, early radical scholars who recognized the link between racism and colonialism either did not link, or were not seen to link, the structure of inner cities and global colonialism. Now, many years later, the work of anti-racism in geography is ongoing, and the social events that anti-racist geography addresses are as urgent and challenging as ever.

## Early Activism in the Discipline

The issue of "race" was high on the agenda of the AAG throughout the 1960s. In 1964–1965, President Gilbert White appointed Saul Cohen, then at Boston University, as the Executive Officer of the AAG. As the Civil Rights movement gained momentum, Cohen toured the historically Black universities of the south, and then applied to the National Science Foundation (NSF) for funding to enhance graduate training in geography with a view to enhancing – or in some cases establishing – Geography curricula in those colleges (AAG Diversity Task Force 2006; Waterman 2002). Cohen received a call from the NSF to say that the funding had been approved on the condition that the program be aimed at students from "small Southern colleges" rather than "black students." The NSF thus funded the Commission on Geography and Afro-America (COMGA), tasked with sending undergraduates from these colleges to Northern universities. Don Deskins, a faculty member in Geography at the University of Michigan, became the first COMGA Director in 1968. Eventually, 20 students (19 Black and one White) were recruited to undertake training at Clark University (Cohen, personal interview, March 12, 2014).

At a 1968 meeting of the AAG Executive Committee (prior to the annual meeting), plans were still being made for the meeting to take place in Chicago, but issues of racism were not absent (*The Professional Geographer* 1968). In addition to the COMGA activities:

> The committee talked about a new commission which would promote teacher training for Negros [sic]. ... The committee suggested that the name be changed to "Commission for Black American Geographers." The Council moved that the question of the name of the committee be deferred until the group has had a chance to assess its responsibilities and opportunities. (Minutes of the AAG Executive, 1968, AAG Archives B188F13)[1]

Alan Pred was appointed as the first Chair of this committee, titled the "Southern College Program," and a place was created in the Washington office for a committee administrator who would organize an inventory of geographers and geography, create a newsletter, and encourage exchanges and curriculum development. The agenda was forthrightly a liberal one, meant to bring a greater diversity of students into the academy, and especially to allow them to enter the public education system. Yet there is no mention of race-related research, or of a role for geographers to address the fundamental social conditions that the Civil Rights movement addressed.

Meanwhile, a series of meetings occurred in Michigan under Deskins' leadership that culminated in a "Proposal for the AAG session on 'The Status of Negroes in American Geography'" (AAG Archives, B188 F13) to recognize that "many of the AAG's current activities are pertinent to recruiting more Negroes into geography, to strengthening and expanding geographic activity in predominantly negro educational institutions, to increasing the interaction between black and white geographers, and to making geography more relevant to the pressing social problems of the nation." This group made a number of suggestions aimed at changing the institutional ethos of the AAG, including "personal commitment," also encouraging new approaches to teaching. Organization occurred using a mailing list containing 41 names – many of them known to this author – among whom 29 were white, male, early-career university professors, 7 were white, male graduate students, and 5 (two of whom were women) of unknown affiliation, who may have been undergraduate students.[2] An evening "open forum" was planned for the annual meeting in Washington in 1967, consisting of three paper presentations: "The geographic dimension of race relations – urgent research" (Richard Morrill); "Recruiting Black American Geographers" (Theodore Speigner); and "The Geographer in the Community" (William Bunge), with commentary from Don Deskins ("Washington Meeting: Special Session on 'The Status of Negroes in Geography'" AAG archives B188

F13). Ron Horvath, then at Michigan State, also presented a position paper in which he wrote:

> America is in crisis. Times of crisis are times for reassessment. It is appropriate that the geography profession reassess its position on the matter of race relations in America. Why? Because we have utterly failed to make any significant contribution to a solution to the racial dilemma facing America today. Most appalling is the obvious fact that we are a segregated community – only a fraction of *one* percent of our membership is Afro-American. Only two black professors were revealed by a survey of geography M.A.- and Ph.D.-granting institutions in this country. Only a handful of geography departments have even a single black major at either the graduate or undergraduate levels. (Horvath 1968: n.p.)

Horvath's data came from surveys conducted in Northern, predominantly white, geography departments with graduate programs, subsequently published (Horvath *et al.* 1969; Deskins and Speil 1971). The initial survey had found only two black geographers with Ph.Ds teaching in Northern established geography departments (Deskins and Harold Rose). Expanding the survey to include non-Ph.D. holders, they found a further 12 individuals teaching in departments with graduate programs in geography, a number that Deskins and Speil (1971: 284) describe as having risen "markedly." They also found that more than half of Southern, historically Black universities offered introductory geography courses, given by between 30 and 40 Black faculty members (few of whom held Ph.D. degrees), although only 12 Black students graduated with geography degrees in the entire U.S. in 1969.

Had Horvath's words been written today, we would be able to point to greater "diversity" in the discipline, but otherwise his observations still ring depressingly true. COMGA became defunct in the 1970s. In retrospect, it seems a half-hearted effort, a small liberal band-aid covering a major wound that could do next to nothing to staunch the existing racism inside and outside of geography. Subsequent surveys of the discipline showed only gradual change during the 1980s (Shrestha and Davis 1988, 1989). Interest sparked again within the AAG with the appointment in 2002 of the AAG Diversity Task Force, chaired by Joe Darden. Its "Final Report" (AAG Diversity Task Force 2006) reiterated many of the issues already identified in the 1960s reports, and called again for renewed attention to anti-racism in the discipline. Throughout those decades, however, anti-racist activism, anti-racist scholarship, and radical geography ran on parallel tracks. To understand this disconnected situation requires an examination not only of the content of radical and anti-racist scholarship, but also of some of the personal relationships that marked these events.

## Early Anti-Racist Scholarship

Until the late 1960s, anti-racist literature was virtually non-existent in the geographical literature. Don Deskins (1969) created a bibliography that found 43 publications plus 15 Master's and Ph.D. theses "on the American negro and racial issues" between 1949 and 1968, almost half of them published between 1964 and 1968.[3] He found that the majority of these were descriptive, many following the tradition of population geography. Only two geographers of color feature in the list: Deskins and Rose; and only one publication, by Richard Morrill, can be described as addressing the problem of racism as a social process. Morrill wrote:

> Inferiority in almost every conceivable material respect is the mark of the ghetto. But also, to the minority person, the ghetto implies a rejection, a stamp of inferiority, which stifles ambition and initiative. The very fact of residential segregation reinforces other forms of discrimination by preventing the normal contacts through which prejudice may be gradually overcome. Yet because the home and the neighborhood are so personal and intimate, housing will be the last and most difficult step in the struggle for equal rights. (Morrill 1965: 339)

For a time during the late 1960s, questions of racism, especially with respect to residential segregation, were debated in the pages of *Antipode* and among practitioners of radical geography. Certainly, segregation was one of the most significant issues addressed by urban geographers, who took their cue from the Chicago School emphasis on the spatial order of the city since the 1920s. But the influence of the Chicago School led to a fixation among geographers for quantifying the spatial order. Most of the radical geographers were becoming skeptical of analyses based on neoclassical economic theories, including Morrill's, however, and the discussions were passionate but short lived, as the issue of class eclipsed race. For two decades the pages of the geography journals were almost completely silent (even during the "relevance debates" of the 1970s) (for a review, see Kobayashi 2014b). For the likes of David Harvey and other Marxian geographers, revolution to achieve social justice was in order (as other chapters in this collection discuss), but social change "does not entail yet another empirical investigation of the social conditions in the ghettoes." Harvey viewed such work as "moral masturbation" on the part of "bleeding heart liberal[s]" (Harvey 1972: 10). Harvey saw spatial phenomena, including residential segregation, as a product of class, thereby shifting attention among radical geographers away from processes of racialization.

Certainly, Harvey did not mince words. His masculinist tone served to entrench the idea of radical geography as a project over which the boys could duke things out. The brash, virtually uncensored tone of *Antipode* in its early days was no doubt part of its appeal, but dissenting voices, or even alternative voices, were effectively silenced. A decade after *Antipode*'s founding there was very little discussion of race in its articles, and an almost complete shift to class. Jim Blaut was an exception, working along with Brazilian geographer Milton Santos to link racism and colonialism at a global scale. Yet this work had more purchase in development geography than it had in the circles of urban geography in North America, a point to which we return later.

Even as Harvey's position won the day, Bill Bunge's "geographical expeditions" became the first major attempt to undertake participatory action research in the inner city. Reading between the lines of the *Antipode* article, Harvey's words were a disdainful dismissal of Bunge's work in Fitzgerald, Detroit, and later in other cities, with a range of collaborators (Katz *et al.*, this volume). Indeed at that time, racialization was not well understood by geographers and almost all of the geographical scholarship involved mapping "race" as a demographic variable and discussions about the problem of residential segregation. There was virtually no engagement with African-American communities. But Bunge (perhaps for these reasons) did not confine his writing to the geographical journals. He wrote this line in a typically sardonic pitch in the African-American journal *Crisis*: "How backward of them [African Americans] for us not to have found them" (Bunge 1965: 495; cited in Donaldson 1969: 26). Relatedly, the anthropologist Hortense Powdermaker wrote that she had: "found the field of geography to be wholly inadequate in its treatment of the Black American. The Black man knows more about whites than they do about him" (Powdermaker 1968: 361). At the same time, as Warren, Katz, and Heynen (this volume) demonstrate, Bunge's approach never transcended its masculinist origins and its intellectual and social influence was limited.

A very small number of black geographers were writing anti-racist articles, the main ones being Harold Rose and Don Deskins, and later Joe Darden, making only limited connection with either the radical geography literature or, later on, the critical poststructuralist literature. Much later, during the 1990s, a rapprochement among various strands of radical geography would emphasize both the construction and the situation of anti-racism, creating openings for theoretically informed geographical scholarship on racism in North America. That ongoing anti-racist project could benefit from a consideration of some of the geographers of color who – if they were considered at all – were never seen as a part of radical geography. Yet, while their theoretical perspectives did not align with the dominant radical paradigm, their views of a society in which overcoming racism was possible were decidedly radical. Thelma Glass was one of those.

## Thelma Glass: Activist (1916–2012)

> When I looked at that bus as it passed my house and nobody was on it, it was a feeling of joy that will be with me forever, ... I had the idea that maybe we were finally going to be successful in getting everybody to cooperate. (Glass quoted in Bernstein, 2012: n.p.)

Thelma Glass, a geography professor at Alabama State University from 1947 until her retirement in the 1970s, was one of the original modern Civil Rights activists long before the radicalization of the discipline of Geography. In 1947, Glass had joined the Women's Political Council, a group of women with strong affiliation to Alabama State College. She was the secretary in 1955 when Rosa Parks was arrested for refusing to give up her seat on a bus, an act of defiance that she had initiated with the support of these women. The night Parks was arrested, December 1 1955, the group stayed up mimeographing 35,000 handbills calling for what became the Montgomery bus boycott, which lasted more than a year, until a Supreme Court decision (Browder vs Gayle) on December 20 1956 ruled segregated buses unconstitutional. Martin Luther King, Ralph Abernathy, and other prominent or soon-to-be-prominent civil rights leaders joined the bus boycott, but it was the Women's Political Council that did the organizing, and the "secretarial" work to keep it going. During that entire time, members of the group helped to transport boycotters around the city by organizing carpools (Robinson 1987), while remaining "invisible" to the wider public (Barnett 1993). Glass was highly regarded in her university and surrounding community; in a survey of Black women who stood out in the Civil Rights movement, Barnett ranked Glass as number 12.

Glass is best recognized for her activities during the 1950s, but they did not stop then. In 1987, she filed a civil rights case against the Montgomery Sherriff's Department for harassment involving a stop in which there was a dispute over the speed of a car in which she was a passenger, along with two other black senior citizens. She alleged that her "civil rights were violated by the manner in which we were approached by these law enforcement officers and that by the pointing of their guns and tone of their voices, that I was being threatened" (Airtel, SAC Mobile to Director FBI, FD 601, 4 May 1987; see also Craven, 2013).[4] The FBI mounted an investigation, which not surprisingly found the allegations to be unfounded, but Glass's protest was a significant expression of an issue of racial profiling and police violence that was only to increase well into the twenty-first century.

The likes of Thelma Glass are seldom mentioned in histories of the discipline or cited in geographical literature, but her role as an activist is well remembered by African Americans, and many are the students whose lives she influenced (George and Monk, 2004). What a radical idea! Rather than her scholarly words, the results of her actions survive to speak volumes about her radicalism. But she worked in a discipline segregated by both race and gender. She "supported" the likes of King and Abernathy, and was virtually ignored by geographers, who apparently failed to make the connection between her work and the project of changing the world.

## Harold Rose: The Flaming Liberal (1930–2016)

Harold Rose spent a career distinctly, even deliberately, apart from the project of radical geography; yet, if only in retrospect, his work articulates a broader definition of the term "radical" aimed at making an anti-racist difference. Rose's career began when the issue of residential segregation was very high on the geographical agenda, although for the most part, especially in such subfields as urban ecology, "race" was a variable to be mapped but not explained. Mainstream geography not only cringed from the prospect of going beyond the bounds of its maps, but also held firmly to the view that spatial science required a detached analysis of external patterns (Rose 1969).

Rose took up a teaching position at the Florida A&M University while completing his Ph.D. at the Ohio State University, then held a post-doctoral position in population studies at the University of Chicago (where he became well acquainted with Brian Berry), before accepting a tenure-track job at the University of Wisconsin, Milwaukee. He told me that he sought the position at the University of Chicago because he was worried that, as a black man, he had to do everything he could to enhance his recognized credentials. Urban geographers at the time drew very selectively from the theories of the Chicago School, ignoring the more humanistic interpretations that have influenced the tradition of symbolic interactionism. As Rose (1978) was to write several years later, a preoccupation with the concept of space had led geographers to ignore the lives of the people who actually occupied what geographers called "space." He disagreed profoundly with the radical geographers of the time, because he did not believe that their theories fit very well with the ways in which American cities were actually occupied. His work progressed over the 1970s from fairly descriptive accounts of emerging African-American communities to more and more passionate statements of anti-racism (Kobayashi, 2014a, 2014b).

Rose was elected AAG president in 1976, and gave his Presidential Address in 1977 (Rose 1978) on the topic of lethal violence against young black men. He seems ahead of his times in recognizing that "Black lives matter"; the major theme of his talk was that there are spatial patterns to violence and that understanding violence goes far beyond mapping the variables of residential segregation. He was adamant, also, in stressing that he was talking about the threat of violence to young black men, and not about some vaguely defined threat to white Americans. Rose was well aware of what he was doing. Within the first two paragraphs he used carefully chosen words to state that (1) he did not intend to dwell upon his own accomplishments; and (2) he did not intend to turn his Presidential Address into a "geography of happiness" when so many lived a geography of despair. At the same time, he bemoaned the lack of what he called "appropriate" demographic training to fit the topic. Notwithstanding his claim that he was not a theorist, he proceeded to debunk most of the extant theories on violence, including the "culture of violence" theory, suggesting that violence occurs in a context of despair created by the formation of ghettoes through segregation and impoverishment. His tone was radical, challenging both established ways of seeing geographically as well as the moral basis of choosing what geographers study, but his methods remained highly conventional. Still, it was another poke at the white, male establishment in the discipline.

It seemed that Rose received considerable respect for his position, but this had little impact on the discipline. I found a total of 24 citations (few for a Presidential Address), nearly all of them focused on the issue of racism. Just three of these used the words "race" or "racism" in their title, and it does not give me great pleasure to say that I authored all them in a direct attempt to situate and bring attention to Rose's work within the discipline (Kobayashi 2014a, 2014b, 2017). We might ask ourselves, therefore, why Rose's work did not have a greater disciplinary impact.

The first and possibly most important reason is that the discipline did not feel comfortable with anti-racist scholarship. In an extended personal interview in April, 2014, Rose expressed to me the up-hill battle he faced throughout his career:

> The atmosphere was pretty negative and ... I persevered ... I was able to do what I wanted to do. I was able to get a post-doctoral fellowship at the University of Chicago in 1960 ... I studied in the population research

center there for a year. I did that primarily to legitimate what I was doing. People saying, 'You don't have any credentials to allow you to do this kind of stuff. Where are your credentials?' So I got the post-doc simply to ward off naysayers and those who were opposed to the kinds of stuff I was doing. ...

... I remember I was giving a talk at the University of Kentucky once and the guy came to take me ... to give the talk. ... There was a little hotel near the campus of the university and ... he said [he was there to] pick-up Professor Rose. They said "that nigger upstairs?" [laughs] I don't know, I didn't know what to expect [at the] University of Kentucky. ... all in a day's work.

... things got better over time. But for the first ten years or so, well, maybe not first ... things began to open up a little bit around 1969, 70. Attitudes began to change a little. And as more people entered urban geography as a specialty, it was hard to overlook the race issue. Pressure against that at the local university was so strong. ... Somebody said I'm gonna walk [out of here] because it's not worth the hassle. Things began to change outside the university environment, then the pressure to change *inside* the university environment became more intense.

... Geographers did not address issues dealing with race in the United States. They might do this somewhere else but not in the United States. And so, but things began to change ... we started the COMGA program. ... What we wanted to do was create a cadre of young black PhDs who could strengthen geography in the black institutions in the southern United States. At least that was the attitude at the beginning. Much of that thinking ... represented the work of Saul Cohen. ... Don Deskins did the legwork but Saul initiated the idea.

... I don't know. Luck probably. ... So I helped the Department of Geography, you know. They knew that. But they were satisfied never to hire another black man. They had their black.

... So geography was very conservative here. I was a burning liberal ... yeah I was a burning liberal in geography in the middle of the road of urban studies. Operating in two different worlds. (Personal interview with Harold Rose, Milwaukee, April 2014)

Rose's life between the two worlds of academia and urban Milwaukee was emblematic of the times. When he received tenure during the 1970s, the first thing he did was to buy a new house in a middle-class suburb. His was the first Black family to move into that development. Within two years of their moving in, White flight had resulted in a complete turnover from an all-White to a predominantly Black neighborhood, an experience that was of immense importance in Rose's own research on the transformation of American cities. But he also related to me that Bill

Bunge was a regular visitor. Bunge's family was still living nearby in Milwaukee at the time; indeed, his father had owned the very land upon which the suburb was built before selling it for urban development. Bunge would arrive on foot and tap on the window of the living room, which also served as Rose's office. Over coffee, Bill would admonish Harold for not taking (what he considered) a more radical stand on questions of racism

This is much more than a simple story about "Bill and Harold." Rose appreciated Bunge's commitment and passion toward anti-racism, but did not share his disdain for the academy or for Rose's desire to live on what Bunge called "the plantation." Rose was committed to *a* radical view of geography that would allow his scholarship to be recognized on the same basis as that of his white colleagues. Unlike Bunge, he had not come from a privileged background that he could throw over as a matter of choice. He believed that the best way to have his say was to do so in a way that would garner established recognition. Of course, we have only Rose's recollections many years after the fact, but he spoke of Bunge with a chuckle in his voice and a twinkle in his eye, expressing a deep friendship for someone he regarded, but largely forgave, as a bad boy of the discipline. He spoke in a similar tone of the relationship between Bunge and certain members of the University of Chicago, as though he considered himself a friend and arbiter to all. But he maintained a firm opinion that Bunge was wrong in his belief that revolutionary social change was required to overcome racism. Much as he disparaged, and even resented, the inability of the disciplinary mainstream to address racism, he had faith that mainstream methods would do the best job.

Rose also told the story of how, after publishing several articles in good journals in the early 1970s (Rose 1970, 1972), he was called into the office of the Head of the Department and told that the university did not expect him to publish such articles, and that he should be more mainstream. Hence his "flaming liberal" comment: Rose was not a revolutionary, but certainly not a bleeding-heart liberal either.

Thus, in different ways, both Rose and Bunge lived their lives outside the mainstream. The discussions between these two men also represented a microcosm of larger discussions in the discipline about the geographer as activist that have only come to light in the geographical journals in much more recent years. Bunge always worked against the grain; Rose attempted to work with the grain but to change it. His understanding of activism was practical and involved committed citizenship: Serving on the commission to address school bussing (which he opposed), on the school board, and on a range of committees to overcome poverty in Milwaukee.

I remain deeply moved by the insight I have gained as a bystander in the relationship between Bill Bunge and Harold Rose, both of whom were deeply and passionately committed to social change, notwithstanding their political differences and different social backgrounds. I think that my analysis, like those of others, demonstrates both the terribly human side of all scholarship and the importance of passion and political commitment in scholarship. We can learn much by documenting those lived moments of scholarship that do not always make it into the pages of the scholarly journals.

The incorporation of anti-racism into the discipline of Geography remains very much an unfinished project. Recent events, such as the violence in North Carolina in the summer of 2017, are all too reminiscent of what occurred five decades ago during the Civil Rights movement. We might therefore ask what radical geography has contributed to the struggle. Yet there is no question concerning the contributions of these two highly charged, passionate, committed individuals. They will be remembered for their path-breaking work to revolutionize not only how we study geography, but also who studies geography: whose faces and ideas represent the discipline.

## Anti-Racism and Radicalism: A Re-Mapping

Long before geographers were inspired by the work of Robert Park in Chicago to map residential segregation, W.E.B. Du Bois began mapping the results of racialization in the city of Philadelphia. It is widely acknowledged that Dubois' *The Philadelphia Negro* (1967 [1899]) was the first in-depth field study of its kind. Dubois was already deeply committed to anti-racism and to addressing the conditions that arose in American cities as a result of the events of the Civil War, and later industrialization and the Great Migration He exerted a strong influence upon the many civic groups that attempted to address the development of slum conditions in Philadelphia and beyond, later becoming the founding President of the NAACP. Beyond this contribution, however, Dubois also recognized the tension between European and Black epistemologies, the sources of conflict therein, and the possibilities for political transformations of modern social conditions. Gilroy (1995: 112) calls this contrast a "politics of fulfillment and a politics of transfiguration," placing it between smug conservatism and radical scholarship. For Dubois, who was well connected with the pan-African movement, there were fundamental connections between conditions in the residentially segregated streets of Philadelphia and other American cities and the global project of colonialism (Dubois 2007[1940]). Yet this connection – despite its fundamental spatiality – went relatively unnoticed in the discipline of geography until quite recently (Wilson 2005).

Park's and Dubois' paths crossed in a variety of ways, particularly over their relationship with Dubois' reputedly more conservative nemesis Booker T. Washington (Dubois 2008 [1903]; Harlan 1998; Walters 2002), for whom Park briefly worked as a research assistant. Park postulated both a "moral order" and a spatial order in the contemporary city, claiming that while blacks were trapped by their low position on the moral order, other groups, including European immigrants, could move through the spatial order (inner city to suburb) to improve their conditions. Yet urban geographers turned to Park (1925, 1939), not Dubois, in order to understand the role of race in the modern city, espousing a particular understanding of Park's work. Notwithstanding Park's immense intellectual influence, particularly on the broader field of symbolic interactionism (Plummer 1991, 1997) geographers seized upon the idea of urban ecology (Entrikin 1980), rescued it from vestiges of environmentalism, and applied it to models of urban development that by the 1970s had become the cornerstone of the sub-discipline of urban geography (Peach 1975; Berry and Kasarda 1977).

It was against just such deterministic and rule-bound interpretations of spatial science and location theory that radical geographers reacted so strongly by the 1970s (see the debate between Harvey 1973 and Morrill 1973; Kobayashi 2014a, 2014b). Yet while radical geographers vilified the urban ecological work exemplified by Berry and Kasarda, they virtually ignored the more liberal approaches of the likes of Rose and Deskins. Indeed, among the acrimonious debates that characterized *Antipode* at that time, Berry (1972: 33) admonished that

> Harvey remains a product of the white European experience; he still has faith in logical rationalism, and views the central city negatively as a problem. Yet bubbling Black ethnocentrism, with an alternative twentieth-century core of theories of social change, provides an alternative value system that provides a fundamental basis for optimism.

Of course, Berry's comment can only be taken as ironic but interesting social commentary, and one cannot help but wonder just what he meant by "optimism": even an instrumental notion of social change does not seem to have permeated his or other urban geographers' approach to treating "race" as a variable rather than understanding racialization as a social process. But Berry's sentiments point to the question of how things *might* have transpired if the idea of anti-racism had not been suppressed by the idea of class in the first decade of North American radical geography, or if the analytical concept of double consciousness that Gilroy so perceptively identified in Dubois (and later in the work of Frantz Fanon) had been taken up in order to uproot – or at least unsettle – geographers' fixation upon spatial pattern. Ultimately, Berry and Kasarda's work is as assimilationist as any at the time, and there is no basis for believing that their project aligns with that of Dubois. Indeed, rather than enter the fray of social activism, they caution against instrumentalism of the sort that would result in social change on the grounds that it would be unscientific (Berry and Kasarda 1977: 429). But we need to look beyond urban geography to see the possibilities of anti-racism in early radical geography.

## James Blaut (1927–2000): Anti-Racist Visionary

Finally, I turn to James Blaut, the scholar who perhaps more than any other early radical geographer insisted upon the salience of racialization, which in his terms was coterminous with colonialism in the formation of the capitalist world. Blaut got the conversation started in the very first issue of *Antipode*, when he characterized the discipline as "full of jingoistic prejudice" (Blaut 1969: 10). He also was one of the first geographers to point out the symbiotic relationship between the discipline of geography and the larger colonial project. Blaut expanded about that relationship in many publications thereafter, sometimes addressing racism more explicitly.

Dubois also eventually received a brief but significant mention, when Blaut (2012 [1993]: 55) recognized the significance of colonial scholars striking back at the European metropole. It is ironic that Blaut should mention him in a list of international scholars, since Dubois was not from the developing world; he was born in rural Massachusetts (the first African-American to receive a Ph.D. from Harvard, in 1895). But then, as Blaut wrote, "No single geographer is wholly free from ethnocentric bias" (Blaut 1969: 10).

This little slip is perhaps not so insignificant. Blaut is most well-known for his work on colonialism and in a global context. He wrote convincingly of the modern, efficient, and powerful course of imperialism, both within the discipline and in the world. But his work was about racism far away, a product of the colonial development ethos. Indeed, he was one of the first geographers from a northern country to engage in activist work outside the northern context. The issue here is not so much, however, with what Blaut thought or wrote. In several pieces, most notably, "The theory of cultural racism" (1992) and his introduction to *The Colonizer's Model of the World* (2012 [1993]) he wrote specifically on the problems of racialized thinking. Moreover, in an article that deviates somewhat from the major themes of his work, Blaut (1983) actually links residential segregation in America to larger processes of colonialism, something that received very little attention at the time notwithstanding the vast numbers of articles by geographers on segregation. If Blaut was ahead of his contemporaries with respect to colonialism – and we could argue that even today his approach is neither well understood nor appreciated – he was so far ahead in terms of linking urban processes in the advanced capitalist city with colonialism that his thinking on this was hardly considered.

## Conclusion

Eric Sheppard (2015) reminds us of the many reasons why we should not treat geography as a club. Rather, we should think geographically and advance activism based on engaged pluralism, transcending both antagonism and exclusion. There were some important anti-racists among the early radical geographers, most notably Jim Blaut, Fred Donaldson, Ron Horvath, Richard Morrill, as well as Harold Rose and Don Deskins. But while their work was certainly not forgotten, it was pushed aside in short order as geographers went on to engage in both antagonism and exclusion when it came to incorporating anti-racism. Radical geographers boisterously and antagonistically challenged racialization, favoring class formation as an explanatory process. They also excluded feminist and anti-racist voices from discussions of urban geography, as much in response to the conservative approach of urban ecology as to any deliberate response to plural voices. They may have denied Robert Park but they ignored W.E.B. Dubois and other writers of the pan-African movement – scholars who belatedly have emerged as significant in anti-racist geography. They discounted or ignored local activism such as that engaged in by the likes of Thelma Glass. But they also engaged passionately, all too humanly, in a cause for which they believed only an engagement with the more brutal forms of capitalism could bring success.

It is certainly too simple to say that analyses of the process of racialization suffered at the expense of attention to class formation, although there is no question that the emphasis on class formation on the part of radical geographers led many away from "race" formation. The fact that analysis of residential segregation was firmly in the grasp of neoclassical economic geography until relatively recently did not help. In hindsight, we can see two things. First, with few exceptions the interests of the few geographers of color did not align well with either school. White radical geographers lost interest in racialization as a spatial phenomenon, as part of an overall turn away from the study of landscape and culture that marked humanistic geography in the 1970s (which, by the way, was even more White).

Black geographers had a diverse agenda. Thelma Glass's mission was to educate Black students and to bring about change on the ground. Her mission was fundamentally geographical, focused on place and relations in place, but she did not publish enough to make a dent on mainstream geographical thinking. Don Deskins wanted to change the face of the discipline, through efforts that have been part of a slow and very partial movement of diversification. Harold Rose turned his political sights not upon the discipline but upon the ways that geographers could contribute to public policy, particularly around such issues as school bussing, residential change, and racialized violence. Today, his work is as relevant – and geographical – as it ever was.

Jean-Paul Sartre, who benefited from many years of sometimes fraught collaboration with Frantz Fanon, once wrote that it was only once the colonized had migrated to the metropole, in the years after de-colonialization (at least officially), and with the creation of new concentrations of oppression such as those found in the *banlieu* (suburbs) of Paris, that scholars in the metropole began to come to terms with what Sartre called the "new racism" (Sartre 1972; Kobayashi and Boyle 2014). All racisms, from the suburbs of Paris, to the inner cities of Philadelphia or Chicago, to the streets of San Juan, Puerto Rico, where Jim Blaut was politically active, are the products of colonialism, capitalism, and the out-and-out hatreds that drive inequality and oppression, but Sartre's point should be well taken in our discipline. These disparate places represent very different geographies, and if it took us a long time to recognize their connections, and to give due credit as a result to early activists such as Thelma Glass, surely making those connections is one of the most important tasks of the geographer today.

In retrospect, geographers of the 1970s occupied distinct and seldom intersecting intellectual worlds. For the most part, Marxian scholars, humanists, and spatial scientists neither agreed with one another, nor overlapped (Kobayashi and Mackenzie 1989). Both development geographers and anti-racist scholars were beginning to emerge (although the latter did not really get going until the 1990s), but in general the racialization wrought by colonialism far away and the racialization of North American cities were not considered as part of a larger global context of racism. Jim Blaut and Bill Bunge did not write on the same page. Notwithstanding the brief florescence of discussions of racism in the first pages of *Antipode*, these issues rode on parallel tracks for some decades. Yet the passion of the early radical geographers has carried forward. We build upon their raucous voices and their silences, their passions about class and colonialism and as well as their blinkered understanding of racialization, positioning us to glance back, listening to the countervailing voices, radical in their own ways, whose perspectives have contributed, albeit many years later, to what we might hope to be a transformative vision.

That vision must surely include the most pressing issues that involve contemporary geographical manifestations of racialization: the effects of enslavement that have carried forward into entrenched poverty and seg- regation in North American cities; the racialized violence that marks ordinary life in cities; the impact of migration throughout the world in further entwining relations between former colonies and the metropole; and the pernicious effects of political relations in which elected and appointed public officials deny racialization and even engage in direct racist attacks upon their citizens (and others). Geographers today are more and more engaged with the anti-racist vision, but they have as much as ever to fill their plates.

## Notes

1   This excerpt comes from a carbon copy of Executive minutes in the AAG Archives, dated only with the year, 1968. The published Executive Committee and Council minutes for 1968 do not contain these lines (*The Professional Geographer* 1968). Curiously, a survey of the minutes of Executive and Council meetings throughout the 1960s and early 1970s (published regu- larly in summary form in *The Professional Geographer*) finds no mention of the events under discussion here.
2   I have tried to trace subsequent information about the five, but have been unable to do so.
3   This bibliography was originally handed out in mimeographed form at the Washington AAG Annual Meeting.
4   The FBI documentation on this case is publicly available as a result of a Freedom of Information Act request made by Shawn Musgrave of *Muckrock*. Available at: https://www.muckrock.com/foi/united-states-of-america-10/ thelma-glass-fbi-1862/ (accessed October 25, 2017)

## References

AAG Diversity Task Force (2006). Final Report: an Action Strategy for Geography Departments as Agents of Change. Available at: http://www.aag. org/galleries/default-file/diversityreport.pdf (accessed October 25, 2017).
Barnett, B. M. (1993). Invisible southern black women leaders in the Civil Rights Movement: The triple constraints of gender, race, and class. *Gender and Society* 7 (2): 162–182.

Bernstein, A. (2012). Thelma Glass, Ala. Teacher at the forefront of civil rights activism, dies at 96." *The Washington Post* 26 July. Available at: https://www. washingtonpost.com/national/thelma-glass-ala-teacher-at-the-forefront-of-civil-rights-activism-dies-at-96/2012/07/26/gJQA0VseCX_story.html?utm_term=.9b48d91ca749 (accessed October 25, 2017).

Berry, B. J. L and Kasarda, J. D. (1977). *Contemporary Urban Ecology*. New York: Macmillan. *Browder v. Gale*, 142 F. supp. 707. 1956.

Blaut, J. M. (1969). Jingo geography (part I). *Antipode* 1 (1): 10–15.

Blaut, J. M. (1983). Assimilation versus ghettoization. *Antipode* 15 (1): 35–41.

Blaut, J. M. (1992). The theory of cultural racism. *Antipode* 23: 289–299.

Blaut, J. M. (2012 [1993]). *The Colonizer's Model of the World: Geographical Diffusionism and Eurocentric History*. New York City: Guilford Press.

Bunge, W. (1965). Racism in Geography. *The Crisis* 72 (October): 494–449.

Craven, J. (2013). 'My civil rights were violated' Thelma Glass's FBI file. *Muckrock* 28 March. Available at: https://www.muckrock.com/news/archives/2013/mar/28/thelma-glass-police-harassment-allegation-investig/ (accessed October 25, 2017)

Deskins, D. R. Jr (1969). Geographical literature on the American Negro, 1949–1968: A bibliography. *The Professional Geographer* 21 (3): 145–149.

Deskins, D. R. and Speil, L. J. (1971). The status of Blacks in American geography: 1970. *The Professional Geographer*, 23 (4): 283–289.

Donaldson, F. (1969). Geography and the Black American: The white papers and the invisible man. *Antipode* 1 (1): 17–33.

Dubois, W. E. B. (2007) [1940]. *Dusk of Dawn: An Essay toward an Autobiography of a Race Concept*. Oxford: Oxford University Press.

Dubois, W. E. B. (1967) [1899]. *The Philadelphia Negro: A Social Study*. New York: Schocken Books.

Dubois, W. E. B. (2008) [1903]. *The Souls of Black Folk*. Project Gutenberg EBook available at: www.gutenberg.org/ebooks/408 (accessed November 22, 2018)

Entrikin, J. N. (1980). Robert Park's human ecology and human geography. *Annals of the AAG* 70 (1): 43–58.

George, S. and Monk, J. (2004). Teachers and their times: Thelma Glass and Juanita Gaston. In J.O. Wheeler and S. Brunn, eds, *The Role of the South in the Making of American Geography: Centennial of the AAG, 2004*, pp. 327–342. Columbia, MD: Bellwether Publishing.

Gilroy, P. (1995). *The Black Atlantic: Modernity and Double Consciousness*. Cambridge: Harvard University Press.

Harvey, D. (1972). Revolutionary and counter revolutionary theory in geography and the problem of ghetto formation. *Antipode* 4 (2): 1–13.

Harvey, D. (1973). A comment on Morrill's reply. *Antipode* 5 (2): 86–88.

Horvath, R. J. (1968). What can be done? Toward a Kerner Report for Geography. AAG Archives B 188, F13, 9.

Harlan, L. R. (1998). *Booker T. Washington in Perspective*. Jackson, MI: University Press of Mississippi.

Johnson, H. (2008). 1968 Democratic convention: The bosses strike back. *Smithsonian Magazine* August. Available at: https://www.smithsonianmag.com/history/1968-democratic-convention-931079/ (accessed October 25, 2017).

Kobayashi, A. (2014a). The dialectic of race and the discipline of geography (Presidential address). *Annals of the AAG* 104 (2): 338–347.

Kobayashi, A. (2014b). Neoclassical urban theory and the study of racism in geography. *Urban Geography* 35 (5): 645–656.

Kobayashi, A. (2017). Race and racialization. In D. Richardson, N. Castree, M. Goodchild, A. Kobayashi, W.-D. Liu, and R. Marsdon, eds, *The International Encyclopedia of Geography*, pp. 5515–5524. Oxford: Wiley Blackwell.

Kobayashi, A. and Boyle, M. (2014). Colonizing colonized: Sartre and Fanon. In A. Bakan and E. Dua, eds, *Theorizing Anti-Racism: Linkages in Marxism and Critical Race Theories*, pp. 184–204. Toronto, Buffalo and London: University of Toronto Press.

Kobayashi, A. and Mackenzie, S. (eds) (1989). *Remaking Human Geography*. London: Unwin Hyman.

Morrill, R. (1965). The Negro ghetto: Problems and alternatives. *Geographical Review* 55: 362–381.

Morrill, R. (1973). Socialism, private property, the ghetto and geographic theory. *Antipode* 5 (2): 84–85.

Park, R., Mackenzie, R. D. and Burgess, E. (1925). *The City: Suggestions for the Study of Human Nature in the Urban Environment*. Chicago: University of Chicago Press.

Park, R. and Thompson, E. T. (1939). *Race Relations and the Race Problem: A Definition and an Analysis*. Durham, NC: Duke University Press

Peach, C. (1975). *Urban Social Segregation*. London: Longman.

Plummer, K. (1991). *Symbolic Interactionism, Vols I & II*. Aldershot: Edward Elgar.

Plummer, K. (1997). *The Chicago School, Vols I, II, III, & IV*. London: Routledge.

Powdermaker, H. (1968). *After Freedom*. New York: Atheneum.

Robinson, J. A. G. and Garrow, D. J. (1987). *The Montgomery Bus Boycott and the Women Who Started It: The Memoir of Jo Ann Gibson Robinson*. Knoxville: University of Tennessee Press.

Rose, H. M. (1969). The origin and pattern of development of urban black social areas. *The Journal of Geography* 68: 327–332.

Rose H. M. (1970). The development of an urban subsystem: The case of the Negro Ghetto. *Annals of the Association of American Geographers* 60 (1): 1–17.

Rose, H. M. (1972). The spatial development of black residential subsystems. *Economic Geography* 48: 43–65.

Rose, H. M. (1978). The geography of despair. *Annals of the Association of American Geographers* 68 (4): 453–464.

Sartre, J.-P. (1972). Le Nouveau racism. *Le Nouvel Observateur*. December: 18–22.

Sheppard, E. (2015). Thinking geographically: Globalizing capitalism and beyond. *Annals of the Association of American Geographers* 105(6): 1113–1134.

Shrestha, N., and Davis, D., Jr. (1988). Status of minorities in geography: A national report. Washington, D.C.: Association of American Geographers.

Shrestha, N., and Davis, D., Jr. (1989). Minorities in geography: Some disturbing facts and policy measures. *The Professional Geographer* 41 (4): 410–421.

*The Professional Geographer* (1968). 1967–1968 council meeting (final), August 16–17, 1968, Washington D.C. *The Professional Geographer* 20 (6): 412–414.

Walters, R. (2002). *Du Bois and His Rivals*. Columbia, MO: University of Missouri Press, 2002.

Waterman, S. (2002). Scholar, manager, mensch: Saul B. Cohen. *Political Geography* 21 (5): 557–572.

Wilson, B. M. (2005). Race in commodity exchange and consumption: separate but equal. *Annals of the Association of American Geographers* 95 (3): 587–606.

# 2

# Myths, Cults, Memories, and Revisions in Radical Geographic History

## *Revisiting the Detroit Geographical Expedition and Institute*

Gwendolyn C. Warren, Cindi Katz, and Nik Heynen

John Henry was a steel driving man, or at least so goes the legend. This "tall tale," the only North American myth to feature the life of African Americans, is useful for setting up our chapter on the Detroit Geographical Expedition and Institute (DGEI). The power of myth to take on important political meaning while at the same time obscuring embodied historical geographies lurks everywhere. Too often, we are seduced by and abide mythological narratives because we do not know any better and lack the necessary insight to challenge them. Other times they offer us convenient connections between places and politics that may otherwise be opaque. And yet, just as often, the complicated histories unmasking myths would force us to run the risk of destabilizing traditions thought too precious to disrupt, regardless of the emotional violence they might sustain. Given the intellectual and material harm centuries of masculinist and racist geographic thinking has fostered, we feel even the most precious radical histories require our best collective efforts at setting stories straight or at least unbraiding some of their knots when opportunities emerge to do so. The return of Gwendolyn Warren, the former co-director of the DGEI, to the field of Geography

*Spatial Histories of Radical Geography: North America and Beyond*, First Edition.
Edited by Trevor J. Barnes and Eric Sheppard.
© 2019 John Wiley & Sons Ltd. Published 2019 by John Wiley & Sons Ltd.

offers us an opportunity to reflect on the DGEI and the legacy of a small Detroit neighborhood called Fitzgerald, which is often thought of as the epicenter of the DGEI.

*Steel Drivin' Man*, a book by historian Scott Reynolds Nelson, offers a path for unpacking politically powerful myths that have been constructed through racialized and gendered power relations. Nelson (2006) recounts the "true" story of John Henry, the black man, who was persecuted though Virginia's white supremacist "Black Codes" at a time when convict labor was coerced in the U.S. South during Reconstruction. John Henry was imprisoned in the infamous Richmond Penitentiary, and forced to labor on the mile-long Lewis Tunnel for the C&O railroad. After John Henry died, through the brutality of his capture, his story began to build until a song was written about him. The first printed score of the song was copyrighted by blues legend W.C. Handy, and it is now the most recorded song in U.S. musical history[1].

The mythic status of John Henry, when mobilized by Pete Seeger for instance, was used as a symbol for labor struggles across the U.S. Given the positive portrayal of his racialized might and power, so rarely valorized in mainstream U.S. culture, John Henry's strength and perseverance were mobilized symbolically in the freedom marches of the Civil Rights Movement. The phrase "how's your hammer hanging" took on political and cultural significance, and still has resonance for expressing solidarity in many African American communities. But this expression of political solidarity was also sexualized, so that John Henry's "hammer" accumulated layers of masculinist meaning as the myths about him traveled and evolved. This gendering of John Henry's "vitality" is especially important when juxtaposed with the entry of Polly Anne, "John Henry's woman," at the end of the song. Polly Anne introduces other important layers to the myth. Although Polly Anne is presented to be as powerful as her mate, she is rarely discussed, and has received scant attention across U.S. cultural history. The myths of John Henry, and relative absence of Polly Anne in these stories, offer guidance for unpacking narratives of the Detroit Geographical Expedition and Institute (DGEI).

Drawing on Barthes's (1972) discussion of "myth as depoliticized speech," Melissa Wright (2006) makes an argument that is productive for our resituating and examining the DGEI and Fitzgerald as we parse their mythical character in relation to how we can understand this moment with fresh eyes and new historical geographical insight. Wright (2006: 3–4) suggests, "myths are vehicles for foreclosing discussions of politics as they use fantastic characters and situations that depict hierarchical relationships broadly believed to have bearing on 'real life' without having to explain these relationships." About the power of history and myth more specifically, Barthes (1972: 142) said,

> What the world supplies to myth is an historical reality, defined, even if this
> goes back quite a while, by the way in which men [sic] have produced and
> used it; and what myth gives us in return is a *natural* image of this reality.

Given this understanding, it seems important to interrogate the myths
that shape our intellectual and political lives and give rise to larger-than-
life disciplinary figures.

Focusing on the DGEI, an extraordinary collective endeavor of
community research and education in the late 1960s and early 1970s,
we discuss the double movement of "mansplanation" within the history of radical geography. Our chapter shows how myths about radical
praxis can play tricks with history and geography, wherein some people
and places acquire cultish status while others are eclipsed with profound impacts on our understanding of the discipline *and* its community
engagements. As the history of the DGEI was recovered in the 1990s
and 2000s (see Barnes 1996; Katz 1996; Merrifield 1995; Heyman
2007; Heynen 2013; Barnes *et al.* 2011), the attention was largely
focused on William Bunge, "the man behind" the DGEI, and Fitzgerald
"the place behind" it, given that it was the title of Bunge's 1971 book
(also see Bunge 1974). The building of the DGEI and Fitzgerald
mythology occurred at the same time as the co-director of the DGEI,
Gwendolyn Warren, went largely unrecognized (see Warren and Katz
2014). Warren was mentioned in most accounts, but few sought her
out to hear her perspective, whether based on her role in the DGEI or
her long career as a working community organizer. Removed from the
field of geography, she did not realize that the DGEI had become one
of the more romanticized and inspiring episodes in the history of radical geography, with Bunge its charismatic leader. We intend to recover
part of the lost narrative of the DGEI through the auto-ethnographic
insights of Gwendolyn Warren, while at the same time thinking about
the role of race, class, and gender in the histories, narratives, and
futures of our field.[2]

The DGEI was a revolutionary project of geographical knowledge
production and exchange whose perils and potentials were fading from
memory by the 1980s, a forgotten alternative of how geography might
be done and learned. More accurately, its forgetting was willed – its history erased, at multiple layers, not just by a profession and university
apparatus that exiled one of its co-founders, but even by the Geography
left who saw it as a wildly utopian moment all the more impossible as
they/we got absorbed in our own professionalization and the accommodations it required. In other registers, the DGEI was left in the dust of
its "empirical" or "applied" concerns by the poststructuralist turn in
Geography with its interests more in theorizing difference than what is

common, which now captivates the intellectual and political imaginations of so many. In the course of the remembering and forgetting, the DGEI and many of its key participants were mythologized in Geography and beyond.

## Radical Geographic *HIS*tory

We will first focus on the basics of the popular, mythological, version of the DGEI and Fitzgerald as has been articulated within radical *HIStory*. As the story goes, the DGEI arose in the aftermath of the 1967 uprisings and brought together academics and neighborhood residents to engage in collaborative geographical research and produce a "pipeline" to higher education that was radical in every way. The DGEI was established in Fitzgerald, a neighborhood in northwest central Detroit. It aimed to reroute the school to street to factory pipeline into a school to street to university pipeline. That ambitious project really shook things up on the streets and in the university. As the DGEI took shape in Fitzgerald, Bunge claimed it was "not a nice geography" or a "status quo geography." It was rather a community-defined *practice* of Geography, which drew on the skills and tools of Geography with a twist. It was neither Geography's mid-century descriptive ideographic regionalism, nor was it a Geography of the "quantitative revolution." And while it took aboard some of the tools of empire – the expedition, the map, the survey – it turned them on their head so that the power of the "expedition" was in the hands of the people "explored," and what was explored came of experience and spurred consciousness of power's effects in place (see Merrifield 1995; Katz 1996).

This version of the DGEI's story is about the activist-intellectual work of a band of geographers with revolutionary aspirations who in the 1960s mapped rats and words and children's pain. Their productions and exchanges of knowledge were part of an effort to organize politically. Their work spurred geographical imaginations in action that mobilized memory as infrastructure and history as possibility. Recalling past geographies and revisiting their sedimentations in place were tools for building alternative geographies of everyday life and mapping future alternatives. The Detroit Geographical Expedition and Institute, as it was formally known, the "Institute" to mark its educational ambitions, was a collective practical engagement with the city that aspired to document, analyze, interrupt, and reroute the multiple trajectories of "slum formation" that were ongoing in the late 1960s as white flight and the decentralization (suburbanization) of the automobile industry, which began in the 1950s, accelerated.

The decentralization of the automotive industry was impelled in part by union activism and the successes of various strikes – wildcat and otherwise – and enabled by the automobile itself and the enormous infrastructure that supported it. The state-sponsored development of the interstate highways advanced sprawl, and directly and indirectly destroyed much of the urban fabric of the U.S. in general and Detroit's in particular. The shifting tax base to the suburbs further decimated the infrastructure and public services of Detroit.

The DGEI, which started in 1968, was a community-based collective of "folk geographers," academic geographers, young people, whether students – high school, undergraduate, and graduate – or school leavers, and residents. At its best, it was a community of practice that shared knowledge, skills, memories, and imagined futures. Its ambitious and exhilarating project was to work the grounds of one square mile of the city after the 1967 riots to understand the very fibers of its existence so that the political economic historical geographical processes engulfing it might be deflected, rejected, undone. As Bunge (1971) wrote in the recently republished keystone text (2011), *Fitzgerald: Geography of a Revolution*, "they" saw the Fitzgerald neighborhood as "America," as containing the story of the nation itself. Just as a drop of blood can tell the health of a patient so too could this mile square plot. "Their" overall intent was to produce an analysis of the spaces of nature, humankind, and machines at five scales – the neighborhood, the metropolis, the nation, the continent, and the planet.

The DGEI was co-directed by Wayne State geographer, William Bunge who with his family had taken up residence in Fitzgerald in the early 1960s, and a militant high school leaver from the neighborhood, Gwendolyn Warren. They were a formidable, unlikely, and somewhat unholy alliance; more unholy as their narrative evolved. Guided by community participants, mostly youthful, Expedition members mapped land use of the most elemental kind. They mapped their rage at the crumbling they saw around them, and trained and energized a phalanx of militant and well-informed geographers to do the same. Among the brilliantly mundane phenomena they documented and mapped in the vicinity were what a child touches, that is, what comprised the tactile surfaces of the neighborhood's open space. They asked basic and deeply political geographical questions about 'slum formation' reckoned in such things as "machine space" versus play space, traffic accidents involving children, childless spaces, hard boundaries like walls and for-tressed exteriors, the presence of rats, rat bites, broken glass, and dead trees, numbers of residential sales, racial and class composition of the residential and commercial areas, and the words that punctuated the landscape. All of which were mapped, charted, and analyzed.[3]

Their work was intensive and revealing. In one schoolyard, for instance, participants combed one square yard and counted 59 pieces of glass and other jagged objects at or near the surface, weighing 6.5 ounces. Multiplying these findings by the size of the schoolyard they made the startling revelation that it contained about 820,690 jagged objects weighing 2.8 tons; and this was not the worst yard in the neighborhood. Such spaces may well now be the grounds of the booming urban agriculture in the "vacancies" of Detroit. Landscape and landscape histories matter in countless ways though they are often unremarked. As the provocative epigraph/dare at the start of one of the DGEI's *Fieldnotes* publications put it: "...if you went down 12th Street or down Mack, or any such place, and you saw that street, what would we be able to read in that landscape that you couldn't?" Plenty as it turns out.

The project was one that revolved around knowledge-action, their meticulous field research was always already part of a strategy for taking a community-based stand against depredations. Depending where you stood, Fitzgerald was scripted and narrated as somewhere between a slum and a thriving community. The Expedition asked questions like whether and how the neighborhood might be regenerated *forever*. The DGEI did not see this as a utopian question. At the heart of its project was an educational imperative: it was the expedition *and* institute. The latter came more from the community while the former was spurred by the academics but was rerouted by community participants. The idea, now a cornerstone of Participatory Action Research, was that the community would do its own research and construct theories about the issues it faced in order to shape and inform practice. To that end, it developed an extension program in neighborhood education, initially through University of Michigan and then Michigan State, both of which were soon destroyed because of the program's deeply radical nature (see Horvath 1971). Among its features were that faculty worked through tithing their time; there was open enrollment for students who did as much community as classroom work; tuition was free or paid for through the tithed salaries of faculty; the program was administered by the community participants; and the programs were based on the campuses of the state's best universities with an insistence that those institutions' resources be available to community-based action researchers. Their model was the extension education programs for rural youth that were part of the land-grant university system, which included Michigan State University. Somehow, though, when extension education was provided for non-white radicalized urban youth it was not seen in the same way by the state (Horvath 1971).

The educational program drew on a case-study method of instruction, which created multilayered communities of practice (see Lave and

Wenger 1991). The first case focused on school desegregation and decentralization, and was drawn on by then State Senator and future Detroit Mayor Coleman Young to analyze redistricting possibilities. The DGEI's plan, which was derailed by those in power, would have redressed some of the most pressing equity and social justice issues facing the local black community. It did not appear coincidental to anyone that its development coincided with Michigan State's dismantling of the program in favor of accepting individual students to matriculate at the university. Also, not coincidental, at around the same time, Bunge was denied tenure at Wayne State. This denial and its aftermath have long been central to his mythical status; being oppressed by the system, while doing so much good for the community, made him into a kind of academic martyr.

Though the Institute was not without its problems, it was destroyed by, among other things, an academic administration that refused to forge, let alone nurture, any meaningful link between the intellectual resources of the state university system and the needs and interests of poor people of color in Detroit. We trace the historic contours of the DGEI to show how it sparked alternative radical and creative modes of geographic scholarship, but also how they were actively undermined and destroyed by those in power – university administrators, state and municipal representatives, and real estate investors, among others. Remembering the work (and failures) of the DGEI, and telling its *HIStory* has produced a political imaginary, a research agenda of community-based scholarship now deeply inscribed within the DNA of radical geography. The DGEI created a field of possibilities as much as possibilities for a field. Its work provided "exemplary suggestions," to use a phrase of Peter Linebaugh quoted in Kristin Ross's (2015) *Communal Luxury*, for thinking and acting that both framed a period and glimpsed another. Our recounting of this well-known *HIStory* is not meant to just frame and narrate this period of radical geography, but to open up possibilities for delving deeper into the myths of the DGEI as a cautionary tale for contemporary radical geographers.

The prevailing narrative of the DGEI reinvented the field of geography in many ways because it offered a new mode of inquiry that differed in radical ways from the two dominant strains in the field. One was the long-standing descriptive tradition that focused on regional and local variations of the earth's surface looking for common patterns of land use and the unique attributes of place. The other was a quantitative approach to explain spatial difference and structure, which arose in the 1950s in part as a scientific critique of the ideographic tradition that held sway through the middle of the century (see also Barnes and Sheppard, this volume). The so-called quantitative revolution sought laws that would explain spatial organization and variation, fastening, perhaps not surprisingly, on the role of distance in the location of various resources. It drew on increasingly sophisticated modeling techniques, along with spatial statistics that were later enabled by digital information science. The DGEI offered a radical break with these positivist traditions and interrupted the complacency of geographical inquiry to examine relational questions on the ground with no pretense of "objectivity" in the positivist sense, what Sandra Harding (1992) would later call, "weak objectivity." Theoretical in the deepest sense – that is in relation to the historical geographies of lived experience – the DGEI looked at the uneven geographical distribution of resources, the histories of land occupation and exploitation, and the everyday effects of racial capitalism in place and space.

As we hope we've made clear to this point, the *HIStorical* narrative of the DGEI is one in which its participants valued participatory research and collaborative writing all the way down. While there have been other narratives; stories that noted problems with the DGEI's division of labor and their mode of theory building, which retained some vestiges of positivism and a masculinist attitude (see, Katz 1996; Warren and Katz 2014), little of this thread was woven into the *HIStorical* record of the DGEI. All of this notwithstanding, the DGEI and its academic participants were instrumental in spurring the shift to radical Geography in the late 1960s associated with the founding of the Union of Socialist Geographers and *Antipode: A Radical Journal of Geography* in 1969, which blossomed into a thousand forms of radical and then "critical" geography over the years. But the myths of the DGEI are inevitably altered in the encounter with *HERstory*.

## Radical Geographic *HERstory*

The received narratives of the DGEI center on the neighborhood of Fitzgerald. This geography runs into some historical problems when we realize how little of the DGEI took place in this community, and that with the exception of Bunge, Frank Truesdale, and Gwendolyn Warren, no other community members from this neighborhood were involved in the DGEI. Because so little is known about how the Co-Director of the DGEI, Gwendolyn Warren, joined the project, we'll start with how she and Bunge met and began working together.

In 1966, the students of Detroit's Northern High School boycotted the 1966–1967 school year after the school newspaper was censored for noting both the school's inferior learning environment and that students experienced police harassment around the school and their bus stops. In Spring 1967 the students from Fitzgerald who participated in the boycott, calling themselves the Infernos, picketed the expanding sex industry and burlesque clubs in their neighborhood. As a result of this organizing, a group of teenagers from the school met with the Fitzgerald Community Council. Bunge and Warren first met at this meeting. Several weeks later Warren was in Esquire Delicatessen on the corner of 6 Mile Road and Livernois. Bunge happened to be there too, and approached her and attempted to strike up a conversation. Because she did not recognize him and thought he looked out of place, she did not talk to him and went about her business. A few days later Bunge went to Warren's house and asked her father's permission to speak to her.

Warren and Bunge sat on the front porch and he told her about his project, the Detroit Geographic Expedition (DGE). He said that he had graduate students working with him to map the history of human spaces in the city of Detroit. He explained that in order to understand better the natural flow of the community he needed a local interpreter, a guide, someone who could interpret what the community saw and experienced, especially through the eyes of its children. He seemed to know all about Warren and the student boycott. For Warren, the words Bunge used were exciting; change was possible, but the thing that stood out the most was the disapproval he expressed toward whites. It was ugly and much worse than the statements made by blacks about other blacks. At this initial meeting, and frequently thereafter, Bunge expressed himself in ways Warren didn't understand or appreciate, particularly the racialized values he assigned to different groups of people.

Bunge saw and represented himself as street smart. In what he seemed to think was a way of "acting black," Bunge would do things like call himself a "nigger," which was upsetting until Warren realized he did not understand the correct use of the word. As she put it, he thought it made him tough and soulful as opposed to stupid and dangerous. For him, whites – especially white men – were immoral, Jews were dishonest, and black women were better in bed than their white counterparts, because they could take pain and suffering in ways white women could not. There was always an underlying sexual and sexist stream that ran through his essentializing narratives about gender and race, inflecting his stereotypical ideas about how people interacted.

After several hours of discussion on the porch, Warren agreed to meet with Bunge and his students. On a number of occasions, Warren listened to Bunge talk about his project, the DGE, and why he moved to Fitzgerald to serve a more deserving people. She thought he was "crazy as hell." But eventually they joined forces, and traveled around Detroit meeting with city, union, and civil rights leaders primarily with Bunge asking for opportunities to serve as the research arm of the movement.

He would list what he thought were his radical credentials of trustworthiness, which included introducing Gwendolyn Warren and her work on the school strike, followed by showing the newspaper article he kept in his wallet that named him as a Communist.

Soon it was clear to Warren that the interests of the DGE were very different than her own, which included finding a job. It was a new experience to talk to whites who expressed disdain for their own people and seemed to truly prefer her culture. Bunge and friends talked about all kinds of things from white self-hatred to how their community could manage the outcomes of Warren's community's future. It seemed to Warren, however, that when Bunge and his students were done doing research in and around Fitzgerald they would go home, and reap the benefits of their work with little regard for the toll it took on her or the members of the communities they were trying to help. After listening a while, the fascination wore off as the picture became clearer. After making the rounds with Bunge, Warren saw no long-term benefit of spending more time with him, so she decided not to return to his home and project, and instead chose to spend more time with her friends and looking for other opportunities.

Several weeks into this hiatus Bunge dropped by Warren's house excited to tell her what he had been doing. He had been meeting with several of the Fitzgerald Community Council members and church leaders, and they had come up with a solution to assist with jobs and recreational outlets for the youth in the community. The Council would help sponsor youth dances if the young people in the community agreed to manage the activities by rules set by the church. Bunge went on to say that the proceeds would go to the youth community organization – the Infernos – to do what it liked until its members could find jobs or other sources of income. After Warren shared this news with the rest of the Infernos, everyone was excited and anxious to get started. The dances were overwhelmingly successful: the line to get in wrapped around the block, and everyone was on their best behavior. In about two months, the Infernos had saved over five thousand dollars.

With the help of Bunge and other adult members of the community, the Infernos sponsored the teen dances and eventually opened their own restaurant on Puritan Avenue. The restaurant was called the Infernoburger, which did not sit well with the Fitzgerald Community Council; they didn't think the incendiary sounding name was appropriate. Nevertheless, most of the skilled labor – electrical, plumbing, and finished carpentry – was provided by the parents of their middle class, largely white friends who lived in Marygrove, an area in the northern part of Fitzgerald. Warren and her friends referred to these kids, whom they'd hung out with for a long time, as the Marygrove Gang.

While most of the start-up labor came from adults in the community, the kitchen equipment – stove, burners, refrigerators, tables, and chairs – were paid for by the Infernos from the dance proceeds. Once the restaurant was opened, all of the kids north and south would go there. Warren recounted, "We sold hamburgers and fries to those who could afford to buy them, mostly adults who came to observe, and we gave away the rest free to any kid that asked. You would have thought by our sense of ownership and pride that we were running and managing a major business enterprise. The restaurant was closed twice a day for 45 minutes for cleaning. Members as well as kids who ate there for free would clean the restaurant thoroughly, burners, floors, bathrooms, windows, everything."

The Infernos received citywide exposure for their youth-coordinated dances and restaurant. As Warren recalled, "I believe we had at least two grand openings, and a series of special events." By keeping the menu simply hamburgers, cheeseburgers, fries, drinks, and assorted candy, the operational cost was low. All dance proceeds went to support the restaurant; various supporters came by to assist with the effort. The *Detroit Free Press* covered these activities in a positive and upbeat manner, which encouraged people to come by the restaurant and see for themselves. Attendance at the dances had reached capacity with overflow crowds, and "we even had a small savings set aside, which we kept in the refrigerator next to the fries."

About four months after the restaurant opened all hell broke out. The local police had determined that the Infernos were a violent youth gang and that every crime in the community was committed by them. They were accused of everything from rape, burglary, grand theft, and strong-arm robbery; you name it, and they did it according to the police. The guys were constantly being harassed, followed, arrested, and roughed up. The police would come into the restaurant, gun in hand, threatening to kill anyone if they moved. The community stepped up and demanded that the harassment stop, but it didn't. It intensified. With the closure of the restaurant and the conclusion of the dances, the relationship between the DGE and the Infernos came to end. Of its members, only Warren and Frank Truesdale maintained a relationship with the evolving DGE.

The tensions between young people in the community and the police escalated. One day, Warren was walking along on the way home and was stopped by the police. It was dusk dark and no one was around. After a few words one of the officers got out of the car and poked her hard in the stomach with his billy club. She fell to the ground and thought she was about to be killed right then and there. After the cops left, she lay on the ground crying in pain, shaking in fear, and trying to catch her breath. Laying there she thought about the guys; if they had been here someone would have died. Things between the police and the community were tense in ways that have sobering resonance with the current climate more than 50 years later.

During the summer of 1967 the Detroit race riots broke out. They were one of the most violent urban revolts on U.S. soil during the twentieth century. The uprisings were sparked after Detroit Police Vice Squad officers raided an afterhours "blind pig" (an unlicensed bar) on the corner of 12th Street and Clairmount Avenue in the center of the city's poorest black neighborhood. There was a party going on at the bar celebrating the return from Vietnam of two black servicemen. Officers had expected a few patrons would be inside, but when they came upon the party they arrested all 82 people attending. Shortly after the arrests started, an empty bottle was thrown into the rear window of a police car by crowds gathered outside. By mid-morning, looting and window-smashing spread out along 12th Street. Police officers began to report injuries from stones, bottles, and other objects that were thrown at them. And that was just the beginning.

Fires started, spreading rapidly in the afternoon heat. By the end of the second day, fires and looting were reported citywide. Michigan Governor George Romney ordered 800 State Police Officers and 8,000 National Guardsmen to address the situation. They were later augmented by 4,700 paratroopers from the 82nd Airborne Division, ordered there by President Lyndon Johnson. In the five days and nights of violence, 33 blacks and 10 whites were killed, 1,189 were injured, and over 7,200 people were arrested. The first death report during the riot was a 16-year-old black youth.

The uprising represented a turning point for most city residents. White flight doubled to 40,000 in 1967, and doubled again in 1968. Construction of the city's freeways, newer housing, and threats of further integration due to the demolition of the city's two main black neighborhoods, Black Bottom and Paradise Valley for 'urban renewal,' caused many whites to leave for the suburbs. Virginia Park, a predominantly Jewish neighborhood, rapidly became a predominantly black neighborhood by 1967. Twelfth Street became the center of black retail.

In *Fitzgerald* Bunge argues that the difference in intensity between the riot on 12th Street and in Fitzgerald could be attributed to the community's efforts to create a "good community," suggesting that years of caring made a difference. He went on to say that their implementation of "anti-slum" techniques saved the day. Based on Warren's experience in the community, in congratulating Fitzgerald's less extreme experience of the riots, Bunge failed to explain why of the 34 stores that were looted, all but seven were in the poor sections of Fitzgerald, or to note that 12th

Street was the heart of Detroit's black community where the uprisings might be expected to be more intense.

A different interpretation of the riot and its effects through *HERstory* can be found in *Field Notes No. 3: The Geography of the Children of Detroit*, in a discussion paper by Frank Truesdale, Roderick Shepherd, Sharon Evans and Debra Hampton, "Everybody was Eating Back Then." According to this account of the uprisings, "Some people had a backyard full of furniture; they would put a couch on top of their cars and ride down the street holding them by hand. A&P on Puritan (a local grocery store) was looted by families along San Juan Drive, Prairie, and Tuller Streets. Living rooms were covered with food." Their account suggests that for many people, "looting" enabled "everybody to eat" better.

Bunge did not discuss the fact that no individual homes were targeted in the riot or that the property of the church and the wealthy (white), including in Fitzgerald, were protected by federal troops with tanks. The poor section of Fitzgerald was actively involved in the riots and looting. It was common to enter homes in this part of the community and find the floors covered with items taken during the looting. Food, clothing, shoes, and household goods were frequently on display. There was also a concerted effort by many to retrieve loan paperwork that was on file at finance companies, and personal items located in the local pawn-shops. The items taken by residents were dictated by individual family needs. Families with small children took diapers and baby food; others just took food – mostly canned goods – some stole shoes to wear. At the time, there was a sense that the seven shops that were located in the northern wealthier section of Fitzgerald were looted by their white business owners for the insurance. Most of the businesses in that section that were damaged never reopened again.

In the same way that Bunge declared himself a "nigger" because he stayed in a hotel in the black community for three weeks while partic-ipating in the Civil Rights Movement in Chicago, he declared himself successful with his Fitzgerald social experiment without asking the residents how they felt or even whether they had participated in the riots. He regularly suggested his dismissal from Wayne State University had to do with being a "nigger lover," versus the version of the story that folks of the time will recall. Bunge had asked black male students in his class if a white female student, who was reluctant to go in the field out of fear, "was rape-able." When he was dismissed for his behavior, he declared that "nigger lovers" made poor university pro-fessors, claiming that he was "a nigger" from Fitzgerald. Tellingly, he did not appear to recognize the racism or sexism of his remarks in the classroom.

As a teenager, Warren had limited exposure to white people other than those in positions of authority like teachers, police, and shop owners. With the exception of a few character roles, everyone on TV was also white. Bunge was the first white male adult with whom she had extended communication. Several weeks after the riots, Bunge came to her house again, this time to ask her to attend a meeting on school decentralization with him at State Senator Coleman Young's office. As this was one of Warren's political concerns she was immediately interested and said that she would attend this time, but noted again that she did not want to participate in the DGE.

During the meeting with Senator Young, Bunge, along with community leaders and union organizers, argued that the community should have significant input and control of the newly proposed School Decentralization Plan. Bunge argued that their group was the only community-based effort that would be represented by public school students, local unions, community associations, and university faculty dedicated to political empowerment and community involvement. He noted that the state's land-grant university system was based on the model of community involvement and research, and the educational community had historically worked with communities that lacked the skills to solve issues of this magnitude and importance.

After leaving the meeting, Warren asked more about land-grant universities and the development of a skills exchange bank. She said she would help Bunge, but only if he helped her and her friends. The Infernos needed income and learning opportunities that didn't include the military or a factory. After having spent time with Bunge's students talking about their plans for the future, Warren felt confident that young people in the community would be successful if given the opportunity to attend college. They were intelligent and motivated but lacked access to higher education. Warren and Bunge discussed what would have to happen for her to commit to working with him on an expanded DGE (key would be adding an educational component, which became the Institute), and to feel comfortable working with him as a community organizer. She agreed to work with the community to ensure the highest level of participation and he agreed to provide the technical and academic support necessary to achieve the desired outcome, which was community control of all research.

Additionally, she demanded that the project could not be located in Fitzgerald. It had to be in an environment where the whole community might have greater control and access. Warren felt that to be truly successful, the community needed more comprehensive access to the university system not just attention from a small group of Bunge's graduate students. There was a community sentiment that the graduate students could help study the problem, but they could not lead the fixing of the problem, because it was not theirs to fix. Bunge agreed to these terms. He also seemed to understand that he was becoming a distraction, and that he needed to both reduce his visibility in the community and assume a more traditional role in this effort. He could no longer be a "brother from another mother"; he had to be a real academic and "act white" since he very clearly was.

During the summer of 1969 with the assistance of a newly recruited citywide team the principles of the DGEI were established. Bunge wanted to do community research and to be an agent of change, but he was growing more unacceptable to the black community. Warren wanted to make a change in her community, but she didn't have the requisite skills to make a real impact. Bunge understood the world of academia and Warren understood what motivated members of her community. They agreed to expand the DGE (Detroit Geographical Expedition) by the creation of the Institute, which necessitated a name change for the organization to the DGEI. Bunge was to manage the research component of the Expedition until a suitable replacement could be trained, and Warren would run the Institute. They agreed on core management principles for their combined efforts (see Horvath 1971):

1  Any Detroit resident could participate in the project as long as their contribution was productive.
2  Every participant would receive a full scholarship and be eligible for college credit if they met the requirements for the successful completion of each class taken. And with the completion of 45 college credits with a grade of C or better they would be eligible for admission to Michigan State University as a sophomore, regardless of their educational background.
3  Community control of all research projects that were to be conducted in the community's name.
4  Research projects would use only a case study form of instruction, and the projects would be selected and approved by the students or community. Relevance of cause was critical for people confronting skill limitations and/or poor educational experiences. The problems addressed would have to be more important to the individuals participating than their fear of past negative experiences.
5  They required the best facilities and staff; the resources used had to dignify the seriousness of the issue at hand. Students were required to challenge themselves and perform at a level of excellence necessitated by the project, which included challenging any skills deficits. The facilities and staff should mirror the community's commitment and only bring their best effort.

The division of labor between Bunge and Warren worked well. Ron Horvath, Edward Vander Velde, and Charles Ipcar took the lead on everything related to faculty recruitment from Michigan State University (MSU) and other state universities. Their commitments included playing a critical role in managing Bunge's interactions with the students and with Warren (see Horvath 1971, 2016). Horvath's involvement in the project was crucial to its success. Warren recruited students who were committed and energized, and Horvath recruited faculty from various universities who served with excellence. The DGEI team that participated in the development of the program's principles and operational guidelines for the Institute were folks Warren met during the Northern High School walkout and other community activities. Frank Truesdale was the only other Inferno who participated in the DGEI. Everyone else joined the project after the Institute was conceived and it became operational.

The agreement that the faculty tithe their time to teach meant only those committed to change and excellence volunteered. The faculty members selected were as committed to the greater good as anyone involved, and quickly adapted to the needs of their students. They were special, and it showed. They met the students where they were, sometimes hungry, scared, and non-believers. But the faculty kept coming, sometimes acting as faculty recruiters themselves. Program offerings went from one class with 40 students in the summer of 1969 to 31 classes and over 500 students by the spring of 1970, with 50 new classes and instructors ready to begin in the fall. These numbers represent classes taught by MSU faculty only. Some of the students who enrolled were high school graduates taking advantage of free tuition in order to accumulate college credit toward their degree. Others enrolled because of their interest in community change. Several thousand city residents attended classes and orientations of the various research projects. And once the course requirements were met, almost a hundred students enrolled in other colleges or universities in the state.

The student recruitment effort was extremely successful. Students came from every corner of the community – factory workers, prostitutes, drug dealers, and grandmothers and grandfathers – they all came. The opportunity to share what they knew while simultaneously controlling and managing the intended outcomes was addictive. Giving the students the resources to rediscover their strength and find their moments of greatness was profound by itself; no additional accomplishments needed to be achieved to experience the sense of individual success. But there were abundant other accomplishments.

During the winter of 1969 the DGEI was awarded the distinct honor of working on the new school decentralization plan for the city of Detroit.

Collecting data, making maps, and writing the report represented its first real success as a research and educational program. The subject matter and case-study method approach to the project created a firestorm in the classroom. The final exam for two of the geography classes and an English class was based on material relating to the presentation and acceptance of the report by the community.

Like the rest of the students involved in the DGEI, Warren suffered from negative educational and life experience. By the time she left the Detroit Public School System at 16 years old, her family had moved 21 times and she had attended over 12 different schools. Her basic skills had suffered greatly, but her mind was nimble and agile, and, like most 18-year-olds, her energy was boundless. She enrolled in classes at the University of Michigan Extension Program located on Wayne State's campus in Detroit while at the same time also recruiting students and interacting with the community leadership on the status of the study. It was difficult for her to take classes with the rest of the students, while maintaining control of the student body and acting as the DGEI's co-director. Her first semester at Wayne State included classes in social stratification, statistical methods, and technical field research. The faculty teaching these classes were not aware of the agreement with Michigan State University, nor were they participating faculty in the Institute.

Every extra moment was spent reading and re-reading the class textbooks and related material. Warren transferred 68 semester units from the University of Michigan Extension Program to Michigan State University at the end of her first academic year. She received straight A's with the exception of a geography class, where she received a B. Go figure! The winter of 1970, she was accepted as a full-time student at Michigan State University, graduating with high honors in the winter of 1971. Warren was accepted in graduate school at Michigan State University in the spring of that year.

Once she moved to East Lansing to attend Michigan State, Warren rarely saw Bunge unless they had to discuss DGEI business. Her concerns about him continued however. For instance, while in San Francisco attending an AAG annual meeting in 1970, Bunge was interviewed by the *San Francisco Chronicle*, and said, "The world is divided into spaces – Homosexual spaces and Heterosexual spaces. The heterosexual spaces are where you have parks and schools, and hospitals, and homosexual spaces are where you have prisons, armies and jails." Warren's distrust of Bunge grew after reading the interview. While his politics of exclusion, misogyny, and racism were masked by the good things he made possible through the DGEI, they were made visceral in statements such as these, and Warren found herself wanting to form an all-women's army to help the children of Detroit (see Warren 1971: 8). The *HIStory* of the DGEI has failed to capture these tensions and the difficulties such sentiments and ideas created for many Bunge claimed to be working to help.

Bunge's biases would go on to create long-term problems for Warren. While at school in East Lansing, she received a call from a close friend of both Bunge and herself whom she had met while working in Detroit. Warren was happy to hear from him because he had never telephoned her before, and she missed talking to him. Warren asked how he was doing. He said well but he had something he thought she should know. He went on to say that Bunge was telling anyone who would listen that she was a lesbian and no longer capable of leading the project because of it. Being outed in this way after all of her hard work and dedication created immense feelings of hurt, betrayal, and disappointment for Gwendolyn Warren. She was disappointed not because Bunge thought she was a lesbian, but because he was attempting to hurt and undermine her. Until that point, she had never felt rejected or disapproved of by him, but this was very different. As a young woman she couldn't make sense of it until much later in life. Although while she was living in California a few years later, Warren received a letter from Bunge that attempted to explain his behavior, they never spoke again.

The last project that the original DGEI project staff participated in was *Field Notes Discussion Paper No. 3 "The Geography of the Children of Detroit."* Most of it was transcribed from earlier recordings and for the most part was not edited prior to being printed. Needless to say, the participants were disappointed, not in the content – which remains compelling and original – but the presentation. Its flawed production suggests the ways that the DGEI had been stretched thin by 1971, and in retrospect signaled the beginning of the end of the project. Beyond the research activities, managing the needs of the new students that were accepted each semester by the university fell primarily to Horvath and Warren, who were at the same time receiving a lot of pushback from Michigan State.

One night after a series of meetings regarding the future of the DGEI on campus, Warren received a home visit from Dr. Robert Green, Dean of Urban Affairs; Joseph McMillian, Director of the Equal Opportunity Program; Lloyd Cofer, Director of Special Services; and Thomas Gunnings, Assistant Director of Minority Counseling Services. All were high-ranking minority administrators on campus. They decided to stop by Warren's house and have a "brother to sister" private talk with her about the DGEI. At first Warren was taken aback by their presence at the door, and just stared at them while trying to gather her thoughts. After she invited them into the house, Dr. Green said that they had come to her home because they needed to talk by themselves without the others being around, "because this was Black folks" business. The administrators explained in some detail how long they had worked to build and expand their role and relationship with the University. And now with the hiring of its first Black president, Clifton Wharton Jr., things had to be perfect or we would never have an opportunity like this again. Talking with these four Black administrators about the appearance and communication skills of other Blacks took Warren back to a place she didn't like. The administrators inferred that they were embarrassed by the presence of so many students from Detroit. It seemed that the students enrolled in the university through the aegis of the DGEI were represented to the University as Detroit's best. But the administrators did not see it that way. They wanted what they believed to be the true representatives of the Black community at the University, not the people they perceived to be its burnouts. They thought it was important to compete grade for grade with the white community by recruiting merit scholars not factory workers.

The group of administrators indicated that they wanted Warren to work with them, and offered to help her with a graduate assistantship or a job. It seemed that they were fine with a few students from the inner city enrolling in the university, but not the hundreds that came as a result of DGEI. One of the administrators stated that he believed the program was operating as a front for white left-leaning faculty, and that the project couldn't be community controlled because Warren lived on campus and no longer in Detroit. Warren tried to defend the work of the DGEI by talking about the political impact of the project in Detroit. How by bringing the students to school as a group it avoided class separation and enhanced their power base in Detroit. She made clear that attending the University collectively made it easier for these students to adjust to the cultural environment in this new community. But she was told that she sounded like an "emotional baboon." The administrators suggested that Warren had been brainwashed by white people. They said they would not support additional resources for the group, not even if the faculty worked for free.

The next day Robert Ward Jr., another DGEI team leader, and Warren went to see the new university president at his office. To their surprise, they were granted an appointment right away. President Wharton was very gracious and welcoming. He was aware of the issues, and listened patiently as Warren and Ward explained their concerns. He inquired about their experience at the University and what he could do to improve things. In retrospect, Warren and Ward concluded that he managed them like kids. They left the meeting honored to have met him, but with no commitment on his part to change a thing.

Robert Green, Director of the Center for Urban Affairs, and his staff continued to insist that things change or no additional resources would be committed to the project. The major problem came from the University's insistence that students enrolled in classes held in Detroit, rather than the main campus in East Lansing, pay full tuition rates. The DGEI had free instructors who turned their salaries back to the University to pay for their students' tuition, and all administrative tasks were performed by DGEI project staff. They only required that the University take care of recordkeeping and grade processing. Even the space for classes was donated by universities and secondary schools in Detroit, but Michigan State University (MSU) wanted full tuition rates regardless (Horvath 1971).

During the 1970 fall semester several student demonstrations were held[4] on the MSU campus in support of the DGEI (Figure 2.1).

## DGEI rally

The Detroit Geographical Expedition and Institute stages a rally at the Administration Bldg. in October to protest the cutting of funds for the program that allows inner city students to take courses in Detroit that prepare them for a chance at a college education.

**Figure 2.1**  Rally protesting cuts to the Detroit Geographical Expedition and Institute. Gwendolyn Warren with bullhorn. *Source:* © *The State News*, Michigan State University. Reproduced with permission.

Supporters from Detroit and other student organizations participated in the demonstrations. Rubin Barrera and Richard Santos from MECHA (Movimiento Estudiantil Chicano de Aztlán), the Chicano student organization, also spoke at the rallies to support the DGEI. Barrera said during his presentation, "The University must accept the responsibility of going into the barrio, the ghetto, and the reservation," adding, "The people is where the problem is." "Human needs come first," Santos said (Saddler 1970). Carolyn Ramsey, a representative from the League of Revolutionary Black Workers, said the League supported the DGEI 100%, and added that they would bring all their supporters from Detroit if necessary. The University minority administrators continued to express concerns regarding the lack of support from the other universities, noting that the program had grown larger than they anticipated (Saddler 1970).

On October 11, 1970, the University issued a final Position Statement on the DGEI. The statement was issued by Dr. Robert Green and his staff at the Center for Urban Affairs. The statement addressed "control" of the program as a concern, as well as the competing interests and needs of other minority students and programs. He went on to talk about Michigan State University's coordinated effort to recruit and admit minority students nationwide. Further, they believed that the DGEI and its representatives had misrepresented the University's commitment to the program both orally and in writing, and that it was their responsibility to distinguish between "experimental" programs like the DGEI and carefully conceived long-range minority programs with proven success. Their final opinion was that

> the DGEI should not move past the experimental stage until evidence of long-term success and benefit could be documented and that no expansion of the program should be made before success was demonstrated.

Dr. Green went on to say that

> most experimental efforts lead to frustration and confusion if they make commitments that cannot be honored. The Black community is well aware of claims made by organizations and institutions for which no delivery can be made. We believe that the difficult struggle in which minority people are engaged to improve their general status is not facilitated by expedient and piecemeal efforts.

During the last meeting with the University, the administration presented their final terms for future program offerings. No tithed faculty would be allowed, the University would handle all administrative activities, financial support for the program would not be guaranteed, and money for ten classes would be allowed only for the next three-quarters (not the 50 classes the Institute had lined up) (Horvath 1971). As Ron Horvath (1971: 84) stated, "It is ironic that a place like MSU, where the concept of the land grant college was pioneered (although on the idea of service to rural Michigan), rejected a program based on the same ideas when it came to servicing the poor of urban Michigan."

The DGEI program located at the East Lansing campus was under increased attack. Non-tenured faculty members were harassed, and their jobs were threatened. On several occasions, tenure was denied, and graduate assistantships were rescinded. The DGEI staff was harassed, and one professor was even charged with stealing food from the faculty lounge to feed DGEI students. She was later denied tenure. In the meantime, Warren had been accepted to graduate school at Michigan State University, Department of Sociology and Howard University School of Law, starting that winter and had to decide between them.

Horvath and Warren called a group meeting of faculty and students to discuss the status of the program with the University. They had over 450 students who were taking classes, and some were waiting for final acceptance to the University. After talking over the situation with the group, Warren decided to prioritize the new students that were coming to campus and avoid any new conversations or conflicts with the University until the acceptance letters went out. She decided it was best to stop the DGEI once that occurred, because they could not sustain the fight politically. After all the students who had been accepted were registered for the fall, Horvath and Warren met with the University representatives to let them know their decision. The administrators seemed surprised and taken aback. Horvath and Warren thanked them for their support and left the meeting. Warren spoke later with President Wharton to let him know that it was an honor meeting him and attending his school. Wharton took Warren's hand and said the honor was his, and that he was proud of her for graduating with high honors and being accepted into to an excellent law school. He added that he wished he had more students like her. They both got the irony of his statement at the same time and smiled. After Warren moved to Washington D.C. to attend law school the program ran its course and faded away, although inspired by the DGEI, and learning from it, other Expeditions were started in Toronto and Vancouver (e.g., Stephenson 1974; Bunge and Bordessa 1975; Merrifield 1995; Blomley and McCann, this volume; Peake, this volume). Decades later, but spurred by prior Expeditions, a group in New York City received funding from the Antipode Foundation to launch one of their own. While each "expedition and institute" has had its own place-specific concerns, and also foundered on its own mix of issues, the resonance of the DGEI across space and time suggests how inspiring its aspirations were and remain. Its collaborative productions and exchanges of knowledge made a difference at a variety of scales, including perhaps most significantly at the level of individual. In reviewing the sprawl of the DGEI's ambitions as well as its stumbles and falls, we mark its still exciting relevance to the practice of radical geography, but also the important lessons it offers for those engaged in community-based and participatory action research (see Warren and Katz 2014).

## Conclusions

There are important and lasting spatial dimensions to myths, their production, evolution, and diffusion (see *Antipode* 2017). As was the case in the story of John Henry and Polly Anne, it is important to pay critical attention to the way gender, race, and sexuality are used to propagate some myths and allow them to evolve into naturalized truths, while leaving other stories to die on the vine. Radical geographers have the capacity to do a better job of unearthing the racialized and patriarchal histories and geographies that have shaped the discipline just as we have the capacity to call out homophobic and other forms of persecution and discrimination. Discerning between *HIStory* and *HERstory* is an effort to do just that, which is why we have focused on providing this counter-narrative on the DGEI to the ones in general circulation.

In working through and challenging the main myths about the DGEI, it is important to recognize that Bunge came to Detroit and worked with passion to change the composition of its geography by educating black working class and poor young people, which was an important contribution and went against the grain of his upbringing and class position (Morrill 2010). At same time, his personal flaws – his anger, his sexism, which included his making sexual advances on young black women in the neighborhood, his racism and tendency to call himself a "nigger," and his essentialist thinking, among other things – worked at odds against his serious and radical ambitions. He was a white wealthy straight man of his time who was dedicated to working toward civil rights (see Bunge 1965). But, he undid himself time and time again. Our heroes and heroines are not immune to human frailties and failures. Knowing this, we do not want to undermine the inspiration the DGEI has brought to many radical geographers in thinking about community-based research, nor disregard that while their relationship was fraught and problematic, Warren is the first to say that she was also inspired in crucial ways by Bunge, and would not have accomplished some of what she has in life without his influence. Indeed, Bunge has been a profound inspiration and influence on all three of us.

So, at the same time as we work to capture bigger, more complete and nuanced narratives, and unbuckle some of our cherished myths, we recognize that this task and its revelations can be painful. We understand and value the ways that knowing and sharing the accomplishments of particular projects and practices, even when they become mythological, is important. But it is also important, perhaps more so, to poke at and complicate the myths, remembering that they hold sway in part because they flatten competing narratives and remove their object or practice from its history and geography (Barthes 1972; see Katz 1992). Remembering the DGEI opens a path to revitalized geographies of practice and possibility in the discipline and more broadly, but it also speaks to the political potency of the construction and reconstruction of a field's history, of what retrieved and revivified histories of collective practice mean for the current moment, of what the radical geographies they made possible mean for producing space now, in multiple places. Places that might include Ferguson, Baltimore, Charleston, New York City, or Oakland, as well as Detroit.

Warren's coming into the DGE and helping create the DGEI led to more community participation, which required and called forth an expanded exchange of values and skills. Her insistence that the community must control the questions being asked and the information being developed in their name, and oversee explaining the benefits of participation to the community is a lesson from which many radical geographers can continue to benefit (Warren and Katz 2014).

In recounting Warren's *HER*story it is not our intent to turn off radical geographers and others who were inspired by the DGEI, but rather to add to and amend the record and to make it more inclusive so that people can see a bigger picture of what was happening, and better grasp the uneven practices by which knowledge is produced and exchanged in the work of social justice, both in the moment and over time. Digging deeper into myths is about locating accountability for the product, the people involved, and the process; it is a way to resituate their stories in history and geography. Given how inspired current activist-scholars are when they learn about the DGEI, how do we alter its stories without losing them? We three have discussed this at length, and feel the weight of the burdensome contradictions that inhere in presenting competing narratives and complicating cherished myths.

Ultimately, however, we think the future of radical geography is better off if we recognize and take account of the bruises, slights, and misrecognitions of its past; to move beyond its fictions – even cherished ones. It always helps to remember, and remember again, that when we do community-based research, we not presume to know what's what or talk at community members, but rather recognize and engage the questions of power that riddle these engagements, and be in conversation and collaboration working together to share access to greater resources and skills. At the same time, it is also important to recognize the multiple voices and inequalities of power in the academy; together – inside and out – we comprise multiple communities of practice (Lave and Wenger 1991; see Smith 2003) and potential action. The DGEI's *HER*story teaches us not to see communities as projects, but as opportunities for solidarity and the production and exchange of knowledge and skills. Recounting Gwendolyn Warren's story of the DGEI, and learning about her life's work following her participation in it, helps us to see that these processes are ongoing and their effects unending and sometimes astonishing in their radical potential.

# Notes

1  While various versions of *The Ballad of John Henry* circulated during the early 1900s, it evolved as an African American work song, and its origins are unclear and somewhat controversial. Through its long history, there have been many different versions of the song. For reference, see lyrics for Pete Seeger's popular version here: http://songmeanings.com/songs/view/3530822107858949536/[songmeanings.com]."

2  As a co-authored text, we have generally referred to Gwendolyn Warren in the third person, but we use quotation marks in a few places to note Gwendolyn's individual voice. Much of this text is derived from a three-day convening of the three of us in New York in August 2017. In the course of long and winding, difficult, pleasurable, and amazing conversations, all of which we recorded, we stitched together the story we tell here. Much of it is in Gwendolyn's voice, but always in conversation with Cindi and Nik who wove their questions, narratives, recollections, and concerns into the collective fabric. This story, like all stories, is partial, and we are collectively working on a longer more autobiographical piece that draws more fully on Gwendolyn Warren's recollections and experiences.

3  One of the biggest oversights of *Fitzgerald*, for all the geographic inspiration it has created, is that Robert "Snoopy" Ward, who made all the maps for the DGEI and *Fitzgerald*, is not credited for all of his cartographic skill and creativity.

4  http://archive.lib.msu.edu/DMC/state_news/1970/state_news_19701013.pdf

# References

*Antipode* (2017). "Symposium – The Detroit Geographical Expedition and Institute Then and Now: Commentaries on 'Field Notes No.4: The Trumbull Community'" Antipode Foundation. https://antipodefoundation. org/2017/02/23/dgei-field-notes/

Barnes, T. J. (1996). *Logics of Dislocation: Models, Metaphors, and Meanings of Economic Space*. New York: Guilford Press.

Barnes, T. J., Heynen, N., Merrifield, A., *et al.* (2011). Classics in human geography revisited: William W. Bunge's (1971) *Fitzgerald: Geography of a Revolution* Cambridge, MA: Schenkman Publishing Co. *Progress in Human Geography* 35(5): 712–720.

Barthes, R. (1972). *Mythologies*. New York, Hill and Wang.

Bunge, W. (1965). Racism in geography. *The Crisis* 72(8): 494–497, 538.

Bunge, W. (1971). *Fitzgerald: Geography of a Revolution*. Cambridge, MA, Schenkman. A second edition was published in 2011 by University of Georgia Press with a Foreword by N. Heynen and T. Barnes.

Bunge, W. (1974). Fitzgerald at a distance. *Annals of the Association of American Geographers* 64: 485–88.

Bunge, W. and Bordessa, R. (1975). *The Canadian Alternative: Survival, Expeditions and Urban Change*. Toronto, Canada: Department of Geography, York University, Geographical Monographs.

Harding, S. (1992). Rethinking standpoint epistemology: What is "strong objectivity?" *The Centennial Review* 36 (3): 437–470.

Heyman, R. (2007). "Who's going to man the factories and be the sexual slaves if we all get PhDs?" Democratizing knowledge production, pedagogy, and the Detroit Geographical Expedition and Institute. *Antipode* 39(1): 99–120.

Heynen, N. (2013). Marginalia of a revolution: naming popular ethnography through William W. Bunge's *Fitzgerald*. *Social & Cultural Geography* 14(7): 744–751.

Horvath, R. J. (1971). The "Detroit Geographical Expedition and Institute" experience. *Antipode* 3(1): 73–85.

Horvath, R. J. (2016). Pedagogy in geographical expeditions: Detroit and East Lansing. In S. Springer, M. Lopes de Souza, and R. J. White, eds, *The Radicalization of Pedagogy: Anarchism, Geography, and the Spirit of Revolt*, pp. 101–124. London: Rowman and Littlefield.

Katz, C. (1992). All the world is staged: Intellectuals and the projects of ethnography, *Environment and Planning D: Society and Space* 10: 495–510.

Katz, C. (1996). The expeditions of conjurors: Ethnography, power, and pretense. In D. L. Wolf, ed., *Feminist Dilemmas in Field Research*, pp. 170–184. Boulder, CO: Westview Press.

Lave, J. and Wenger, E. (1991). *Situated Learning: Legitimate Peripheral Participation*. Cambridge, Cambridge University Press.

Merrifield, A. (1995). Situated knowledge through exploration: Reflections on Bunge's Geographical Expeditions. *Antipode* 27(1): 49–70.

Morrill, R. (2010). Bunge, William (1928–). In B. Warf, ed., *Encyclopedia of Geography*, p. 301. London: Sage. DOI: http:// dx.doi. org/10.4135/9781412939591.n120 Last accessed October 1, 2017.

Nelson, S. R. (2006). *Steel Drivin' Man: John Henry, The Untold Story of an American Legend*. New York: Oxford University Press.

Ross, K. (2015). *Communal Luxury: The Political Imaginary of the Paris Commune*. New York: Verso Books.

Saddler, J. (1970). DGEI head denounces 'U' for not increasing funds, *State News*, 63 (63): 1–10. Michigan State University, October 13. http://archive. lib.msu.edu/DMC/state_news/1970/state_news_19701013.pdf

Smith, M. K. (2003, 2009). Jean Lave, Etienne Wenger and communities of practice. *The Encyclopedia of Informal Education*, http://infed.org/mobi/jean-lave-etienne-wenger-and-communities-of-practice/ http://infed.org/mobi/jean-lave-etienne-wenger-and-communities-of-practice/ Last accessed September 27, 2017.

Stephenson, D. (1974). The Toronto Geographical Expedition, *Antipode* 6(2): 98–101.

Warren, G. (1971). *Director's Annual Report, Field Notes 3*. Detroit: Detroit Geographical Expedition and Institute: 4–9.

Warren, G. and Katz, C. (2014). "The Detroit Geographical Expedition: What is its relevance now?" Conversation at the Graduate Center, City University of New York. Available at: https://vimeo.com/111159306 Last accessed August 18, 2018.

Wright, M. W. (2006). *Disposable Women and Other Myths of Global Capitalism*. New York: Routledge.

# 3

# Radical Paradoxes
## *The Making of* Antipode *at Clark University*

Matthew T. Huber, Chris Knudson, and Renee Tapp

The radical journal *Antipode* was founded at the Graduate School of Geography (GSG) at Clark University in 1969. It emerged at a transformative moment in history, with an aim to take on and eventually supplant the conservative establishment in geography. Why and how did students and faculty at Clark come to challenge a geographical field dominated by quantitative methods and spatial science when only a few decades earlier Clark was "the last bulwark of environmental determinism to be stormed" (GSG 1968: 31)? In just a few years, Clark morphed from a relatively staid department to a leading center of radical geography in the United States.

In this chapter, we argue that the emergence of *Antipode* was a result of a combination of historical regimes of capitalism, widespread social activism, a contemporaneous shift in leadership at the GSG, and the graduate student community forged in Worcester, MA. Since the GSG missed out on the quantitative revolution of the 1950s and 1960s that occurred in other departments like the University of Washington or Michigan, the department was able to assemble a faculty of radical scholars in the mid-to-late-1960s critical of that revolution, unopposed by the extant faculty. Especially important was when the new director of the GSG, Saul Cohen, increased the number of students and faculty in the department starting in 1965. Many of these new recruits challenged the status quo of both the discipline and society. Unable to find academic journals that reflected their politics, they founded *Antipode*. Their intellectual commitment to the project was important to keeping the journal going, as were the numerous hours spent by students mimeographing,

*Spatial Histories of Radical Geography: North America and Beyond*, First Edition.
Edited by Trevor J. Barnes and Eric Sheppard.
© 2019 John Wiley & Sons Ltd. Published 2019 by John Wiley & Sons Ltd.

binding, and mailing issues. The doctoral students were a reserve of committed labor who maintained the journal, even as other journals of the era, like *The Insurgent Sociologist*, the *Transition*, and the *USG Newsletter*, folded.

*Antipode* changed over time. It began as an eclectic commitment to social justice documenting the inequalities of poverty, racism, sexism, and colonialism, and quickly shifted to theory grounded in Marxian political economy. By the mid-1970s, it featured erudite debates on Marx's theory of rent. Along with this change in content, the journal became more professional, giving it greater legitimacy in the eyes of the neoliberal university. *Antipode* was also marked initially by a heavily male and masculinist discipline, but which was increasingly contested by a dynamic group of GSG female geographers who labored for the journal, wrote articles, and struggled to expand and redefine what counted as "radical" in radical geography.

This chapter contains five sections. First, we provide a brief early history of the Clark GSG from its founding until the early 1960s when it had become intellectually stagnant and politically conservative. Second, we discuss the growth of the GSG under Cohen's directorship linking it to postwar Keynesian capitalism and its financial largesse. Third, we detail the history of the founding of *Antipode* that we set against both outside national social unrest and activism of the second-half of the 1960s, and internal departmental change. Fourth, we turn to the processes by which *Antipode* was materially produced in a DIY fashion using free graduate student labor. We also confront the troubled history of gender in Geography as a whole, and how women within the GSG began to build a radical feminist geography both within *and* against *Antipode*. Finally, we examine the suturing of the radical geographical project to the professional demands of tenure, theory, and publishing in "quality" journals. In the epilogue, we reflect on the creation of the Antipode Foundation and its contribution to the present state of the field.

## "The Last Bulwark of Determinism": The GSG from its Founding Through the 1950s

"During the nineteen-thirties," Professor Samuel Van Valkenburg said to students in a seminar he gave in the late 1960s, "the general movement in the United States had been against determinism as expressed by Ellsworth Huntington and Ellen Churchill Semple. Clark University was regarded as the last bulwark to be stormed" (GSG 1968: 31). Van Valkenburg knew this history personally. He had taught in the department almost since its founding, had been its director for three decades, and was a frequent collaborator with Huntington. Van Valkenburg's point that the GSG was long out of step with mainstream geography can be extended further. Until the late 1960s, the GSG was marked by a focus on graduating as many students as possible, training undergraduates to teach high school, and to propagate a qualitative economic geography. It was this setting that allowed the space for new radical geographers to enter the department and launch *Antipode*.

The GSG's conservative identity can be traced back to its founding in 1921, when its director and President of Clark University, Wallace W. Atwood, strove to build "a great geographical institute" (Koelsch 1980: 573). His first choice for the GSG was Ellen Churchill Semple, who was then at the University of Chicago. Although Semple's environmental determinism was "subject to criticism by certain of her students and colleagues at Chicago," she was also held in high esteem by some parts of the geographical community, as shown by her election as president of the Association of American Geographers the year before (Keighren 2008: 270). Although geography was beginning to turn away from the notion that human societies were products of the environment, Semple's influence on the GSG lasted well past her death in 1932 (Keighren 2008).

Environmental determinism at Clark persisted not only in courses and texts (Semple's 1911 *Influences of Geographic Environment* was used at Clark until 1939), but also through the research and graduate supervision of Semple's student Walter Elmer Ekblaw, hired at the GSG in 1926 (Barnes 1949; Martin 2015). As the only one of Semple's two advisees to supervise students, Ekblaw was at Clark for over two decades, until his death in 1949. During this time, he supervised one-third of the GSG's Ph.D. dissertations, which singlehandedly totaled more doctoral theses than that of any other U.S. geography department, except Wisconsin and Chicago. Although Ekblaw's environmental determinism strongly influenced his students, his impact outside of Clark was circumscribed. Only one of his students, who also stayed on at Clark, ever supervised another dissertation (Bushong 1981). Ekblaw's influence at Clark, though, was considerable, as he was arguably Semple's most ardent proselytizer. One student in his "Human and Cultural Geography" course recalled that "Ekblaw lost his temper [when I tried to discuss possibilist theories] and [he] said my remarks were absurd because 'You can't grow bananas at the Pole'" (quoted in Keighren 2008: 289–290)." Ekblaw also adopted the work of Ellsworth Huntington – a notorious climatic determinist – who lectured and taught summer courses at Clark, even up to the year before his death in 1947 (GSG 1946). Merle Prunty, who received his Ph.D. at Clark in

1942, suggested "a latter-day form of environmental determinism predominated" in the department, with much less attention paid to Hartshorne or Sauer than Semple and Ratzel (Prunty 1979: 45).

Through the 1940s, the GSG was dominated not only by Ekblaw's environmental determinism, but also by Atwood's focus on teacher training. Atwood was an influential educator across the U.S., publishing textbooks and distributing educational films. One of his skills was drawing intricate pictures in chalk using both hands simultaneously (Prunty 1979). Like Ekblaw, Atwood was also a prolific advisor, with the result that the GSG graduated more Ph.Ds in the 1907–1946 period than any other US geography department (Bushong 1981: 203). Because of this great volume, critics of Clark doubted the quality of the Ph.D. students (Martin 2015: 177). Compared to Chicago, for example, where more than two-thirds of its geography graduates went on to teach at major universities, Clark placed less than half of its students at such schools, with many going on to small teacher-oriented colleges (Rugg 1981). The result was that Clark effectively operated as "an advanced teacher training institute for future teachers rather than research workers." Consequently, training in research methods at Clark formed a smaller part of the curriculum than other doctoral-geography programs (Martin 2015: 176).

Atwood and Ekblaw both died in 1949, but the GSG carried on with little change. According to William Koelsch, an M.A. student of the time, "Clark in the 1950s would have been quite familiar to a graduate student in the 1920s and 1930s. Prunty's criticism... would have been germane 20 years later, with few exceptions" (Koelsch 1988: 42). With Ekblaw's death, his editorship of *Economic Geography* passed to Raymond Murphy. Ekblaw, a biogeographer, founded the journal as a student and edited it for 25 years (Murphy 1979: 39). The journal was in poor condition when Murphy took it over. It was behind schedule, it carried a large number of Ekblaw's students' papers, and to save time Ekblaw edited out the footnotes and typically wrote all book reviews himself (Murphy 1979: 40). While Murphy revived the journal during his tenure, it continued to shun quantitative geography until Gerald Karaska took over the editorship in 1969. Murphy recalled that Karaska "felt that the future of economic and urban geography lay in more, rather than less, use of modern statistical techniques. To some, the change was long overdue; to others the magazine lost much of its interest and usefulness" (Murphy 1979: 42). Clark had been so old-fashioned for so long – missing out on the quantitative revolution in newer departments, like Iowa, Seattle, Michigan, and Penn State – that it could leap straight into radical geography. Before it did, however, the department required new faculty and renewed purpose.

## Saul Cohen and Keynesian Capitalism

The rise of radical geography and *Antipode* in the late-1960s emerged in the larger context of postwar American capitalism and the expansion of higher education. Economic shifts from relatively low skilled manufacturing to "knowledge" or "white collar" work required workers with more extensive training in higher education (Boweles and Gintis 1976). This context led to the 1944 GI Bill that guaranteed a college education to veterans after WWII (Geiger 1995: 59). Postwar "Fordism" was characterized by high levels of growth and profits (Harvey 1989: 132), and exceptional marginal tax rates – over 90% – on high-income earners (Linden 2011). Higher revenues spilled into higher education funding (Geiger 1995). Most importantly, federal money came in the form of "military Keynesianism" that not only supported the profits of private military contractors, but also research universities (Mintz and Hicks 1984). After the 1957 Sputnik crisis, the US government funneled prodigious funds into higher education (Geiger 1995: 60). The National Defense Education Act of 1958 expressly aimed to make the US more competitive with the Soviet Union through funding scientific research at universities. Beyond military-based funding, The Higher Education Facilities Act of 1963 and the Higher Education Act of 1965 greatly increased funding for universities to upgrade their facilities to accommodate increased enrolments. All this money changed the face of higher education: "The federal largess, superimposed on mushrooming enrolments and state support, produced an ephemeral golden age in American Higher Education" (Geiger 1995: 60–61).

Clark reflected these trends and during the postwar era it came to resemble the "corporate university" we know today. The GI Bill led to a spike in enrolments from 650 prewar and 1023 by 1947 (Koelsch 1987: 167). Tuition increases became a common trend outpacing inflation: $500 in 1952, $1000 in 1959, $1500 in 1963 (Koelsch 1987: 169, 184). In 1946, the newly hired President Howard Jefferson hired a Director of Public Relations to help attract potential tuition payers (Koelsch 1987: 195). With increased revenue in the 1950s, the university hired a "management consulting firm," which recommended they hire a full-time business manager to report to the President (Koelsch 1987: 195). This replaced the single "treasurer" of the board of trustees who had managed the financial affairs of the entire university (Koelsch 1987: 194). In 1962, Clark appointed Dwight Lee the first Dean of Graduate Research to focus on grant generation (Koelsch 1987: 195). In the same year, the board of trustees was expanded from 18 to 30 members with its members tightly woven with philanthropic elites within Worcester (Koelsch 1987: 197).

In this context, President Howard Jefferson invited Saul Cohen to revive the Geography department in 1965, which was languishing from its dated approaches to research and teaching. Cohen had just taken a year as the executive director of the Association of American Geographers where he "vigorously pursued numerous grant opportunities in various Washington funding agencies and developed a network of contacts that were to benefit the profession and the Association for years to come" (James and Martin 1978: 153). Cohen told us the agencies were "very, very lavish in handing out grants."[1] According to Cohen, Jefferson said, "I'll give you anything you want. There's only one thing I can't give you – money." With Cohen's extensive contacts in the National Science Foundation and the US Office of Education (USOE), Cohen replied, "I don't need money" and he agreed to join Clark.

Cohen marshaled these funds to remake Clark Geography in a very short period. This included major facilities grants that allowed Geography to move into a newly renovated building (now called the Jefferson Academic Center). The grants also allowed an expansion of hiring and graduate education. Between 1965 and 1971 at the GSG, the number of graduate students expanded from 40 to 60, and tenure-track faculty from six to 15 people (Koelsch 1988: 46). During this period Cohen hired Richard (Dick) Peet, James (Jim) Blaut, and David Stea, all of whom became central to *Antipode* a few years later. In a 1972 "progress report" filed to the NSF, Cohen reported that in 1970–1971 Blaut, Peet, and Stea were being paid in full by a NSF Science Development Grant.[2] The following year, another critical figure, Anne Buttimer, was hired and paid by the same grant (The Saul B. Cohen Papers). In an age of civil rights activism, Cohen was also at the forefront of recruiting African-American graduate students into Geography. At a time when the entire discipline of US Geography contained a single African-American faculty (Harold Rose), Cohen helped create the Commission on Geography and Afro-America (COMGA): "to see…what system could be developed to draw talented undergraduates from those [traditionally black colleges] into graduate programs" (Waterman 2002: 562; and also Chapter 1, this volume).

The grant money also provided resources for the GSG to reestablish itself as a center of cutting-edge research. According to Koelsch (1987: 209), "Cohen had a good sense of where geography was going and…the department began to develop new strengths in areas of environmental cognition, international development (particularly in relation to Africa), and environmental hazards." Although Cohen told us that he did not intend to create a radical department, *Antipode* emerged out of the stew. Jim Blaut's USOE funded research on environmental perception featured kindergarteners under the framework of the Place Perception Project

(Mathewson and Stea 2003: 218), but his broader underlying philosophy was informed by a radical critique of imperialism (Mathewson and Stea 2003: 217). Dick Peet was trained as a quantitative geographer at Berkeley – his dissertation analyzed the spatial expansion of commercial agriculture (Peet 1968). He was hired to continue Clark's not-so-radical tradition of lecturing on economic geography, but after coming to Clark he soon began teaching a radical course on poverty.[3]

Saul Cohen also aimed to improve the undergraduate program. Prior to his arrival, Cohen claimed the Geography department "...had no interest in undergraduates which was stupid... I think they had all of 30 majors..."[4] In explaining his priorities, Cohen said:

> The first thing we do is pay attention to the undergraduates here at Clark... We ended up with a hundred plus majors.... One thing we insisted was every faculty member had to teach an undergraduate course. At that time when I came, there were only a half-dozen faculty and they were too pristine to teach undergraduates.[5]

According to Koelsch (1988: 46), "the increased appeal of the department's cognitive and environmental orientations," as well as a freshman cultural/historical course offered by Martyn Bowden, "brought hundreds of Clark undergraduates into the department, yielding six-to-ninefold increases in numbers of majors." Radical geography courses like Peet's poverty course and Blaut's psycho-geography course also attracted undergraduates. Blaut was a "charismatic teacher" who brought his radical critiques to the classroom (Mathewson and Stea 2003: 217). The first semester Dick Peet offered his course "Geography of Poverty," he was shocked to walk into a room overflowing with 150 enthusiastic students (the course expanded and regularly attracted 250 students). As Peet explained to us:

> Maybe one reason [Cohen] was so tolerant [of radical geography] is that the enrolments doubled or tripled in one course. Plus, then [students] came through the rest of Geography looking for other good stuff too. So, it turned out pragmatically a good thing to do.[6]

Anne Buttimer said, "Saul is the one to be given the most credit for turning an old-fashioned department of geography into a graduate school of geography."[7] While not a radical himself – Peet called Cohen "liberally inclined"[8] – he laid the conditions for radical geography to develop. As we will see next, Cohen later developed antagonisms with certain radical geographers, but not necessarily because of their radical politics; for the most part, he tolerated or overlooked the rise of radical geography in the GSG. As Cohen said, "We were an open department and faculty were encouraged to pursue their own ideological directions."[9] Buttimer also confirmed Cohen "appreciated diversity."[10] Thus, it was not only Cohen's money, but also his tolerance that created the material conditions for *Antipode* to emerge.

## "Our Goal is Radical Change": 1960s Politics and the Founding of *Antipode*

The evolution of the GSG from its founding through the 1960s can be seen in miniature through the history of field camp. From 1927–1966 field camp was held in the first three weeks of the fall semester. Students would work in teams to collect physical geography data from a site chosen within a couple hundred miles from Clark. In 1966, in response to incoming students who had a greater interest in human geography, the GSG moved field camp to Barranquitas, Puerto Rico. Jim Blaut (who had taught at the University of Puerto Rico before Clark), and David Stea, who like Blaut was bilingual, played prominent roles in running the camp. Stea's cross-appointment in the GSG and the Psychology department was in part due to his and the department's mutual interests in his involvement in field camp.[11] According to Stea, spending several weeks in Puerto Rico was transformative for graduate students, many of whom had never left Massachusetts before:

> People were exposed to a different cultural setting… We had informal discussions in seminars that I think in general were consciousness raising… We hatched various schemes, sitting around a bottle of rum. I attribute the loosened atmosphere to some of the kinds of things we generated there.[12]

Field camps in Puerto Rico connected concerns with justice and poverty to life in Worcester and centered the GSG as a place to study and fight global poverty, reflecting a broader commitment within the radical geography movement to "connectivity" between places and the scaling up of local, radical politics. The greater awareness of the world that GSG students received at field camp found an outlet in the growing social activism on campus. According to D'Army Bailey, an undergraduate who moved to Clark in 1962 after being expelled from Louisiana's Southern University for his anti-segregation work, the Worcester Student Movement for Civil Rights was "the forerunner of real student activism on the Clark campus" (Bailey 2009: 157). This activism occurred, however, on a campus where even by 1969, only seven out of the 400 incoming students were black ("Black Students" 1969). When the Black

Student Union learned that President Jackson had failed to meet his target of 30 incoming black students, and did not provide more financial aid, they organized a sit-in at the Administrative Center on February 20, 1969. As a result of this action, the Black Student Scholarship Fund was established, and a black student cultural center was set up in a university-owned house. After further recruitment, 33 black students enrolled in the fall.

Geography graduate students were actively involved in community activism, perhaps most prominently in running Your Place, a community center funded by federal grants that facilitated social work in the Puerto Rican neighborhoods of Main South surrounding the Clark campus (Peet 1985). The most prominent activism at Clark involved activities against the Vietnam war. According to Ph.D. student, Ben Wisner:

> In 1969 some 20–30 of us were arrested for refusing to leave the army recruitment center in Worcester – at the time I volunteered with Worcester's Draft Information Center and had largely organized this Clark geography presence at the civil disobedience action. We spent an afternoon in a cell together in the county jail with a number of our professors and instructors [including GSG Professor Bob Kates]. I recall we had a lively "seminar" on war and peace.[13]

After the expansion of war to Cambodia on April 30, 1970 and the shootings at Kent State on May 4, Clark and other schools voted to strike, with the result that about two-thirds of Clark's 1,100 students walked out. What started as a three-day strike was later voted to extend until the end of the semester (Koelsch 1987: 216–17). Clark students and faculty, including those in geography, used the time to engage in "leafletting, opinion polling, and participating in a sit-in at the local draft board which resulted in the arrests of nearly 300 people," mostly from Clark, and the nearby College of the Holy Cross (Koelsch 1987: 217). Kates recalls that there was "little internal conflict over the war" on campus, as students and faculty were largely united in opposing the war.[14]

Clark geographers engaged in activist struggles outside the academy – including the antiwar, civil rights, and environmental movements – but they also saw these struggles as internal to the field of geography itself. The wider debate in geography over the quantitative revolution was a political one within the GSG. Before the founding of *Antipode* and the rise of radical GSG faculty, the department was strong in research on environmental cognition and environmental hazards management that, in contrast to the isotropic plane of statistical spatial methods, took seriously people's perceptions of and their agency within space. Anne Buttimer's arrival at Clark in 1970 inspired students like Ben Wisner to engage in humanistic geography by "approaching the experience of space, life worlds and nature in a phenomenological manner..."[15] For Wisner, these "philosophical and interpretive/ qualitative directions were attractive to *Antipode* founding grad students who [...saw the quantitative revolution in geography as...] a spent force and, to our juvenile, restless minds, too 'establishment' and 'buttoned down.'"[16]

According to Stea, after one of his seminars, "We were standing around chatting, and we said, 'Why don't we start writing – however we can produce it – a little journal of our own that represents our points of view in radical geography?'"[17] They chose the name *Antipode* because for them "Radical geography was not at that point nearly as Marxist as it became. It was just anything that was alternative... This was the opposite pole to the pole of conventional geography."[18] The journal's logo for the first ten years was one drawn by undergraduate Cynthia Briggs. It combined the symbol of the Campaign for Nuclear Disarmament ("ban the bomb") and a drawing of the Earth (Figure 3.1). The journal's broad aims were explained in Stea's introduction to the first issue: "Our goal is radical change – replacement of institutions and institutional arrangements in our society that can no longer respond to changing societal needs... that often serve no other purpose than perpetuating themselves" (Stea 1969: 1). Topics covered in the first issue showed the breadth of what its founders considered radical geography: the New Left (Peet); anti-imperialism (Blaut); low-income housing (Jarrett and Wisner); Black America (Fred Donaldson); and critiques of geography textbooks (Reed Stewart), and research methods (Kates and Jeremy Anderson).

## The Social Reproduction of *Antipode*

### Material production

Early issues of *Antipode* were produced, assembled, and distributed at Clark in cottage-industry fashion. *Antipode* relied on the informal labor of the administrative pool of women and graduate and undergraduate students to handle the physical production of the journal. The graduate students were not paid for their labor, while the administrative assistants had *Antipode* work on top of their daily tasks. Like generations of graduate students before and after them who exchanged labor for sustenance, the Clark students that worked on the early editions of *Antipode* often were compensated with free food and drinks. And like generations of

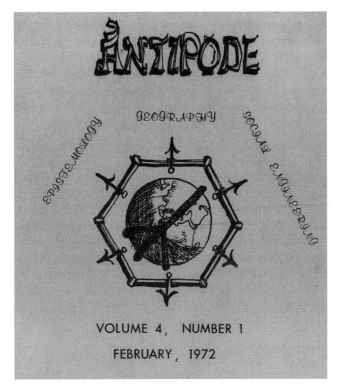

VOLUME 4, NUMBER 1

FEBRUARY, 1972

**Figure 3.1**   *Antipode* cover, Volume 4, issue 1, February 1972.

administrative staff before them, the women's names who typed and mimeographed the journal have not been recorded or remembered.

Dick Peet and students gave the articles – many times hand-written drafts – to the administrative pool and student laborers who would then use typewriters to draft and format the articles for publication before reproducing them with a mimeograph machine, which, in the early days, had to be turned with a crank by hand.[19] In the beginning, "the first two issues were printed on a mimeograph machine and collated, then stapled with the heavy, colored covers in the basement of the GSG building."[20] Predictably, this informal system produced less than perfect final versions of the journal. Phil O'Keefe recalls a particularly disastrous event when he and Myrna Breitbart were mimeographing a red and black cover for a special issue on anarchism that turned out to be pink and green.[21] Although Wisner couldn't recall "where the money came from for paper, etc.",[22] Dick Peet remembers that Saul Cohen allowed the department resources to be used for the first several issues of *Antipode*.[23]

Once *Antipode* began receiving subscriptions and payment, the journal was able to afford to rent a small box at the local post office to serve as its official mailing address. After the first few issues, the journal also eventually moved mimeograph production out-of-house to a local print shop. Ann Oberhauser, who started Clark in 1982 as a Master's student before entering the Ph.D. program, worked with Dick Peet as both a teaching assistant and the subscription manager for *Antipode*. As a subscription manager, she was responsible for keeping track of who paid – either in cash or check – and who owed money, advertising the journal at conferences, and mailing physical copies of the journal to a global network of radical geographers from Massachusetts to British Columbia to Southeast Asia.[24] Paper editions of *Antipode* were also carted to conferences where students like Oberhauser sold subscriptions to individual academics.

Here "place" played a central role in the everyday production of *Antipode* – especially in the built environment in and around the University. Ho Toy, a Chinese-Polynesian restaurant at the corner of Park Avenue and May Street, became a favorite hangout for the *Antipode* crew. Dick Peet remembers how the students would be horrified that he ordered chips (French fries), which they would then quickly eat themselves, leaving him none.[25] A few years later, Mona Domosh remembers filling boxes of printed copies of *Antipode* in the Libby Lounge and being rewarded with pizza and beer.[26] Other bars, like Moynihan's, served as a meeting place for graduate students to debate ideas and discuss radical politics. The DIY approach to the journal's production helped solidify the perception among students that they were working on the edge of the radical margins; they were Marxists laboring for free. *Antipode* not only entailed power relations in regard to labor, but also to gender.

## Gender, power, and the radical project

Clark occupies a significant space of educating and politicizing women geographers, yet the relationship between women and radical politics at Clark is complicated. The rise of second- wave feminism and changing social conventions that occurred in the personal and professional lives of faculty, students, and society, spilled over to *Antipode*. Many people involved in the early history of the *Antipode* project found it impossible to draw clear distinctions between the politics of the journal and the politics of the department. On the one hand, women were instrumental to the early founding of *Antipode*: they contributed to the journal's material construction and marketing, served as editors, and wrote articles. They were, in other words, a source of manual labor for the journal. However, understanding the influence of their theoretical labor on the journal is less clear-cut and the answer to this question reflects longstanding and still unresolved issues about sexism in radical and Marxist geography at large. This section examines the complicated relationship between *Antipode* and feminism at Clark drawing on the voices of women and men who were in and around Clark geography during this time.

Of course, perhaps the most famous female geographer of the early twentieth century, Ellen Churchill Semple, worked at Clark from 1922 until her death in 1932. Yet, according to Keighren (2008: 270), she discovered late in life that she was paid "$500 less per year than her male colleagues," prompting her to disinherit Clark. Her smaller salary was justified "on the basis that Semple was 'without dependents' [...which she deemed...] a 'mid-Victorian argument from a group of modern capitalists'" (Keighren 2008: 270). Although her work is controversial, she voiced a strong sense of injustice at this inequity.

> [T]hough I worked longer hours and made a larger scientific literary output every year than the men professors in my department ... [and had a] national and international reputation [which] equaled or surpassed theirs,' the academy, like the society it reflected, was not a meritocracy (Keighren 2008: 270).

From 1932–1970, Clark's Graduate School of Geography employed no female faculty members despite having produced "over 30 percent of doctoral dissertations in geography [that were] written by women in the United States up to 1969...more than double the number awarded at its nearest 'competitor,' the University of Chicago" (Monk 1998: 15). The void of women in faculty roles reflected trends in the academy, and geography more specifically, where the "discipline of geography has historically been dominated by men, perhaps more so than any other human science" (Rose 1993: 1–2). It fostered a climate of sexual harassment that included,

> leers, jeers, insults, sexist jokes and provocative remarks, forms of behaviour that many women have kept to themselves as a private experience not to be shared least they rock the boat, or disturb the male structures of patronage on which they depend for advancement [...especially impacting...] women students [who] are the most vulnerable because of their relative youth, inexperience and lack of power within the institution (McDowell 1990: 325).

Hanging in the basement of the Geography building at Clark where *Antipode* was assembled, a troublingly gendered poster symbolized the struggle between human and quantitative geography and loomed over

**Figure 3.2** "Qualifactus versus Quantifactus." Poster hanging in the Clark Graduate School of Geography in the late 1960s (Wisner 2015).

students as they engaged in radical politics (see Figure 3.2). The image of Quantifactus became even more charged when, according to Wisner, the caption "The Rape of Geography" was added to the poster.[27] Yet, despite producing a significant number of women Ph.Ds in geography, women were not seen as equals, even within radical geography.

In 1970 Anne Buttimer arrived as a postdoctoral researcher, and, in 1971, became an assistant professor; the first female faculty in the 40 years since Semple's death. Although Buttimer sat on *Antipode*'s editorial board, published articles in the journal, and pursued her own radical humanistic geography project, she had a strained relationship with Marxist geography. With the Marxists she shared a critical position against quantification:

> My position was not only critical of the logical positivist stances of the quantitative revolution. I was also critical of some opponents of the movement, e.g., the Marxist position for their proposals of alternative structures, and of SERGE ("socially responsible" geographers) who acted as though they knew what was best for everyone. We had many [of these].[28]

Buttimer's quest to develop a humanistic geography in the face of positivism influenced the graduate students at Clark who were working to develop a radical politics around *Antipode*. According to then graduate student Ben Wisner, Buttimer "was a big influence on some of us as she was struggling at the time to found what is now known as humanistic geography. She provided 'cover' for some of us uneasy with more conventional approaches [to geography]."[29]

Despite the support she offered radical graduate students at Clark and the contributions she made as an *Antipode* editor, Buttimer and her work were dismissed as counter-revolutionary by those she considered dogmatic Marxists.

> The editor, Richard Peet, was a genuine acolyte of David Harvey, agreeing with every one of his words, often in opposition to my views. For example, Harvey objected to the applied work in which many of my students had become engaged, arguing that all this "busy work" was delaying the advent of the real structural revolution that was needed in society! Yet, in the early years of *Antipode* there was a genuine diversity of critical views. I remained on the editorial board even during my absences during visits to Sweden in 1973 and 1976.[30]

When Buttimer left Clark in 1981, the department hired Susan Hanson in the same year. When she arrived at Clark, Hanson hosted a party at her house for the women graduate students in the GSG. She recalls their surprise at socializing with a faculty member outside of the department, an opportunity that had not been afforded to them as women until her arrival.[31] Susan Hanson recounted that as late as the mid-1990s:

> I would go around giving talks at other universities and I had Assistant Professor women and grad students asking me, 'Do you think it's really safe for me to do work on women?'… Because in their department sometimes it wasn't safe. They didn't feel safe.[32]

Despite the School's established record of educating women, an interviewee recalls GSG as, "definitely a masculinist environment [...] all of the professors were male."[33] Alison Hayford, then a Ph.D. student at the University of Michigan, who became integrated into the Clark *Antipode* network through David Harvey, reflected that there was "a lot of misogyny on the left" in this period overall.[34] She explained that a key part of these power dynamics within radical geography was that most of the men were faculty and most of the women were graduate students. She told us,

> We weren't their equals in terms of power... the problems come... when women start entering into the same rank as you [men], and they have the same right to speak as you, and they are no longer your grateful but still dependent students; they are people who are demanding to be treated as full-fledged members of the group... [men] were fine dealing with them as students... but they had a much harder time dealing with women as colleagues.[35]

A campus wide investigation on gender discrimination among undergraduates at Clark found that the university accepted less qualified men over more qualified women to maintain a student housing ratio (Vollmer 1968). Clark's track record of granting Ph.Ds to women also was somewhat stunted by Saul Cohen's "revitalization" of the GSG. As Monk (1998: 24) explains,

> At a time when Saul Cohen was bringing federal funds into Clark to revitalize the institution, as well as to support training of African Americans (Koelsch 1987), most of the students recruited, including those participating in National Defense Education Act year-long and summer institutes, were men.

This reflects a larger bias by the NSF and other grant-giving institutions toward men (Rossiter 1995). Indeed, when we asked Cohen about how women and the feminist movement shifted the departmental culture at the time, he flatly declared, "It didn't... Actually, when I was there, women played a very minor role. They weren't particularly active in the field at that time. The role of women was not... a focus." The absence of female faculty meant that whatever ideas about feminism were percolating at Clark for generations had come from outside the department or from among the graduate students themselves.

The first record in the Clark student paper, *The Scarlet*, of a Marxist-feminist giving a public lecture on campus was in October 1970 when the anthropologist Evelyn Reed spoke about the institution of marriage and the oppression of women by capitalism (Roboff 1970). The sociology department offered a "Women in American Society" class in spring 1971 (Plette 1971b). While limited professional development opportunities existed in the department, women at Clark built their own movement internally. Katherine Gibson explains the dynamics during her time in the department in the late 1970s to early 1980s.

> In some ways, I felt like my education was less influenced by those professors than by the cohort. It was influenced by the work we did together; the professors were a bit on the side. My supervisor, and Julie's supervisor was Karaska. He took us [Julie and I] on and said you can do what you like... Even though the academic leadership was male, the dynamism was coming from the women.[36]

The strong networks and geographical connections that formed between women, starting with recruitment and extending beyond graduation (Monk 1998), were a vital component to women's role in *Antipode*. When Mona Domosh arrived in the 1970s, she thought all geography departments were made up of strong, supportive, and curious women.[37] Ann Oberhauser recalls how working at *Antipode* formed close friendships that lasted decades; for her it was a very important time in developing a radical community that would shape the careers of those involved.[38]

Many of the women we interviewed for this project – not all listed here – expressed an ambivalent relationship to Peet and orthodox Marxism, but they nonetheless all felt committed to the journal. For them, both *Antipode* and Clark offered opportunities for women not only to pursue questions about Marxist economic theory and women, but also to do academic research in realms traditionally occupied by men. Nevertheless, in the first five years of the journal (1969–1974) 82% of the articles in *Antipode* were written by men, 11% by women, and 6% co-authored between men and women. These women included Anne Buttimer, Alison Hayford, and Doreen Massey who wrote pieces including *The Geography of Women: A Historical Introduction* (Hayford), *Health and Welfare: Whose Responsibility* (Buttimer), and *Towards a Critique of Industrial Location Theory* (Massey).

This foray into a more expansive radical geographic project was short-lived. What appeared to be diversity of content in early issues was in part a result of the new journal trying to have enough material to fill its pages. Editors invited academic friends and colleagues to publish in the journal, with many radical ideas finding outlets during this time. Buttimer recalls,

> [when] Richard Peet, together with a number of other English ex-pats took over [*Antipode*], the general tone became more doctrinaire, no longer critical. I do not understand why other voices were suppressed. But has this not been the typical British way?[39]

As political economy at Clark, *Antipode*, and the discipline solidified around orthodox Marxism, the radical project became narrower, with intellectual and political schisms forming.

Despite Buttimer's clash with Marxism, she also plainly told us she "never emphasized feminist aspects of things." Graduate students during the 1970s confirmed with us that she was not much of an ally when it came to gender issues within the department. The gender divisions within Marxism, *Antipode*, and Clark were evident for Cindi Katz. In the late 1970s, she recalled becoming aware of the divided interests within radical geography:

> [We were a] group of radical students and faculty [that formed] study groups and held political meetings to discuss academic work and a sense of what we should be doing as activists. We met once a month and talked about politics. We made three study groups... value theory, oppressed groups, and something else. All of the women turned up in the oppressed groups one, and all of the men went to the value theory group... and we were like, 'oh my god, what does this tell us?' It was an important part of what we were doing.[40]

Many of the orthodox male Marxists acting as gatekeepers to the radical project were trained in the quantitative geographic tradition, itself a masculinist subfield (Madge *et al.* 1997). As time went on, it became clearer that the journal would support only a narrow, masculinist interpretation of Marxism defined by Dick Peet and other male Marxists engaged in the *Antipode* project. Men carefully guarded the radical terrain, which was particularly hostile to women inside the classroom, within departmental life, and at academic conferences. Anyone who has spent time in the GSG has experienced Peet's sexism – his tendency to call female grad students "honey" and "babe" among other inappropriate comments. Katz told us when she was in a *Capital* reading group Peet made her feel her activism founding Worcester's first rape crisis center was not "radical," but more "like being in the 'ladies' auxiliary.' It was infuriating, but also undermining."[41] She recalled the "oppressed groups" study group had to fight to assert that "the kinds of things we did that were not urban, spatial, Marxist, were also important and radical."[42] Mona Domosh echoed a similar sentiment, that even the idea of studying "culture" was seen as "the most ridiculous thing ever that a Marxist could think about."[43] As a Ph.D. student at Clark, Paul Robbins described what it was like to be a student in Peet's class:

> It was combat...you'd take feminism and lift it up in [Peet's] seminars and then shoot it down. Then you'd go and bring in the post-structuralists and lift them and say "this is very interesting" and then you'd shoot them

down. This was the way he published during the period. The problem is that these were real people, sitting in the room and ostensibly other faculty. That way of approaching feminism and the history of ideas was really unfortunate.[44]

Many male Marxists considered themselves victims of the geography establishment, persecuted for their radical ideas and perpetually in danger of not getting a promotion or tenure because of their intellectual commitment to Marxism (Harvey 2015). Since they were intellectually committed to revolution and justice, it has long been believed that the radical community was immune from sexist behavior. In practice, the belief in the infallibility of the left provided a cover for the pervasive dogmatic and sexist behavior that characterized the new radical domain.

Feminism and (mostly male) Marxism are often seen to be in antagonism with each other in Geography; this antagonism was rooted in precisely this narrow terrain of what counted as radical in the 1970s at Clark and elsewhere. Alison Hayford characterized the period as an Oedipal moment where young scholars attempted to challenge and subvert the establishment of orthodox Marxism.

> Some react to the challenge by remaining open and using the challenge to move themselves forward and some of them become more defensive of their positions and variations in between [...] The people who wanted to do rigid Marxist economics and then the people who simply wanted to bring women in, as many people besides me have pointed out, even raising questions about gender was a revolutionary act. It wasn't allowed within that paradigm.[45]

The struggle for legitimation and recognition within an increasingly rigid project shaped the legacy and subsequent trajectories of radical geography at Clark and beyond. One of the most emblematic feminist fights came within the pages of *Antipode* between Dick Peet and Julie Graham – a Ph.D. from Clark who challenged Marxist orthodoxy and rethought political economy – she wrote:

> The landscape of this region [anti-essentialist Marxism] is very different from that of classical Marxism where Dick Peet dwells. The institutions and rules are different as well. Over the years, the differences between the two regions have contributed to a variety of misunderstandings and even outright battles between their inhabitants. None of these battles has been decisive, leading to the suppression or annexation of the foe. Nevertheless, it is my impression that Peet's region is losing population while mine is experiencing a small but significant gain (Graham 1992: 144).

Paul Robbins recalls that the battle within Marxism spanned professional, personal, and intellectual spheres:

> You'd go to these national meetings and she'd [Julie Graham] get up there on the stage and prominent male Marxists would savage her. She'd get off the stage and [then] their graduate students would attack her. Her project, an alternative radical project didn't get a proper hearing for many, many years because of a direct and continued assault by the traditional male radical left.[46]

Radical feminist geography continued to challenge Marxist dogma within and outside the confines of *Antipode*. The struggle galvanized a generation of scholars into action, a demonstration of "how powerful the insurgency of feminism was."[47] The theoretical, material, and applied work carried out by women at Clark – once decried as "busy work" by male Marxists[48] – pushed into new areas of scholarship, bringing to light a growing divide within "radical" geography. Mona Domosh graduated from Clark to establish *Gender, Place, and Culture*; Katherine Gibson and Julie Graham, eventually writing as J. K. Gibson-Graham in one of the most enduring and widely recognized academic partnerships in geography, offered a feminist rethinking of political economy; Cindi Katz undertook groundbreaking work on children's geographies and resurrected interest in social reproduction theory (see also Mitchell *et al*. 2004 and Meehan and Strauss 2015 for new directions in Marxist-feminist geography); and the arrival of Susan Hanson shaped the work of a generation of urban-economic feminist geographers.

Although the history of *Antipode* at Clark is characterized by masculinist, exclusionary, and sexist practices, Clark is not the only institution where such behaviors appear in the discipline. To move forward, the radical geography project must contend with its history, including how to interpret the legacies of male academics that have behaved in discriminatory and inappropriate ways to women colleagues and students. The present moment, with a renewed interest in the role of women and feminism in politics and the workplace, is an opportune time to examine the sexist histories of the radical past and assert new visions of radical geographies and Marxist-feminist analysis.

## Revolutionary Theory? Tenure and the Professionalization of *Antipode*

The early days of Antipode's formation came with a sense of possibility, but the radical geography project was forced to confront the realities of institutionalized higher education. In the academic year 1970–1971,

Clark witnessed what the student newspaper called a "faculty purge" (Plette 1971a: 1). GSG graduate student Roger Hart claimed some 13% of faculty were fired; among them Jim Blaut and David Stea. David Stea's three-year appointment was simply not renewed due to the conclusion of an NSF grant meant to build a psycho-geography program. Stea also told us that the very prominent psychology department at Clark did not appreciate his research with Blaut on children's spatial learning: "What we were coming up with as results were critical of the orthodox opinions... And the psychology department did not like that at all."[49]

Blaut's tenure case was particularly "agonizing" as Cohen put it in a memo to the faculty.[50] The faculty and grad students were divided on Blaut's tenure case, which went through a roller coaster ride of decisions: the faculty-graduate student committee within the department recommended "no;" the GSG faculty as whole narrowly voted "yes;" and then the Administration decided "no." In the same memo to faculty, Cohen urged that the department accept Blaut's denial "for the sake of University unity and the welfare of the school of Geography" (Saul Cohen Papers). Although this memo paints Cohen as a distanced bystander in these affairs, he was extremely powerful within the administration (a former Dean of the Graduate School until 1970) and likely was instrumental in the administration's negative decision (as seen next, Cohen implies it was "his" decision). Despite Cohen's urging of the department to accept the decision, it was resisted by many faculty and students. There was a massive student mobilization – some adorning "Blaut is beautiful" pins. Students demanded "due process for the professors and student involvement in the decision-making process" (Hart 1971: 5). Roger Hart was involved in organizing the Graduate Student Association's vote of no confidence in Saul Cohen as director of the GSG. Hart reported the vote to Cohen and claims he "turned green" upon hearing the news.[51]

Many believed then and today that "politics" was behind this decision. Wisner *et al.* (2005: 1048) assert, "Security of employment was one price Blaut paid for his activism." David Stea told us that the left-wing faculty "were the ones that got fired – like me."[52] We agree the decision was political, but what kind of politics was at work? It appears these decisions were sometimes less a silencing of radical views, and more about academia-specific but no less political ideologies of professionalism. As early as 1968, Cohen expressed a mounting frustration with Blaut's lackadaisical approach to managing grants. In a harsh letter to Blaut and a collaborator, Cohen expressed exasperation over their failure to submit the final report and fiscal report for a USOE grant. "I cannot further jeopardize the position of Clark (or myself) for failure to produce two simple documents."[53]

Cohen himself acknowledged Blaut was "brilliant" and his scholarship was never in question.[54] We found several external letters unanimously praising Blaut's research. One letter from a prominent scholar exclaimed, "I was astonished to learn that Jim Blaut is just now being considered for tenure; it is something that he should have had long ago."[55] Nevertheless, Cohen told us his reasoning for denying him tenure. "I refused to give him tenure for a simple reason. He walked into a doctoral dissertation defense and started off by saying, 'I haven't read what you've written but' and then went into a diatribe…"[56] Peet confirmed that there were concerns with Blaut's professionalism: "Jim would come a half hour late to classes… just behaved in a totally irresponsible way…"[57] Roger Hart even admitted his advisor was "totally disorganized, the opposite of professional."[58]

It is clear that these faculty were not simply challenging US empire or capitalism, but the buttoned-up hierarchies and behavioral expectations of the University itself. Stea and Blaut were aligned with another fired sociologist, Howard Stanton, who, according to Roger Hart, experimented with letting a 350-student introductory course at Clark "self-grade."[59] All of them were also involved with what they called "mini-versities" based on a principle of democratizing higher education. Stea claimed you could, "have one anywhere" and, in Worcester, they set up one at a store-front to teach about "the real issues – the gradual decay of industrial infrastructure, [issues of] ethnicity, race, etc."[60] While these faculty challenged the very idea of the University, Stea said, "[Clark] alumni…were not happy with this direction."[61]

Out of this purge, there was a need to redefine what radical geography meant. It did not mean compromising one's political views or critiques. It did, however, mean aligning radical politics within the institutional constraints of the University. Peet claimed, somewhat paradoxically, that in contrast to Blaut, "If you're a radical you behave yourself."[62] In this vein, Peet took over the editorship of *Antipode* as, in his words, "the only person really capable [of doing it] … someone who actually does things on time…"[63] Another interviewee agrees that Peet was the "steady, focused manager" that *Antipode* needed at the time.[64]

While *Antipode* began as an eclectic collection of alternative (Antipodean) politics, in the early 1970s it shifted to become a more Marxist theoretical journal. Radical geographers understood that part of Geography's disciplinary marginality was rooted in its idiographic focus on description and its inability to advance theory. Peet recalled a debate during his oral defense at Berkeley in the mid-1960s: "the debate was: is theory possible in geography? I was saying 'of course!' But the majority of people…were saying it's impossible because everything is unique… So just being able to do theory was a struggle."[65]

*Antipode*'s shift to theory with a capital T – called "the breakthrough to Marxism" by Peet (1977: 249) – is often traced to David Harvey's (1972) famous article on "Revolutionary and Counter Revolutionary Theory." In this article, Harvey (1972: 10) denigrates a geography based in empirical description or quantitative statistics: "[revolutionary theory] does not entail yet another empirical investigation of the social conditions of ghettos. We have enough information already..." As Phil O'Keefe recounted, theoretical debates bifurcated into urban-spatial Marxism on one side (e.g., Soja and Hadjimichalis 1979; Harvey 1981) and peasant studies on the other (Blaut *et al.* 1977).[66] These approaches mirrored classic geographical studies into local processes like urban-spatial form or a rural landscape, but the goal was now to situate empirical research within broader spatial and geographical theories of capitalism. Harvey (2015: 4–5) recently reflected on the larger professional context of this shift:

> Survival in the discipline was an issue... And yes I will here offer a *mea culpa*: I was from the very beginning determined to publish up a storm... It was more than the usual publish or perish. For all those suspected of Marxist or anarchist sympathies, it was publish twice as much at a superior level of sophistication or perish... The Faustian bargain was that we could survive only if we made our radicalism academically respectable and respectability meant a level of academicism that over time made our work less accessible.

Despite its shift to theory, *Antipode* remained unhelpful for geographers seeking professional validation. Richard Walker told us one of his colleagues was denied tenure and his *Antipode* article "was used against him."[67] In Dick Peet's own tenure case, the evaluation committee claimed they "did not know what to do with" his editorship of *Antipode*.[68] However, Peet claims he "made sure that his actions answered the [political] questions [about his tenure] because I just believed in getting tenure."[69] Peet also cites his value to the department as a teacher in addition to his more conventional publications on Von Thünen (e.g., Peet 1967). Walker claims his own tenure was contested because: "the reactionaries in my department [went]... after me for being a radical leftist who was destroying the great Sauerian tradition"[70] (see also Chapter 7). The professionalization of radical geography was a political struggle.

All these trends culminated in the late-1970s when Phil O'Keefe and Kirsten Johnson took over *Antipode*'s editorship while Dick Peet was in Australia. O'Keefe recognized academic respectability hinged on theory. "But to be a social science there had to be a theoretical place to it and we all gravitated toward Marx eventually."[71] This is also a period where abstract Marxist theoreticism thrived outside of geography in journals like *Science and Society, Capital and Class*, and *New Left Review*. In a 1979 editorial on "Reorienting *Antipode* for the'80s," O'Keefe (1979: 1–2) wrote, "it is necessary to develop a more rigorous theoretical discourse… Such a theoretical discourse must be rooted in Marxist analysis." In 1981, the editors (now Peet, along with O'Keefe and Johnson) announced formal peer review and a stark break with early iterations of the journal: "In the early days, [*Antipode*] entailed printing articles written at a number of levels of theoretical understanding. This evolutionary phase is now over… This means a sharp improvement in the intellectual quality." (Editors 1981: ii).

O'Keefe recounted the goal was, in a word, *professionalization*. "It massively had to become professional to underpin the efforts of the students publishing in it go get jobs. We had to go the route of impact factor."[72] Peet explained to us,

> You have to keep a status position in academia if you're going to survive as a journal that people want to publish in and get tenure and not have it looked on as a rag and if anything, worse for their tenure case. So, we wanted radical people to get tenure.[73]

Today, *Antipode* has a good impact factor and high-quality publications, but it has become a very different journal than the one launched in 1969.

## Epilogue

The attempts to professionalize *Antipode* did not end with the introduction of peer review, and a focus on theory. After his editorship with *Antipode* ended in 1986, Peet managed the negotiations with Basil Blackwell Ltd to publish the journal. At the time, Peet (1985: 4) explained that this would deliver "the academic respectability provided by a professional publisher." However, initial support from Blackwell was minimal. According to Eric Sheppard, who along with Joe Doherty edited the journal after Peet's tenure: "We labored over everything from the cover, to subscriptions, mailing lists, publicity and editorial support…" (Sheppard 2009). Moreover, Blackwell for a time received payments from sympathetic senior geographers to support the journal (Doherty 2009).

Over time, the *Antipode* Board's finances improved, largely because they had retained ownership rights to the journal. By the early 1990s, the board had negotiated a small percentage of the net revenue from subscription sales. And within two decades, *Antipode*'s growing readership allowed the Board to form the Antipode Foundation in 2011, after John Wiley & Sons, Ltd, which merged with Blackwell in 2007, agreed to give them half of the net revenues from subscriptions. With this funding, the Foundation has been able to support conferences, workshops, summer schools, lectures, and collaborations between academics and non-academic activists, among much else. Through the Foundation, *Antipode* fulfilled the goal of its early editors to provide a respectable journal for early career radical scholars in which to publish. Peet, however, told us he regrets the sale because it relinquished too much control over the "means of production" to Wiley (something Peet reclaimed with his journal *Human Geography*).

However, radical geographers at Clark and elsewhere had long struggled with the double-edged sword of outside validation and support. It was Cohen's federal government funding after all that allowed the great expansion of faculty and students, many of whom, alongside the additional labor of the administrative pool, dedicated their mental and physical labor to *Antipode*.

The pursuit of radical geography within Cohen's GSG came with high expectations for scholars that were unevenly applied. Radical women confronted a highly masculinist discipline often defined by sexist behavior and misogynist attitudes that devalued their contributions and forced them to develop their own professional networks within the discipline. Meanwhile, some men felt they could shun academic professionalism altogether or engage in behaviors that fostered a hostile work environment. While *Antipode* was once used against tenure cases in the 1970s, it is now both a premiere outlet and funder of critical scholarship. Although some things have changed since the journal's founding, *Antipode* remains, for better or worse, radically paradoxical.

## Notes

1  Saul Cohen interviewed by Matt Huber, by telephone, May 4, 2016.
2  Saul B. Cohen Papers, Series 2, Box 5. Clark University Archive. Clark University. Worcester, MA.
3  Dick Peet interviewed by Renee Tapp and Chris Knudson, Worcester MA, April 7, 2015.
4  Cohen, 2016.
5  Cohen, 2016.
6  Peet, 2015.
7  Anne Buttimer interviewed by Matt Huber, by telephone, February 23, 2016.
8  Peet, 2015.

 9  Cohen, 2016.
10  Buttimer, 2016.
11  David Stea interviewed by Matt Huber, by telephone, February 9 and 16, 2016.
12  David Stea, 2016.
13  Ben Wisner, written communication, December 10, 2015.
14  Robert Kates interviewed by Chris Knudson, by telephone, July 25, 2016.
15  Wisner, 2015.
16  Wisner, 2015.
17  Stea, 2016.
18  Wisner, 2015.
19  Peet, 2015.
20  Wisner, 2015.
21  Phil O'Keefe interviewed by Chris Knudson, by telephone, June 16, 2016.
22  Wisner, 2015.
23  Peet, 2015.
24  Ann Oberhauser interviewed by Renee Tapp, by telephone, May 16, 2016.
25  Peet, 2015.
26  Mona Domosh interviewed by Renee Tapp, by telephone, August 18, 2017.
27  Wisner, 2015.
28  Buttimer, 2016.
29  Wisner, 2015.
30  Buttimer, 2016.
31  Susan Hanson interviewed by Renee Tapp, by telephone, September 5, 2017.
32  Susan Hanson, 2017.
33  Anonymous interviewee.
34  Alison Hayford interviewed by Matt Huber, April 24, 2016.
35  Alison Hayford, 2016
36  Katherine Gibson interviewed by Matt Huber, by telephone, April 14, 2016.
37  Domosh, 2017.
38  Oberhauser, 2016.
39  Buttimer, 2016.
40  Cindi Katz interviewed by Matt Huber, by telephone, August 2, 2016.
41  Cindi Katz, 2016.
42  Cindi Katz, 2016.
43  Domosh, 2017.
44  Paul Robbins interviewed by Renee Tapp, by telephone, December 12, 2017.
45  Hayford, 2016.
46  Robbins, 2017.
47  Robbins, 2017.
48  Buttimer, 2016.
49  Roger Hart interviewed by Chris Knudson, by telephone, September 15, 2016.
50  Saul B. Cohen Papers, Series 2, Box 5
51  Hart, 2016.
52  Stea, 2016.

53  The Glenn W. Ferguson Papers, Box B 7-3-2, folder "Blaut, James M. 1967–1971." Clark University Archive. Clark University. Worcester, MA.
54  Cohen, 2016.
55  The Glenn W. Ferguson Papers, Box B 7-3-2, folder "Blaut, James M. 1967–1971."
56  The Glenn W. Ferguson Papers, Box B 7-3-2, folder "Blaut, James M. 1967–1971."
57  Peet, 2015.
58  Hart, 2016.
59  Hart, 2016.
60  Stea, 2016.
61  Stea, 2016.
62  Peet, 2015.
63  Peet, 2015.
64  Anonymous interviewee.
65  Dick Peet interviewed by Matt Huber, Worcester MA, September 14, 2017.
66  O'Keefe, 2016.
67  Dick Walker, written communication, May 5, 2017.
68  Peet, 2015.
69  Peet, 2017.
70  Walker, 2017.
71  O'Keefe, 2016.
72  O'Keefe, 2016.
73  Peet, 2017.

# References

Bailey, D. (2009). *The Education of a Black Radical: A Southern Civil Rights Activist's Journey, 1959–1964*. Baton Rouge: Louisiana State University Press.
Barnes, C. P. (1949). W. Elmer Ekblaw, 1882–1949. *Annals of the Association of American Geographers* 39 (4): 294–295.
Black Students Seize Administration Building. (1969). *The Scarlet*. XVII (14): February 21, 1969, Extra Edition.
Blaut, J., Haring, K., O'Keefe, P., and Wisner, B. (1977). Theses on the peasantry. *Antipode* 9 (3): 125–127.
Boweles, S. and Gintis, H. (1976). *Schooling in Capitalist America: Educational Reform and the Contradictions of Economic Life*. New York: Basic Books.
Bushong, A. D. (1981). Geographers and their mentors: A genealogical view of American academic geography. In B. Blouet, ed., *The Origins of Academic Geography in the United States*, pp. 193–219. Hamden: Archon Books.
The Saul B. Cohen Papers. Clark University Archive. Clark University. Worcester, MA.
Doherty, J. (2009). *Past Editors' Reflections*. Available at: onlinelibrary.wiley.com/journal/10.1111/%28ISSN%291467–8330/homepage/editor_s_past_reflections.htm - Doherty (accessed November 2018).

Editors. (1981). Editorial. *Antipode* 13 (2): ii.

The Glenn W. Ferguson Papers, Box B 7-3-2, folder "Blaut, James M. 1967–1971." Clark University Archive. Clark University. Worcester, MA.

Geiger, R. L. (1995). The ten generations in American higher education. In P.G. Altbach, R.O. Berdahl, and P.J. Gumport, eds, *American Higher Education In The Twenty-First Century: Social, Political, and Economic Challenges*, pp. 38–70. Baltimore, MD: Johns Hopkins University Press.

Graham, J. (1992). Anti-essentialism and overdetermination – a response to Dick Peet. *Antipode* 24 (2): 141–156.

GSG. (1946). *The Monadnock: The Alumni Magazine of the Clark University Geographical Society*, 20(2). Graduate School of Geography.

GSG. (1968). *The Monadnock: The Alumni Magazine of the Clark University Geographical Society*, 42. Graduate School of Geography.

Hart, J. (1971). Dissension appeased. *The Scarlet May* 7: 1.

Harvey, D. (1972). Revolutionary and counter revolutionary theory in geography and the problem of ghetto formation. *Antipode* 4 (2): 1–13.

Harvey, D. (1981). The spatial fix: Hegel, Von Thünen, and Marx. *Antipode* 13 (3): 1–12.

Harvey, D. (1989). *The Condition of Postmodernity*. Oxford: Blackwell.

Harvey, D. (2015). "Listen, Anarchist!" A personal response to Simon Springer's "Why a radical geography must be anarchist." davidharvey.org/2015/06/listen-anarchist-by-david-harvey/

James, P. E. and Martin, G. J. (1978). *The Association of American Geographers: The First Seventy-Five Years, 1904–1979*. Washington, D.C.: Association of American Geographers.

Keighren, I. M. (2008). *Reading the Reception of Ellen Churchill Semple's Influences of Geographic Environment (1911)* www.era.lib.ed.ac.uk/handle/1842/3122

Koelsch, W. (1980). Wallace Atwood's "Great Geographical Institute". *Annals of the Association of American Geographers* 70(4): 567–582.

Koelsch, W. (1987). *Clark University, 1887–1987: A Narrative History*. Worcester, MA: Clark University Press.

Koelsch, W. (1988). Geography at Clark: The first fifty years (1921–1971). In J. E. Harmon and T. J. Richard, eds, *Geography in New England*, pp. 40–48. St. Lawrence Valley Geographical Society.

Linden, M. (2011). *The myth of the lower marginal tax rates*. Center for American Progress. www.americanprogress.org/issues/economy/news/2011/06/20/9841/the-myth-of-the-lower-marginal-tax-rates/

Madge, C., Raghuram, P., Skelton, T., Willis, K., and Williams, J. (1997). Methods and methodologies in feminist geographies: Politics, practice and power. In Women and Geography Study Group, ed., *Feminist Geographies: Explorations in Diversity and Difference*, pp. 86–111. Routledge.

Martin, G. J. (2015). *American Geography and Geographers: Toward Geographical Science*. New York: Oxford University Press.

Mathewson, K. and Stea, D. (2003). In memoriam: James Blaut (1927–2000). *Annals of the Association of American Geographers* 93 (1): 214–222.

McDowell, L. (1990). Sex and power in academia. *Area* 22 (4): 323–332.

Meehan, K. and Strauss, K. (eds) (2015). *Precarious Worlds: Contested Geographies of Social Reproduction*. University of Georgia Press.

Mintz, A. and Hicks, A. (1984). Military Keynesianism in the United States, 1949–1976: Disaggregating military expenditures and their determination. *American Journal of Sociology* 90 (2): 411–417.

Mitchell, K., Marston, S. A. and Katz, C. (eds) (2004). *Life's Work: Geographies of Social Reproduction*. Oxford: Blackwell.

Monk, J. (1998). The women were always welcome at Clark. *Economic Geography* 74 (s1): 14–30.

Murphy, R. E. (1979). Economic geography and Clark University. *Annals of the Association of American Geographers* 69 (1): 39–42.

O'Keefe, P. (1979). Editorial. *Antipode* 11 (3): 1–2.

Peet, R. (1977). The development of radical geography in the United States. *Progress in Human Geography* 1: 240–263.

Peet, R. (1985). Radical geography in the United States: A personal history. *Antipode* 17 (2–3): 1–7.

Peet, R. (1968). *The Spatial Expansion of Commercial Agriculture in the Nineteenth Century: A theoretical analysis of British import zones and the movement of farming into the interior United States*. Doctoral Dissertation, Department of Geography, University of California-Berkeley.

Peet, R. (1967). The present persistence of Von Thünen Theory. *Annals of the Association of American Geographers* 57 (4): 810–811.

Plette, E. (1971a). Students seek reform. *The Scarlet, April* 30: 1.

Plette, E. (1971b). "Women in America" Course Begins. *The Scarlet 53*, February 19.

Prunty, M. C. (1979). Clark in the early 1940s. *Annals of the Association of American Geographers* 69 (1): 42–45.

Rose, G. (1993). *Feminism & Geography: The Limits of Geographical Knowledge*. Cambridge: Polity Books.

Rossiter, M. W. (1995). *Women Scientists in America: Before Affirmative Action, 1940–1972. Vol. 2*. John Hopkins University Press.

Rugg, D. S. (1981). The Midwest as a hearth area in American academic geography. In B. Blouet, ed., *The Origins of Academic Geography in the United States*, pp. 175–191. Hamden: Archon Books.

Roboff, S. (1970). Marxist speaks on women's liberation. *The Scarlet* October 9, 51 (4): 4.

Sheppard, E. (2009). *Past Editors' Reflections*. onlinelibrary.wiley.com/journal/10.1111/%28ISSN%291467-8330/homepage/editor_s_past_reflections.htm - Sheppard

Soja, E. and Hadjimichalis, C. (1979). Between geographical materialism and spatial fetishism: some observations on the development of Marxist spatial analysis. *Antipode* 11 (3): 3–11.

Stea, D. (1969). Positions, purposes, pragmatics: A journal of radical geography. *Antipode* 1 (1): 1–2.

Vollmer, D. (1968) Sex discrimination; does it exist here? *The Scarlet* March 8, 90 (16): 2.

Waterman, S. (2002). Scholar, manager, mentor, mensch: Saul B. Cohen. *Political Geography* 21: 557–572.

Wisner, B., Heiman, M. and Weiner, D. (2005). Afterword: Jim Blaut, scholar activist. *Antipode* 37 (5): 1045–1050.

# 4

# A "Necessary Stop on the Circuit"

## Radical Geography at Simon Fraser University

### Nicholas Blomley and Eugene McCann[1]

In the mid-1970s, the Department of Geography at Simon Fraser University (SFU) in Burnaby, a suburb of Vancouver, emerged, almost overnight, as a hotbed of radical exploration, education, and praxis. Ron Horvath, who came as a visiting professor in 1974, suggests that it constituted the biggest mass of radical geographers in the English-speaking world at the time.[2] SFU was to play an important role in the development of Anglo-American radical and critical geography, yet this story is easily overlooked, as it has not been documented.[3] Indeed, a 1990 survey of the Department's history makes only passing reference to the moment (Hayter 1990). Our most immediate goal in this chapter is to describe what happened at SFU, particularly from the perspective of those who were active participants. While not seeking to romanticize the radical era, we believe it deserves better recognition. Memories and interpretations differ: we have tried our best to be even-handed in our account. This is not intended as a history of the department during the 1970s (human geography at SFU during this formative period in its history was also more than radical geography), nor are we able to provide a full accounting of all those who participated in the "radical convergence" or were peripherally affected.

Our motivation in documenting this story also reflects a desire to understand the history of an institution in which we work. Indeed, for

*Spatial Histories of Radical Geography: North America and Beyond*, First Edition.
Edited by Trevor J. Barnes and Eric Sheppard.

one of us (Nick), the radical 1970s still cast a residual shadow when he arrived at SFU in 1989, particularly in light of its subsequent unraveling and suppression. Mention of "theory," for example, was met with a muttering under the breath from some quarters, given its supposed association with radicalism. Michael Eliot-Hurst, the leading faculty member involved with the moment, was in poor health, and had been marginalized, retiring in 1989 and dying a decade later in relative obscurity.

Beyond documenting this moment, we ask a number of questions. Most immediately, we center on the distinctiveness of the radical moment at SFU, asking who was involved, what happened, and how it was lived, tracing the importance of a number of radical moments of learning, praxis, and politics, specifically the Vancouver Local of the Union of Socialist Geographers (USG) and the Vancouver Geographical Expedition (VGE), while also noting the role of early feminist geography. SFU was not, of course, an immaculate conception. As we show, it was shaped by its institutional context, both in relation to SFU's creation as a new university in the mid-1960s and also to its particular departmental culture. Of, course, in turn, this new context provided a physical space in which various actors coalesced to produce knowledge in and through certain power structures. We also consider SFU as one node in a wider circuit of radical geography, particularly within Canada and the United States. Thus we also ask: How was the radical moment at SFU shaped by its socio-spatial relationships in this network? What came to SFU through these connections, how did SFU shape the network, and with what consequences?

## The Convergence

The radical moment at SFU was short and intense, with few obvious precedents. From around 1973 to 1976, radicals at SFU Geography engaged in a period of avid exploration, engagement, and creativity. Len Evenden, a sympathetic if non-self-declared radical faculty member at the time, described the emergence of what he called a "swirl" of ideas and activities[4]. A radicalized group of graduate and undergraduate students propelled the swirl, working with some younger and visiting faculty. It revolved around particular initiatives (the VGE and the Vancouver Local of the Union of Socialist Geographers, in particular) and benefited from either the direct engagement of some more senior faculty members (particularly Eliot-Hurst, Chair from 1971–1975) or the tolerance of others, notably cultural geographers Ed Gibson and Phil Wagner, and social geographer Bob Horsfall.

It was the students who were most directly active, however. Bettina Bradbury, a History graduate student at the time, remembers the experience of exchanging radical ideas with graduate students from all over the world as incredibly exciting, noting her and her partner John Bradbury's radicalization happened through engagement with other students, rather than faculty, learning Marxist analysis through marathon pub conversations.[5] Colm Regan, also a graduate student at the time, noted that the internal strength and inner resolve of the graduate students was crucial in maintaining radical energy, even after the end of Eliot-Hurst's Chairship.[6]

The energy and vibrancy of the period was palpable, as participants read, debated, and organized. Bob Galois, another graduate student, prefers the term "critical mass" to describe the manner in which a crucial threshold of people and energy was reached as people came together.[7] There was lots of work, "but you just got wrapped up in it" noted Galois, although graduate theses often suffered as a consequence. Susan Williams, an undergraduate, notes that while it consumed evenings and weekends beyond her already heavy student load, "we were excited about the idea that change was going to happen." It was grueling (*Capital Volume II* nearly "killed me," she recalls), but there was a sense of camaraderie and activity driven by a sense of higher purpose, and a "feeling that you were on a … journey together" to change the world.[8] As we note next, many participants described the experience as formative and life-changing.

The motivations that drew participants into this convergence were clearly varied. Some arrived with a radical background, while others came from conservative educational and cultural settings. Susan Williams describes being drawn in through her experiences in Eliot-Hurst's classes, contact with the radical TAs attached to these classes, and engagement with the Geography Student Union, which also was progressive in orientation.[9]

Radical energy was directed at geography as a discipline, but also beyond. Several participants noted the importance of a group mentality. For Colm Regan,

> there was a sense of us-and-them, there was a sense of challenging the establishment … challenging people like [Richard] Hartshorne, and people like that, establishing that geography had been in the service of imperialism… There was a lot of piss and vinegar politics.[10]

Regan also recalled interminable conversations as to whether people identified as "radical" or "socialist," noting that "we were all branded as socialists by the outsiders."

For many, the goal was to confront the "establishment," including its disciplinary manifestations. One manifestation of this critique – and a clear statement of a binary between the establishment and its alternative – comes from a 1973 *Antipode* paper by Eliot-Hurst, entitled "Establishment geography: or how to be irrelevant in three easy lessons." He characterized "establishment geography" as a delusional, pervasive, and quasi-religious sect of devout believers, shored up by such powerful infrastructures as textbooks, academic meetings, graduate training, and departmental structures. He lamented how its long-standing roots in imperialism and capitalism had been now enhanced by the embrace of quantification and "the deification of prescriptive scientism" (Eliot-Hurst 1973: 41). In the new geography, we "find the *rigor* needed to guide the 'planes to North Vietnam, but alas also, the *mortis*" (Eliot-Hurst 1973: 42, his emphasis). In its place, he called for an "antiestablishment geography" that debunks and demystifies establishment geography, develops alternatives to the positivist orthodoxy, engages with human equity and wellbeing, politicizes students, and lives its theory.[11]

Yet the radical moment ended almost as quickly as it began. In 1975, Michael Eliot-Hurst suffered what appeared to be a nervous breakdown, thought to be related to issues in his personal life. He went on medical leave just before his term as Chair ended. The new Chair, a physical geographer, was no sympathizer of the radical project. The positions of some radical visiting faculty were not renewed. Eliot-Hurst's graduate students, central to the swirl, were assigned new advisors. When Alan Mabin arrived in late 1975 as a graduate student, he described a sharpening environment of conservative constriction, driven by budget cuts and an institutional desire to "shut things down".[12] Unsurprisingly, when Fran Walsh, who had been an MA student in 1970–1971 returned to SFU in 1978, it was to what he called a "pacified" department.[13]

## The Union of Socialist Geographers

The convergence crystallized around a number of often overlapping nodes of radical learning, praxis, and politics. The Vancouver Local of the USG played a pivotal role. As Bettina Bradbury described it, the Local was "central to everything that was going on… it was the intellectual life of us and our fellow graduate students… it was a framework through which to think everything, and a heart of a kind of sociability".[14] For Colm Regan, the USG aimed to demonstrate that radical geography was more than a theoretical construct, but was relevant to the struggle by "building a practice of socialist geography".[15]

The wider emergence of the USG, and its international scope, are more fully documented by Linda Peake (this volume). Committed to the radical restructuring of society in accord with the principles of social justice, the USG played an important role in the radicalization of the discipline within the English-speaking world. It was formed at a gathering in Toronto in 1974, associated with the Toronto Geographical Expedition. Many SFU people attended this first gathering (the sign-up sheet lists 34 people, with nine from British Columbia: SFU faculty Eliot-Hurst and Ron Horvath presented at the associated symposium). Reportedly, Eliot-Hurst provided the departmental van to allow SFU participants to cover the 4,300 kilometers from Vancouver. SFU participants appear to have been active in shaping the terms of reference for the USG, including suggestions that one of the goals of the organization should be agitating for staff and student parity in departmental decision-making.

Notwithstanding its Toronto origins, "SFU was prepared to put in work to build it," noted Regan.[16] A Vancouver Local was established at a meeting on September 10, 1974. A discussion ensued on the focus of the USG: should it be confined to the academy, or engage in agitation within the community? Eliot-Hurst suggested that the two were inseparable if the USG were truly socialist, as the goal for members would be revolutionary change as human beings first, then as geographers. Debate ensued concerning the "bourgeois nature of geography, the need for a socialist alternative within the discipline, [and] the development of revolutionary geographic theory."[17]

The Vancouver Local was many things, but three distinct foci can be identified: reading groups; producing the USG Newsletter; and developing conferences that would be run in parallel with more conventional academic meetings. This initiative of the Vancouver Local became a mainstay of USG organizing, piggybacking paper sessions onto CAG and AAG conferences (while maintaining an important distance from such "establishment" events). Fortuitously, the Canadian Association of Geographers annual conference convened in Vancouver in May 1975, where a parallel USG gathering was organized with papers on labor, imperialism, the Toronto Geographical Expedition, and anarchism. Many SFU radicals participated, including Colm Regan, Fran Walsh, Ron Horvath, John Bradbury, Michael Eliot-Hurst, Nathan Edelson, Alison Hayford, and Peter Walsh, as well as visitors from outside, including David Harvey and Jim Blaut. The Vancouver Local also organized a five-hour field trip of "working class Vancouver," with an accompanying 14-page guide written by USG and VGE members. Fran Walsh returned to Vancouver for the gathering, and experienced the meeting as one of intense discussion, debate and "a great sense of enthusiasm and camaraderie."[18] While one contemporary observer noted an potentially exclusive "in group" attitude, he relished the manner in which the USG sessions drew growing numbers of participants, while the rest of the CAG "dabbled in jargon and 'old boy network' name dropping" (Gerecke 1975: 9). There was lively discussion over the focus and scope of the USG. An Annual General Meeting (which reportedly extended over several nights) included extensive, albeit unresolved examination of logistics and organization: "Perhaps in a year … we will elect someone to travel the continent, disseminating information about the USG (a full-time traveling minstrel). Until that time, just talk to the person on your left" (Anonymous, 1975a: 17).

Such parallel conferences played a crucial role as sites for maintaining and invigorating the USG network, as we note next. In this pre-Internet age, maintaining connections beyond such gatherings required other forms of communication. Crucial in this regard was the development of the *USG Newsletter*. The *Newsletter* was initiated at the Vancouver meeting in 1975, with the Vancouver Local electing to take the lead. An SFU editorial collective produced five substantial issues of the *Newsletter* before the McGill Local (including many former SFU students) took over in late 1976. Under Vancouver editorship, the Newsletter came to provide summaries of annual USG meetings, updates on the activity of the Vancouver Local, bibliographies on particular themes, satire, course outlines with a socialist approach, and, particularly early on, free-standing articles that included an exegesis on Marx's theory of circulation. Hopes for "ideological eye opener" sections, in which extracts of geographical writing were to be subjected to ideological critique, did not take off. In its last issue, the editorial group reflected on the difference between the Newsletter and the journal *Antipode*. While complementary, the former was seen as offering a space for shorter articles and more avowedly Marxist in orientation. *Antipode* was described as offering a "radical" political perspective, whereas the Newsletter was "explicitly for socialist geography" (Anonymous 1976a: 1). Nevertheless, the Vancouver Local also edited a volume of *Antipode* (published in 1976), including papers by local participants Bob Galois, John Bradbury, Jim Overton, Peter Usher, Colm Regan and Fran Walsh, and Phil Wagner.[19]

The Vancouver USG Local was much more than a newsletter editorial board, however. It became a focus for shared learning, engagement, and socialization, drawing in a large number of SFU participants, mostly graduate and undergraduate students, with the former taking the intellectual lead. There were monthly meetings, and people were reportedly frowned upon if they didn't show up. Minutes reveal a remarkable array of activity: discussions of course proposals, local politics, engagement with other organizations, planning for the Newsletter or the special issue of *Antipode*, beer purchases, discussions of Chomsky on alienation, prison seminars, correspondence from radicals outside Vancouver, and Hallowe'en party planning.

Of crucial significance was the role that the USG Local provided in allowing participants to engage in forms of collective learning. It is easy for us, 40 years on, to take for granted the availability of a corpus of radical social theory from which to draw. This was not evident in the mid-1970s, given the relatively recent radical turn in geography (Harvey's influential *Social Justice and the City* only appeared in 1973). While some faculty were sympathetic, the graduate program at the time was far from sophisticated, and self-education through organized reading groups became crucial. For Lee Seymour, a Teaching Assistant: "We learnt from each other".[20]

Colm Regan noted that the corpus of "radical geography" was very eclectic, at least to begin with, but became increasingly theoretical and rigorous.[21] However, participants note a variety of other threads and themes, including an interest in imperialism. One product was the massive *Study Papers on Imperialism*, an 87-page document prepared for a mini-course on Marxist perspectives on geography at the 1976 AAG meetings. Eliot-Hurst also became preoccupied with Maoism. According to Susan Williams: "someone was always reading the *New Left Review*".[22] Baran and Sweezy, Poulantzas, Althusser, Mandel, and Kropotkin were in the air.[23]

Participants describe the ethos of the Local, including its reading groups, as internally supportive and communal, without a competitive spirit. They describe "hot-housing" student's theses if they were struggling to complete them, and convening "dry runs" of paper presentations to ensure the maintenance of quality to the "external" community. People worked together at a radical bookstore, and some shared accommodation, notably in a house at 2057 Napier Street on Vancouver's Eastside.

## The Vancouver Geographical Expedition

The Vancouver Geographic Expedition was a manifestation of earlier Expeditions in Toronto and Detroit (Heyman 2007; Heynen and Barnes 2011). Merrifield (1995) notes that the expedition movement emerged from the turbulent era of the late 1960s, motivated by a desire for social relevance and a rejection of establishment geography. Bill Bunge developed the idea with Gwendolyn Warren, establishing the Detroit Expedition in 1969. It drew from geography's positivistic tools, but with an insistence on bringing geography to the scale of people's lives through an anti-racist and anti-capitalist lens, coupled with a conscious subversion of the "expedition" label from nineteenth-century colonialism (see Warren *et al.*, this volume). Theory and practice were mobilized, with the use of field manuals and data reports, and the promotion of community activism and local empowerment in the "base camp community" (Horvath 1971: 2016).

Bunge moved from Detroit, establishing an Expedition in Toronto in October 1972 (Stephenson 1974), expressing greater optimism concerning the radical potential of Canadian students: "Youth is crushed in the United States, unlike Canada where some spirit remains" (Bunge 1977a: 1). Bunge had first visited SFU in 1969 (Bunge 1969: 29), returning in Spring 1973 to outline the Expedition ethos to a larger audience of 200 students.[24] He expressed its desire to document and contest the expansion of "machine space" (Horvath 1974) at the expense of plant, animal and human life, particularly that of children.[25] In August, 1973, six SFU students went to observe and participate in Toronto, and exploratory work in Burnaby, New Westminster, and East Vancouver began in the autumn of that year. It was decided to focus on the Grandview area of East Vancouver, because of its already existing political organization, its proximity to SFU, and its experience of urban change. The focus would be on transport, housing, and recreation (including documenting traffic flows and accidents, and developing children's journey to school routes to reduce accident rates). A manifesto was outlined in January 1974, advocating an emphasis on engaging the city through physical presence and residency rather than abstract forms of visualization, also calling for attention to the dangers of machine space upon "life" (notably that of children), and a desire for community engagement (Anonymous 1974: 1). The overall goal was ambitious:

> [w]e are geographers committed primarily to the betterment of living conditions in a particular region of the City of Vancouver. We believe that local residents should have a decisive level of control over the destiny of that community and that changes adversely affecting those residents without providing significant benefits to the community should be opposed. As social and physical and historical scientists we have a commitment to the general extension of theoretical knowledge. But as citizens of Vancouver our primarily [sic] interest must lie with the betterment of its people's living conditions (Edelson 1974: 7).

By February 1974, some 20 undergraduate students had become involved. Supportive professors encouraged students in their classes to engage with the project, and offered advice.[26] Graduate students

**Figure 4.1** Nathan Edelson, Suzanne Mackenzie, Colm Regan – VGE 1975. *Source:* SFU *Graduate Studies Calendar* 1975–1776, 63. Used with permission of Simon Fraser University.

also participated, offering assistance (Figure 4.1). In April 1974, the Expedition moved its HQ from the SFU Burnaby campus to the house on Napier Street (which was in in the Grandview-Woodlands neighborhood). Several expedition members expressed an interest in becoming permanent residents; the plan was that the house would become a meeting, working, and resting place to help develop local contacts. One outcome was a report on children's recreation prepared by Susan Barry, Gwen Robbins, and Tom Phipps (copy with authors). It provided systematic observations of children's use of the neighborhood, also documenting the many physical impediments to children's play (such as the lack of accessible play areas), as well as crucial resources for children (Figure 4.2).

The Vancouver Expedition had disbanded by the spring of 1975, however, ushering in an extensive post-mortem regarding its merits and weaknesses that revealed some of the fault-lines within the radical convergence more generally. An anonymous autocritique argued that it had outlived its usefulness and lacked revolutionary rigor, describing it

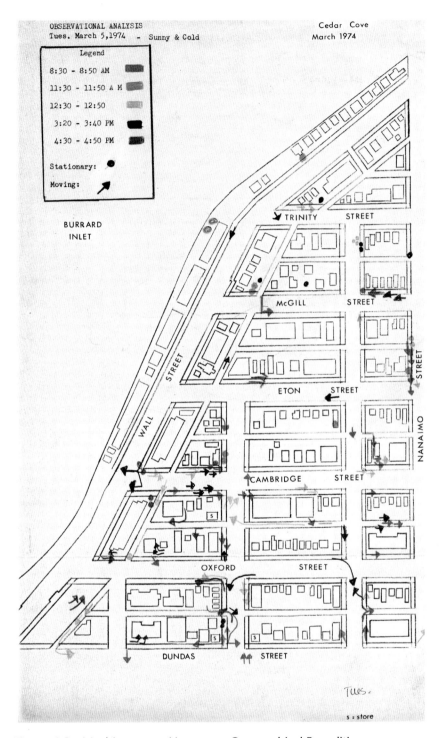

**Figure 4.2** Machine space, Vancouver Geographical Expedition. Observations of children's movement on March 5 1974 at different times of the day. *Source:* unpublished VGE report by Susan Barry, Gwen Robbins and Tom Phipps. Reproduced with permission.

as a loose coalition of "politically and theoretically inexperienced left liberals, socialists and even conservatives" (Anonymous 1975b: 14), lamenting that "we were unable to contribute significantly to class struggle" (14) in the Grandview area. Yet it was argued that the data gathered in the project had been used by community groups and tenants' organizations when responding to the housing crisis, wage controls and the tightening of social services. VGE researchers had provided information to Grandview and East End activists concerning urban property ownership, had written newspaper articles, participated in seminars concerning proposed plans to make Vancouver an "Executive City," created the first computerized landlord data bank available to Vancouver residents, and helped demystify some of the expertise surrounding urban data and decision making. While the Expedition's work on tenant organizing "will not solve the housing crisis, ... it can effectively provide a number of people with both a window revealing the underlying processes of capitalist development and the organizational experience necessary for its radical transformation" (Anonymous 1975b: 14).

A positive spin was also provided by a later commentator, who noted the valuable role of students in the development of social struggle: "[s]tudents and profs have to get over this 'academic guilt complex' and get serious about the skills they have and how they should put them into practice" (Arnold 1976: 32). Conversely Nathan Edelson and Colm Regan argued that time could have been better spent by students on the hard work of reading theory: "Being a student – being a Marxist student – is serious political activity, it requires a good deal of time and energy" (Edelson and Regan 1976: 33). They also criticized the focus on the neighborhood scale:

> [T]hrough an overconcentration on issues of consumption (housing, light, facilities etc.) we often avoided dealing with the place of exploitation in the production process itself. Hence much of our work, despite our intentions, never really successfully integrated the place and functions of a given community within the larger context of a city under capitalism (33).

Foreshadowing her later interests Suzanne Mackenzie, then an undergraduate, argued that the problem was a failure to see capitalism as an integrated social system that encompasses both "community" and "work place issues" (Mackenzie 1976: 33).

Bunge visited the VGE Expedition house in the winter of 1975. It was not a happy meeting. Participants lamented Bunge's overbearing personality. As Susan Williams put it: "Bunge was 'a bit of a wingnut. It was his way or the highway."[27] In a letter in the USG Newsletter Bunge noted that "We were mutually appalled by each other" (Bunge 1977b: 19). He noted Lenin's directive to be self-critical, no matter how gleeful it leaves our enemies, and then declared that the VGE was doomed from the start, lamenting its lack of community control: "The VGE gave the impression that the intellectuals talked a lot and the community either said nothing or the intellectuals never got within ear shot" (Bunge 1977b: 42). Moreover, Bunge criticized the adoption of a permissive politics that did not match what he defined as community mores and norms, given the presence of gay participants in the VGE: "[t]he atmosphere [of the Expedition] suggested that it was bad manners not to be liberal towards homosexuals... [However] any behaviour that seriously insults local mores is unacceptable ..." (Bunge 1977b: 19). The Vancouver *USG Newsletter* editorial collective (Anonymous 1977) pushed back, pointing out a number of errors in his analysis, while also advocating for a more general discussion on sexuality and socialism[28].

## Feminist Geography and the Radical Project

The radical project at SFU was largely driven by men, both as faculty and students. As Bettina Bradbury describes it, in a retrospective commentary (2015: 263), "Geography as a discipline was a tough nut for women to crack." She notes that when she arrived at SFU there were no female professors in Geography or her home discipline of History:

> And in those early 1970s, one male professor thought it completely acceptable to complain to me ... over a beer, that having just remarried he had no idea how to relate to women students any more as he could no longer sleep with them (263–264).

Yet Bradbury, along with several other determined women, began to carve out an important space for women's voices and feminist perspectives more generally, within the radical convergence at SFU. There were some important precedents for this. The SFU chemist Maggie Benston had published a highly influential paper in 1969 in *Monthly Review* that was among the first to apply a Marxist analysis to the oppression of women. Moreover, in 1973, the department ran a one-off class in the Geography of Gender (one of the first in Canada), coordinated by faculty member Bob Horsfall. While it was not repeated, and reportedly generated a "few hackles and a few guffaws" from male faculty (Lebowitz *et al.* 2008: 179), student interest in the class was useful in lobbying for the successful formation of the Women's Studies department at SFU, one of the first for-credit programs in Canada.

Yet, emergent feminist perspectives were clearly on the margins of the radical project in SFU Geography. For example, a session on "alternative perspectives in geography" was organized at the CAG meetings in 1975, which included a paper on indigenous peoples in the Canadian north and a presentation by Alison Hayford entitled "The Geography of women," exploring the roles of women within capitalist society and noting that Geography had excluded women from the field of research. The USG disassociated itself from the session on the grounds that there was no guarantee that its papers would be socialist, while condescendingly commending the authors for challenging "Establishment Geography" (Anonymous 1975c: 15).

Nevertheless, a socialist women's reading group was formed within the USG Local later that year, including Suzanne Mackenzie, Bettina Bradbury, Lee Seymour, Barbara Horvath, Mary Ogilvie, and Susan Williams. Participants noted their desire to engage with feminist theory through a socialist lens, while creating a space for women's voices to be heard. As Lee Seymour notes, "we got pissed off with the men... We were trying to find our voice"[29]. Concerned that men were having all the space, there was a desire for a women's-only reading group. There was no feminist literature within geography, obviously, so participants remember reaching out to identify relevant material, such as that of Maggie Benston, or sociologist Dorothy Smith. This folded over into personal relationships, Lee Seymour noted: "We all began to challenge the men and demand more in the relationship we had with them."

Relations with the broader group were not always easy. For Lee Seymour, there was some resistance to feminist ideas within the USG Local, given the centrality accorded a class analysis. However, she noted that some were more sympathetic to a feminist analysis, and that any disagreement was amicable, also noting the galvanizing presence of many "strong women".[30] Yet participants clearly struggled to be counted. Bob Galois noted that many male Marxists were "unenlightened" regarding gender issues, also remarking that "lots of hormones were flying around at the time." At the USG meetings that paralleled the 1976 CAG meeting at Laval, Lee Seymour, Suzanne Mackenzie and Alison Hayford presented papers at a poorly attended panel session: One of the participants reported at the time that "the fact that we were women and socialists resulted, we suspect, in many people not taking us seriously," it being noted that

> Women's success or the success of women studying women in all disciplines is still defined by men, on their terms and according to their criteria. In Geography in particular, the study of women by women is still treated as a joke (Anonymous 1976b: 10).

Yet the participants had the last laugh. Bettina Bradbury suggested that the experience was important, shaping her future career as a historian who considered herself both feminist and socialist. While sadly we do not have her own testimony, it is worth considering the influence of the SFU experience on Suzanne McKenzie, who went on to an important, though tragically curtailed career, as "the first women geographer in Canada, and one of the first in the world, to develop a substantive research programme in geography and gender" (Anonymous 1999). Her socialist-feminist analysis and her belief that feminist geography was 'a living, pulsating tool for analysis into social change" (Peake 2015: 258) were surely influenced by her deep involvement in the SFU convergence.

## Swirling Outwards: SFU's Radical Moment in Context

Despite Edelson and Regan's argument for a more studious and less locally-engaged approach to radical geography, we have suggested that the radical moment at SFU was deeply embroiled in its local contexts, within SFU and in Vancouver. Yet, when considering the radical moment, it is important to step back from the daily ferment of reading groups, thesis-writing support groups, internecine politics, conference organizing, and community expeditions to place it within the larger context of higher education in Canada and the United States, and within the development of radical geography across the two countries. In the following sections we turn to the institutional context of SFU as a new 1960s university, with its own vision of what might be called "mainstream" or "liberal" radicalism, exploring how this influenced what was happening among its geographers at the time. We will then scope out further to discuss SFU Geography's place in the circuit of traveling people and ideas that, through the 1970s, disseminated radical geographical ideas across the US and Canada. In this context, it is important to consider how the radical moment at SFU was shaped by its relationship with this network and by what came to SFU through these connections, as well as how SFU geographers shaped the network.

### A mainstream radical campus

Simon Fraser University's original campus was built in only two years (1963–1965) to serve the growing demand for university education among a suburbanizing baby boom population that had, since the end of

**Figure 4.3**  Radical architecture. *Source:* SFU *Graduate Studies Calendar* 1975–76, 2. Used with permission of Simon Fraser University.

the Second World War, expanded in settlements eastwards down the Fraser Valley from the original urban core of Vancouver. Its architecture, as Figure 4.3 suggests, was a form of modernist brutalism – "weird and futuristic," according to Alan Wallace, one of the radical students (quoted in Peake, this volume).

Its locally based architect, Arthur Erickson, deliberately intended that SFU contrast with the University of British Columbia (50 years SFU's senior) in both form and function. Whereas UBC sat on the far Western edge of Greater Vancouver, SFU was built to the east, on a largely undeveloped mountaintop in Burnaby. UBC's campus was as traditional in its organization as it was spectacular in its ocean-side location. Departments and faculties at UBC were housed in their own separate, self-contained buildings, spread out across a sprawling campus, which Erickson found "appalling [with] … [n]o place for students." For him, its architecture consisted of "a jumble of cacophonous, rambling personal statements…fortress[es] set up against the intrusions of rival disciplines – jealous enclaves with no exchange" (Erickson quoted in Stouck 2013: 184).

In contrast, SFU was to be a modernist, idealized reinterpretation of the academies of Greece, defined by interconnected spaces that would encourage as much interaction among students as possible. Its rough concrete structure was unitary in design, with its sections all connected into a single low-slung complex – a "single-structure campus," as *Architectural Forum* put it in a favorable review at the time (quoted in Stouck 2013: 196). This ideal resonated with the vision of Gordon Shrum, a former UBC administrator charged by the British Columbia provincial government with establishing the new university. His long, sometimes tense, affiliation with UBC (Fotheringham 1965) motivated him to distinguish SFU through innovations in pedagogy and administration. SFU was to be a campus for students and, ideally eventually a campus by students, in the sense that its spaces would encourage interaction and new ideas, nurture the energies that emerged from random meetings and unusual combinations of students and faculty, facilitating deeper learning and ongoing (re)thinking about higher education itself. As we will see, however, this ideal was soon tested (and found wanting) by the political radicalism of students and faculty.

Fifty years later, SFU is branded as a "radical campus" (Johnston 2009), with references to political radicalism domesticated and combined with nods to Erickson and Shrum's depoliticized radicalism by the University's marketers. Yet, in SFU's first decade, back before the concrete had fully set, as Bob Galois evocatively put it, the two visions were in distinct tension.[31] This tension was perhaps best highlighted by the concerns of the radical geographers. Their efforts at community engagement in the VGE and their internationalism through the USG contrast with the attractiveness Shrum found in the Burnaby Mountain location – a place he saw as "largely removed from the normal world" (Johnston 2009: 43).

## Political radicalism shaking SFU's foundations

Contemporary worldly concerns motivated intellectual debate and political activism on campus, nonetheless: not just in Geography but other departments as well. For example, one of the University's experiments in interdisciplinarity was to create a single department of Political Science, Sociology, and Anthropology (PSA), whose founding Head was the Marxist sociologist Tom Bottomore. Tensions within the department and between the politically left-leaning portion of its faculty and the University's administration, around academic freedom, student involvement in decision-making, and control of hiring decisions, created a crisis. Over 100 students who had occupied the administration building in late 1968 were arrested, a prolonged faculty strike ensued in 1969, and PSA subsequently was abolished and replaced with more traditional, separate departments: Archaeology, Sociology and Anthropology, and Political Science. Five years after the University's founding, the concrete was beginning to harden.

The decision to break up PSA had consequences for the Department of Geography, toward which a number of the PSA students gravitated, seeing it as a haven of left radicalism at a time of University. This no doubt added to the momentum of "the swirl." Perhaps as a result of the various influences that converged in the department's classrooms, offices, and hallways at this time, political radicalism was varied and dynamic in SFU Geography in the late 1960s and first half of the 1970s, as the examples discussed above emphasize. Another aspect of the story, however, is how these left political radicalisms were protected and nurtured in an institution that, despite the lofty goals of its founding, had proven itself willing to implement top-down hiring decisions, surveillance of what was being taught in individual classrooms, and the wholesale dismantling of departments that did not toe the line.

Michael Eliot-Hurst was a key catalyst in harnessing the institutional resources of a mainstream radical institution to help develop left radical geographical praxis. Hired as a mainstream economic and transportation geographer in 1965, Len Evenden recalls that Eliot-Hurst had remained peripheral to early debates among the Geography faculty over the intellectual focus of the department, until he was politically radicalized during a 1969 sabbatical at California State University, Northridge, where he experienced state surveillance of campus activities.[32] Wendy Eliot-Hurst, Michael's partner, played an important role.[33] A PSA grad student, she was an active member of the USG Local, and may have influenced Michael's radicalization.[34]

Eliot-Hurst influenced the radical swirl in three ways. First, his published work, including critiques of mainstream geography (Eliot-Hurst 1973), gained the attention of colleagues elsewhere and attracted radical graduate students to SFU. Second, he was a remarkable, innovative undergraduate teacher in a department where professors taught many courses, large and small. His introductory human geography course became legendary – always over its circa 500 student capacity, with 10–12 Teaching Assistants, unregistered students sitting in the aisles, and, at one point, two technicians assigned to manage its multiple audio-visual aids (film, tapes, slides). Looking back, Ron Horvath[35] identifies him as the "fertilizer for radical ferment" among the students, Bob Galois recalls his "all singing, all dancing, multimedia" lectures as "amazing, pedagogically," and Nathan Edelson calls them "magnificent." When Fran Walsh, coming from a conservative Irish background, attended Eliot-Hurst's class, "the earth moved." For Galois, teaching was Eliot-Hurst's main strength.[36]

Third and perhaps most important to both the rise and demise of the radical moment in SFU Geography, was Eliot-Hurst's role in departmental governance. He became heavily involved in administration upon returning from California, exhibiting skill and innovation, or Machiavellian guile depending on one's perspective, at a time when room for maneuver was lessening. He certainly seemed to value concentrating power – he was simultaneously department chair and chair of the Graduate Studies Committee – as well as opening departmental decision-making to undergraduate and graduate students, who were appointed with voting rights to all departmental committees. With Eliot-Hurst in charge, departmental space and resources were available to all students, but particularly the radical group – who Evenden[37] dubbed the "red guard." Edelson remembers the large amount of departmental space afforded to the students, including a large space that acted as a "student union" with library, seminar, meeting, and study space.[38] With Eliot-Hurst as benefactor to the radical students and faculty, resources also were available to bring in speakers, produce the USG Newsletter and the issue of *Antipode*, while vans and a departmental credit card were available to fund conference trips, including to the founding meeting of the UGSC in Toronto, mentioned above.

Fundamentally, Eliot-Hurst's time as chair was both intellectually and materially crucial to the development of the radical swirl. Yet he was neither the intellectual center nor the daily driver of radical thinking within the group. As Colm Regan argues, "He created this strand of energy … but never really controlled it … Michael created the space that we were able to fill in."[39] For Edelson, Eliot-Hurst's way of operating was a learning moment: "I learned a lot from the way he handled power," particularly regarding how institutional resources could be turned to support radical visions.[40] Yet Eliot-Hurst's way of handling power was also divisive, and he eventually overstretched and broke the alliances among the faculty that allowed his practices to flourish. His decision to put students on the salary review committee, with the ability to vote on whether faculty would receive salary increases, is often cited as the final straw for many, although both Evenden and Horsfall[41] say they had no problem with the practice. Bob Galois sums up Eliot-Hurst as "slightly chaotic, creative and divisive."[42]

## Class consciousness: Students learning and creating radical geography

If Eliot-Hurst created the space for many radical activities, other continuing and visiting faculty, including Bob Horsfall, Ron Horvath, and Jim Overton, supported students as they combined their intellectual interests and political convictions. Eliot-Hurst's secretary, Barbara Shankland also played an important role. Hailing from a Scottish coalmining family, she supported the radical moment. The undergraduate students were also crucial figures. They were special in many ways, with an average age higher than elsewhere. Alan Mabin recalls that he was younger than the average age of his students when he taught a class in 1977.[43] Like students across the campus in its early years, the geography undergraduates had life and work experience and, for the most part, working class backgrounds. They were socially conscious, high quality, and worked extremely hard, Evenden remembers, especially in their upper-level research and field-based courses.[44] They created an exciting milieu; both former faculty and graduate students comment on how much they learned from the undergraduates.

The undergraduate student group in the department was called the Geography Student Union. This name, which persists today, speaks to the politics of the students in the University's early years. The GSU produced a Newsletter, the first issue of which (GSU 1972) called for enhanced student input in departmental affairs and proposed an "anti-calendar" (a means of "exposing worthless courses and professors"), which was eventually produced. The GSU Newsletter also offered an extensive critique of the "true role of the capitalist university," identified as the promotion of bourgeois ideology, with the imperative that students not be exposed to revolutionary socialist consciousness. Geography, as an academic discipline, was presented as deeply implicated in this agenda: geographers were the most conservative of social scientists, due to their "poor intellectual caliber …, the stultifying effect of geographic training on the recipient, the basically boring content of geography as presently taught and the political naivety of geographers" (GSU 1972: 5).

According to a number of our interviewees, these were the words of students influenced by a particular form of Western Canadian radicalism that grew out of the province's largely rural landscape of resource-extractive industries, mixing environmentalism, anarchism, and the Wobblies's trade unionism with the experiences of feminists, American draft-dodgers, and religious non-conformists. Thus, Suzanne Mackenzie, who received her undergraduate degree in 1976 and was a strong force among the radical students at SFU, came from a family that was heavily involved in the union movement in the Kootenay region of British Columbia.[45]

## Convergence and Connections: Radical SFU Geography, In and Of the World

The local confluences and connections that produced radical SFU Geography – the gravitational pull attracting students who had been shaken loose by disruptions across campus or were looking for perspectives not offered by their own departments, the radical openness that attracted undergraduate students to Eliot-Hurst's and others' classes, the undergraduate and graduate interaction and common purpose – were all entwined with places and people further afield.

### Radical convergence

Numerous people had come to SFU from elsewhere in the world. One group was comprised of political exiles of one sort or another. Some came from the Republic of Ireland, frustrated by the Irish university system's conservatism. Fran Walsh had been an early migrant, but returned to University College Dublin to complete a Ph.D. after his SFU M.A., from where he encouraged local students Colm Regan and Alan Wallace to travel to SFU. Walsh then took a job at Maynooth, establishing a critical presence in geography that has persisted. Leaving Ireland was an attractive opportunity for Regan. SFU, where he stayed from 1974–1976, was the most attractive option because of its reputation, offering an escape from what he and others saw as a conservative regional historical geography approach in Ireland that had nothing to say about the situations in Northern Ireland, Southeast Asia, and elsewhere. Alan Mabin escaped the circumstances – and the draft – in South Africa. He already had learned of SFU as a result of reading Eliot-Hurst's publications, and communicating with him and David Harvey. While he moved first to Ohio State, he quickly decided that the west coast would be a better fit and picked SFU. Funded by Ohio State to attend the CAG/USG meeting in 1975, he stayed with Galois and met many of the other radicals. SFU's openness at that time hit those from more conservative locations "like a wall of water," as Evenden put it.[46] This new radical environment energized them tremendously: a "cataclysmic" change as Regan put it.[47]

A second group of incomers gravitated toward Burnaby Mountain, to a great extent as a result of SFU having been noticed by radical geographers elsewhere (in part through Eliot-Hurst's published commentaries, as Mabin's story suggests). David Harvey played an instrumental role in the moves of Bob Horsfall, hired as an Assistant Professor at SFU in 1970, and Nathan Edelson, who arrived in 1972 to work on a Ph.D. with Eliot-Hurst. Harvey knew them at Johns Hopkins and the University of Pennsylvania respectively and wrote recommendation letters for their applications. Bettina Bradbury, the historian who had come to SFU from New Zealand with her geographer husband John, recalls how this convergence created an "incredibly exciting time," with graduate students from all over the world exchanging radical ideas.[48] John Bradbury researched the fate of British Columbia's rural and remote resource towns and was politically active both in founding SFU's Local of the USG and in an unsuccessful attempt to prevent the building of a Chevron oil refinery at the foot of Burnaby Mountain. According to Galois's obituary of John (Rose and Galois 1988: 174), the Bradburys were central figures among the radical group in those "heady disputatious times." Bettina remembers that she spent a great deal of time with the geographers and was radicalized by other students, not the faculty. For her, it a "transformative experience" that laid the basis for an academic career grounded in socialist and feminist analysis.[49]

The travels of these individuals and others like them were also framed by the conjunctural context of the time, as Ron Horvath notes. This was a time when radicalism was appearing in multiple and differentiated ways – some muted, some more vibrant. It was a time that witnessed a

> general growth of rebelliousness among young people … associated with flower power, hippie and pop culture, drug use, the peace and protest movements (anti-Vietnam War, civil rights, women's rights) etc. [and the] massive growth of the international socialist movement among students in 1968.[50]

Alan Mabin argues that SFU's emergence as a radical center must be thought of in the context of similar "new peripheral universities, like University of Western Cape in South Africa and Nanterre in France".[51] Their important role in struggles over social change was based, he contends, on their self-image as special, different, non-elite institutions that attracted leading radical intellectuals, like Tom Bottomore and Michael Eliot-Hurst at SFU. Bob Galois echoes this point when describing the attraction of SFU in 1969: it was fresh, it was new, it had lively radical politics. It was "in tune with the times," but also with the place: it was very definitely not the United States.[52] For Colm Regan, "the States wasn't an option. Clark or Berkeley didn't figure on our map at the time."[53]

## "A necessary stop on the circuit"

SFU was clearly on the map for radical geographers in North America. It was part of a network that, like all such networks, was produced and reproduced through the mobilization of ideas and information being communicated through the USG Newsletter, the flows of graduate students and junior professors, but also the travels of major figures in this emergent field. Harvey, Bunge, Blaut, Milton Santos, and Yi-Fu Tuan – radical geography's "Johnny Appleseeds" (Akatiff 2011: 10), or traveling minstrels, as the USG Newsletter had it (Anonymous 1975a: 17) – each visited SFU, some more than once. In May 1975 the front page of the SFU student newspaper, *The Peak*, ran the headline "Socialist geographers come to SFU" (Figure 4.4). The accompanying article anticipated the upcoming second meeting of the USG and the arrival on campus of Blaut and Harvey. It also noted that SFU hosted the largest USG local, formed the year before. Drawing on an interview with Colm Regan, *The Peak* announced that, "[e]stablishment geography has always been incapable of explaining or solving most of the problems in modern capitalist society" whereas, the USG was "intent on reconciling geography and reality" (*The Peak* 1975: 1).

The role of conferences as "convergence spaces" and in the transfer of ideas (and ideals) is well known (Temenos 2016). Yet it is important to emphasize that they also play the important role of developing and nurturing "weak ties" (Granovetter 1973): the loose bonds of acquaintance and trust that connect geographically dispersed groups of people together into communities of common purpose. In the context of radical praxis, the productive role of conferences is not merely granted by the powers that be, but is taken and (re)made. As Regan notes,

> This was about saying, "feck this," we can organize our own conferences … If you wait for the establishment to give you permission, you will never get it. You get your permission with making your own space. We also wanted to work through our own arguments, our own debates, create an alternative site of energy.[54]"

This networking work was crucial to Susan Williams's experience. She was heavily involved in producing the USG Newsletter and attending the CAG/USG meetings, also presenting at the conference. For her, it was a somewhat "scary" experience, but this was tempered by the communal work in advance, where the group edited and practiced one anothers' papers.[55]

The USG, at SFU and beyond, was a group on a mission: "wrapped up in providing … mutual aid (spiced with great parties and fierce arguments) across multiple [radical] traditions" (Harvey 2015) at conferences that linked them up with fellow travelers from elsewhere. As Williams recounts, they "we were exposed to students from across North America through USG. We felt like we were part of a network, with a strong

Figure 4.4   Socialist geographers come to SFU. *Source: The Peak* [student newspaper], May 22, 1975.

**140**     SPATIAL HISTORIES OF RADICAL GEOGRAPHY

presence at CAG and at the AAG in New Orleans and New York"[56]. Similarly, for Regan these meetings, especially the Chicago AAG,

> gave us a sense of coherence. ... Meeting other people, gave strength and energy to go on. Gave a sense of a movement, not just autonomous sites of energy. But now there was a newsletter, partnerships, exchanges, not just *ad hoc* engagement. Cross fertilizing was important.[57]

These connections lasted beyond students' sojourn at SFU: Thus, Susan Williams moved on to McGill, where she built on her USG contacts to maintain "lots of connections with Clark and to Laval, via another CAG meeting".[58]

Conferences were only part of the circuit, however. Other crucial elements included the travels of leaders in the field, visiting various campuses across Canada and the United States to give talks and hold seminar discussions. By 1973–1974, Horvath says, SFU was the biggest mass of radical geographers in the English-speaking world and a "necessary stop on the circuit" for leading figures.[59] They brought differences in perspective, in how they engaged with faculty and students, and in how they conveyed their ideas, triggering different impacts among the group at SFU. According to Horvath, Bunge had a tense relationship with many of the students he met at SFU (as discussed above), whereas Blaut was the most effective interpersonally and able to persuade numerous students of his ideas on development geography and dependency theory. Harvey, more analytical and academic than Blaut, was also influential. His visit triggered the yearlong (1975–1976) *Capital*/Marx reading group, involving approximately 25 participants. In short, the networking work of students and faculty at SFU tied the department into a wider community, but also fundamentally shaped how radical geography developed.

## Down from the Mountain: Legacies and Limits

The most sustained and recognized influence in an academic discipline involves ongoing publishing in the academic literature. David Harvey (2015) has recently reflected on this:

> Survival in the discipline was an issue [at the time]. Having pushed the door open we had somehow to keep it open institutionally in the face of a lot of pressure to close it. ... I was from the very beginning determined to publish up a storm and I did emphasize to my students and all those around me who would listen that this was one (and perhaps the only)

way to keep the door open. It was more than the usual publish or perish. For all those suspected of Marxist or anarchist sympathies, it was publish twice as much at a superior level of sophistication or perish. ... Many of my colleagues in the radical movement, those with anarchist leanings in particular, did not care for that choice (for very good reasons) with the result that many of them, sadly, failed or chose not to consolidate academic positions.

The radicals at SFU were distributed on both sides of this line. Some published significantly throughout their academic careers, while also teaching, mentoring, and so on. Alan Mabin, for example, became a productive scholar, working in a variety of development fields. He notes the powerful importance of the SFU experience on his writings in the 1980s (e.g., Mabin 1986).[60] Among those who remained in academia, Fran Walsh, argues that he was "a different person" when he left SFU.[61] Bettina Bradbury, says her "work would not have been what it was without" SFU, finding her path difficult thereafter. She was rejected for a position in History at McGill because "they said I was 'too feminist' and 'too socialist.'"[62] She became a renowned feminist historian at York University. SFU undergraduate Suzanne Mackenzie went on to have an important career as a socialist feminist:

> Over more than two decades, her numerous books and articles exploring how women's lives shaped and were shaped by their social environments have become a beacon for and an inspiration to a whole generation of feminist geographers. She has influenced profoundly the ways that gender is now understood in Canadian universities (Anonymous 1999: 105).

Nonetheless, as Nathan Edelson[63] notes, SFU did not produce many radical Ph.Ds in the end, making it "an asterisk" in the received radical history of the discipline.

Others chose a hybrid route, moving beyond academic publishing by choice or otherwise. Walsh, while maintaining an academic career, became involved in an Irish NGO focusing on foreign ownership of natural resources, which shaped legislation to increase taxation rates and controls on such ownership, and became active within the Workers' Party. Bob Galois remained an academic, engaging in important research on colonialism, for example, but in a teaching position at UBC. Colm Regan also became involved in the NGO world. Alan Mabin returned to apartheid South Africa, and became involved in setting up an NGO working for poverty eradication. Lee Seymour, one of those who pioneered feminist geography at SFU, set up Marxist-feminist study groups in St. Johns, Newfoundland. She helped to found the Newfoundland

Association for Full Employment, worked with a local women's center and transition house, and worked for Oxfam Canada and Amnesty International. SFU "had a formative, catalyzing effect," she said. "I've been active all my life... [My time there] was totally formative in terms of my value system ... A huge legacy. Huge."[64] Thus, SFU's legacy exceeded the academy as Seymour and others went on to live radical geography and make radical geographies in their own no less significant ways.

As with any such movement, its legacy is defined both by the paths it clears and the limits it fails to transcend. The question of limits must also be acknowledged. Thus, Galois notes the general silence among other SFU radical geographers regarding settler colonialism. Imperialism was a focus, he notes, but there was little recognition of the history of the land upon which Vancouver sits and of the tension of holding radical conferences on unceded native land.[65] Horvath notes that the new environmentalism didn't figure greatly in radical discussions.[66] Terry Simmons, an early member of Greenpeace, remembers being reprimanded for engaging with "fuzzy-wuzzy environmental issues."[67] Nathan Edelson points to how the group missed the unraveling of Fordism and the emergence of neoliberalism, even as these unfolded around them in the mid-1970s.[68] The critical analysis of race among what was an overwhelmingly white project was also a significant silence (see Kobayashi, this volume).

While it is important to document this story, it has relevance for radical geographers today. Indeed, learning this history was significant for several current graduate students at SFU (mostly in Geography, as well as Urban Studies), who formed the interdisciplinary Place+Space collective in 2016. While the motivations for this collective are broad, its members nevertheless note the influence of SFU's radical moment, learning from its absences, and omissions (e.g., its whiteness, or the need to ensure continuity beyond the presence of individual members), while being inspired to fill its remaining gaps. Like the earlier movement, the current collective is student-led, and like the USG, both outward-facing, building solidarity with other groups, and inwardly supportive:

> While there might not be any direct line between the two "moments", knowing SFU geography's radical history has definitely helped some of us feel an extra sense of institutional belonging .... There are conditions of radical possibility that exist at SFU, and we think that it is no coincidence that a collective like ours formed here. For some, it was very exciting to find out about the radical past of the Geography department at SFU, and it made us feel super enthusiastic about the possibilities for us to challenge conventional structures[69]

In that sense, they note, "we like to think that the legacy of the radical movement is still 'alive' within us."

## Notes

1  We gratefully acknowledge the willingness of our many interviewees to share their memories of this important moment. Thanks to Len Evenden, Lee Seymour, Alan Mabin, and Proinnsias Breathnach for their comments on an earlier draft, which also benefited from the editorial eye of Eric Sheppard. Earlier versions benefited from feedback at presentations at AAG (San Francisco 2016) and at SFU (2016). Various documents relating to this episode, kindly shared by participants, have been archived (http://atom.archives.sfu.ca/union-of-socialist-geographers-2), hopefully to be made available in digital format soon.
2  Ron Horvath, interviewed by McCann (by phone), Australia, July 2015.
3  Why the story has not been properly documented is an open question. Some suggest its deliberate erasure from collective memory. Others, comparing it to radical schools such as Clark, point to the fact that there were few who went on to tell the tale within the discipline. Many of those who participated went on to important forms of activism and political engagement, but not within Geography. While it did foster some significant and productive scholars, two in particular – John Bradbury and Suzanne McKenzie – died tragically young.
4  Len Evenden and Nathan Edelson, interviewed by Blomley and McCann, Burnaby, August 2015.
5  Bettina Bradbury, interviewed by Blomley (by phone), New Zealand, July 2015.
6  Colm Regan, interviewed by Blomley (by phone), Ireland, June 2015.
7  Personal communication, October 11, 2016.
8  Susan Williams, interviewed by Blomley (by phone), Newfoundland, September 2015.
9  Williams, 2015
10  Regan, 2015.
11  For Bob Galois, an Althusserian ideological critique of the discipline was a motivation. He was intent on uncovering Geography's hidden politics, and recovering anarchist alternatives, notably the work of Reclus and Kropotkin (interviewed by Blomley and McCann, Vancouver, May 2015).
12  Alan Mabin, interviewed by Blomley and McCann, Burnaby, October 2015.
13  Fran Walsh, interviewed by Blomley (by phone), Ireland, July 2015 (Fran Walsh subsequently switched to using the Gaelic version of his name, Proinnsias Breathnach).
14  Bradbury, 2015
15  Regan, 2015.
16  Regan, 2015

17  Minutes of meeting of Vancouver local of Union of Socialist Geographers, September 10, 1974. Copy available from authors.

18  Email from Fran Walsh to Linda Peake, March 24, 2011 (copy with authors).

19  Galois's article on Kropotkin is particularly significant, being one of the first English language articles on Geography's anarchist past, seeking to rediscover "alternative traditions in geography that have either been eliminated from the discipline or emasculated beyond recognition" (1976: 13).

20  Lee Seymour, interviewed by Blomley (phone), Halifax, September 2015.

21  Regan, 2015.

22  Williams, 2015.

23  The geographical dimension of the convergence was somewhat diffuse. While the goal was disciplinary in focus for some (such as uncovering Geography's subaltern histories, like nineteenth-century anarchism), many respondents were hard-pressed to explain the specifically spatialized characteristics of their critique, and noted that much of their reading (especially Marx) was more general in orientation. Indeed, Eliot-Hurst (1980) would insist that by privileging space as a category while ignoring its fetishistic qualities, radical geography was inherently diversionary and bourgeois.

24  These notes derive from the unpublished comments of Nathan Edelson (1974).

25  A local SFU trained college instructor, Jim Sellers, engaged in an interesting precursor to the VGE, deploying geography students to engage in "grass roots" research (Sellers 1973).

26  Notably Bob Horsfall, who 'enforced' an organizing session at his home based on his own community organizing experience in Baltimore (interviewed by Blomley and McCann, Burnaby, July 2015).

27  Williams, 2015.

28  This proved an important thread in subsequent years: Eliot-Hurst, who came out shortly before the end of his Chairship, became an early defender of gay rights, helping to establish a 'gay caucus' at the AAG in 1976, seeking to support gay geographers in the process of coming out, or facing discrimination, and advocating for 'agit-prop by gay socialist geographers' (Eliot-Hurst, 1978, 35).

29  Seymour, 2015.

30  Seymour, 2015.

31  Personal communication, October 11, 2016.

32  Evenden/Edelson, 2015.

33  Terry Simmons, interviewed by Blomley, Vancouver, November, 2015.

34  Proinnsias Breathnach, personal communication, August 20, 2017.

35  Horvath, 2015.

36  One illustration is his urban geography textbook (Eliot-Hurst 1975), a highly innovative and creative reader that combined poetry, politics, and analysis.

37  Evenden, 2015.

38  Nathan Edelson, interviewed by Blomley and McCann, Vancouver, March 2015.

39  Regan, 2015.

40  Edelson, 2015.

41  Evenden/Edelson, 2015; Horsfall, 2015.

42  Galois, 2015. Indeed, he remains a lightning rod even now. While many celebrate the role he played in making space for the radical project, others remain hostile. One senior faculty member described him, in conversation with one of the authors, several years ago, as leading a quixotic "children's crusade." For others, the concern was said to be less a matter of political differences than a concern that he would detract from the reputation that the new department was keen to establish.

43  Mabin, 2015.

44  Evenden/Edelson, 2015

45  Alan Mabin points out the importance of the GSU, noting that it was a "very significant conduit, cause, and carrier" of the radical project. Locally grounded, and often mature, the undergraduates provided both weight of numbers, and political solidity (Mabin, personal communication, August 20, 2017).

46  Evenden/Edelson, 2015.

47  Regan, 2015.

48  Bradbury, 2015.

49  Bradbury, 2015.

50  Proinnsias Breathnach, personal communication, August 20, 2017.

51  Mabin, 2015

52  Galois, 2015.

53  Regan, 2015.

54  Regan, 2015.

55  Williams, 2015.

56  Williams, 2015.

57  Regan, 2015.

58  Williams, 2015,

59  Horvath, 2015,

60  Personal communication, August 20, 2017.

61  Walsh, 2015

62  Bradbury, 2015.

63  Edelson, 2015.

64  Seymour, 2015.

65  Galois, 2015.

66  Horvath, 2015.

67  Simmons, 2015.

68  Edelson, 2015.

69  Email: June 19, 2018, Samantha Thompson and the Place + Space collective.

# References

Akatiff, C. (2011). "A great day in Toronto": The founding of the Union of Socialist Geographers. Presented at Association of American Geographers meeting in Seattle. Copy available from authors.

Anonymous. (1974). *The Vancouver Geographical Expedition, January 17 1974.* Copy available from authors.

Anonymous. (1975a). Annual General Meeting. *USG Newsletter* 1(1): 16–17.

Anonymous. (1975b). Socialist practice. *USG Newsletter* 1(2): 14–15.

Anonymous. (1975c). Alternative perspectives in geography. *USG Newsletter* 1(1): 15–16.

Anonymous. (1976a). Editorial comment. *USG Newsletter* 1(4 and 5): 1.

Anonymous. (1976b). Socialist perspectives at the CAG. *USG Newsletter* 1(4 and 5): 10.

Anonymous. (1977). Dr. Bunge and the Vancouver Geographical Expedition: Expletives undeleted. *USG Newsletter* 2(2): 18–21.

Anonymous. (1999). Obituary. *Gender, Place and Culture* 6(1): 105.

Arnold, B. (1976). Socialist practice – Grandview revisited. *USG Newsletter* 1(3): 31–32.

Benston, M. (1969). The political economy of women's liberation. *Monthly Review* 21(4): 13–27.

Bradbury, B. (2015). Twists, turning points, and tall shoulders: Studying Canada and feminist family histories. *Canadian Historical Review* 96(2): 257–285.

Bunge, W. (1969). Detroit Geographical Expedition, Field Notes. *DGE Discussion Paper* 1.

Bunge, W. (1977a). The Canadian-American Geographical Expedition, Field Notes. *DGE Discussion Paper*, 2.

Bunge, W. (1977b). The Vancouver Geographical Expedition. *USG Newsletter* 2(2): 18–20.

Edelson, N. (1974). *History and Philosophy of the Vancouver Geographical Expedition.* Copy available from authors.

Edelson, N. and Regan, C. (1976). Socialist practice – Grandview revisited: Reply (1) *USG Newsletter* 1(3), 32–34.

Eliot-Hurst, M. (1973). Establishment geography: Or how to be irrelevant in three easy lessons *Antipode* 5: 40–59

Eliot-Hurst, M. (1975). *I Came to the City: Essays and Comments on the Urban Scene.* Boston: Houghton-Mifflin.

Eliot-Hurst, M. (1978). The gay caucus, Association of American Geographers. *USG Newsletter* 4(1): 34–35.

Eliot-Hurst, M. (1980). Geography, social science and society: towards a de-definition. *Australian Geographical Studies* 18(1): 3–21.

Fotheringham, A. (1965). How an old boy got back at UBC. *MacLean's* 78(October 16th): 67–68.

Galois, B. (1976). Ideology and the idea of nature: the case of Peter Kropotkin. *Antipode* 8(3): 1–16

Granovetter, M. S. (1973). The strength of weak ties. *American Journal of Sociology* 78(6): 1360–1380.

Gerecke, K. (1975). Vancouver: socialist geographers organize. *City Magazine* 1 (5&6): 9–10.

GSU (SFU Geography Student Union). (1972). *Newsletter.* October 31. Copy available from the authors.

Harvey, D. (2015). *Listen, Anarchist!* http://davidharvey.org/2015/06/listen-anarchist-by-david-harvey/ (accessed December 16, 2017).

Hayter, R. (1990). Geography at Simon Fraser University. *Yearbook of the Association of Pacific Coast Geographers* 52: 191–207.

Heyman, R. (2007) "Who's Going to Man the Factories and Be the Sexual Slaves If We All Get PhDs?" Democratizing knowledge production, pedagogy, and the Detroit Geographical Expedition and Institute. *Antipode* 39(1): 99–120.

Heynen, N. and Barnes. T.J. (2011). Forward to the 2011 Edition: Fitzgerald then and now". In W. Bunge, ed., *Fitzgerald*. Athens: University of Georgia Press.

Horvath, R. J. (1971). The Detroit Geographical Expedition and Institute experience. *Antipode* 3(1): 73–85.

Horvath, R. J. (1974). Machine space. *Geographical Review* 64: 167–188.

Johnston, H. (2009). *Radical Campus: Making Simon Fraser University*. Vancouver, BC: Douglas and McIntyre.

Lebowitz, A., Newcombe, H. and Kimball, M. (2008). Women's Studies at Simon Fraser University, 1966–76: A dialogue. In W. Robbins, M. Luxton, M. Echler, and F. Descarries, eds, *Minds of their Own: Inventing Feminist Scholarship and Women's Studies in Canada and Quebec, 1966–76*. Waterloo: Wilfred Laurier University Press, pp. 178–187.

Mabin, A. (1986). Labour, capital, class struggle and the origins of residential segregation in Kimberley 1880–1920. *Journal of Historical Geography* 12(1): 4–26.

Mackenzie, S. (1976). Socialist practice – Grandview revisited: Reply (2). *USG Newsletter* 1(3), 34.

Merrifield, A. (1995). Situated knowledge through exploration: Reflections on Bunge's "Geographical Expeditions". *Antipode* 27(1): 49–70.

Peak, The. (1975). *Socialist geographers come to SFU*, 30(4): 1.

Peake, L. (2015). The Suzanne Mackenzie Memorial Lecture: Rethinking the politics of feminist knowledge production in Anglo American geography. *The Canadian Geographer* 59(3): 257–266.

Rose, D. and Galois, R. M. (1988). John H. Bradbury, 1942–1988. *Antipode* 20(3): 173–179.

Sellers, J. B. (1973). "Making human geography serve people," A junior college example. *British Columbia Geographical Series, Malaspina Papers* 17: 124.

Stephenson, D. (1974). The Toronto Geographical Expedition. *Antipode*, 6(2): 98–101.

Stouck, D. (2013). *Arthur Erickson: An Architect's Life*. Vancouver, BC: Douglas & McIntyre.

Temenos, C. (2016). Mobilizing drug policy activism: Conferences, convergence spaces and ephemeral fixtures in social movement mobilization. *Space and Polity* 20(1): 124–141.

# 5

# The Life and Times of the Union of Socialist Geographers

## Linda Peake

Through a story of the life and times of the Union of Socialist Geographers (USG), a specific socialist and scientific "centre of calculation" (Jöns 2011; Latour 1987), this chapter aims to complicate the commonly accepted 1969 origins story of radical and critical geography as emerging from a singular place and institution: Clark University and *Antipode*. An independent socialist organization, formed in Toronto, Canada, in 1974, the USG was to disband just eight years later, morphing into the Socialist Geography Specialty Group (SGSG) of the Association of American Geographers (AAG). While the USG constituted a fully fledged formal organization (Akatiff 2011), it is best understood as a movement based on a culture of transnational solidarity, albeit an uneven and contested one. Placing this movement in its historical geographical context allows for a tracing of the networks of solidarity and praxis within which the seeds of Marxist but also feminist, anti-racist, and queer geographies emerged and traveled, linking the emergence of its agents and institutional dynamics of geographical knowledge production to the synergies of the times and places from which they emerged. Although little known about by current geographers the legacy of the USG speaks directly to current modes of institutional organizing and radical knowledge production in the Trump-Trudeau era.

The USG emerged from the social movements of the New Left, which provided the context for the emergence of radical countercultural geography and its activist orientation.[1] With the explicit turn to socialist geography through the formation of the USG, a more traditional scholarly

*Spatial Histories of Radical Geography: North America and Beyond*, First Edition.
Edited by Trevor J. Barnes and Eric Sheppard.
© 2019 John Wiley & Sons Ltd. Published 2019 by John Wiley & Sons Ltd.

orientation gradually developed centered on Marxist theory. The USG was, with *Antipode*, the most important node in the development of socialist geography, operating within a complex of networks across North America and beyond, its growth nurtured by the growing preponderance of socialist geographers in geography departments across the world. Recounting the formation and influence of the USG – through the circulation of its publications and of its members between locals, meetings, and countries – serves to recalibrate the geographical imaginary of the origins of radical geography from a singular point of departure to that of a network of politicized academic geography communities forming across and in relation to each other. The chapter concludes with an account of the demise of the USG, which speaks to the post-structural turn and the gradual fragmentation of socialist geography into one of many critical geographies, as well as an assessment of the USG's role in radical geographical practice and the lessons its existence has for critical geography today.

## Setting the Stage: North American Radical Geography in the 1960s and 1970s

If the role of the discipline of geography prior to WWII was to manage an empire, in the context of the liberal post-war consensus it was concerned with charting a course of professionalization, dominated by a descriptive and quantitative orthodoxy. The ideological consensus that underlay the scientific method as the *modus operandi* of the discipline was to come under increasing scrutiny in the 1960s, with the relevance of mainstream geography being challenged on a scale that was to eventually change it. In the context of the protest-laden 1960s and the rise of the New Left, which spoke to a broader range of issues than the class-dominated interests of earlier leftist movements, the ontological vacuousness of the establishment geography of spatial science paved the way for oppositional tendencies and the turn to radical geography. In both Canada and the U.S., movements played out in the context of civil rights, peace, student, and anti-Vietnam War and environmental protests (Greenpeace was founded in 1971), gay and lesbian activism (epitomized by the Stonewall riots of 1969), and second-wave feminism, as well as moves to regional separatism (with the Front de libération du Québec in Canada) (Peake and Sheppard 2014). Each was to produce radical turns in the social sciences and humanities, including geography, in which the civil rights movement was to have the most immediate impact. To a large degree, however, the majority of the (white and male) practitioners of geography – a field that had also attracted government employees with military and intelligence connections (Akatiff 2016) – were out of tune with the 1960s, failing to connect in any serious way with the emerging radical left. This was much to the frustration of those few left-leaning practitioners who found themselves within a field they considered to be intellectually bankrupt (Alan Wallace, email 2010). From the early 1960s, a few geographers, constituting no more than a minority in any department, were calling for a more relevant geography, characterized by three dimensions: radical theory; engagement in activism both on and beyond university campuses; and radical praxis that encompassed teaching, learning, research, and work toward social change. This call is one that for many radical geographers today – from early career scholars to those approaching retirement and beyond – characterizes their experience in the academy.

## The 1960s and 1970s: Transforming establishment geography

A number of African-American geographers located in the southern historically black colleges and universities had been directly involved in the civil rights movement in the 1950s,[2] but white geographers' connection to this movement was first made via the AAG in 1963 with a petition against holding meetings in racially segregated hotels (Horvath 2013). This was followed by organizing politically charged AAG sessions such as "The Status of Negroes in Geography" in 1968 (Horvath *et al.* 1971). Perhaps best known was the relocation of the AAG meetings in 1969 from Chicago to Ann Arbor in tacit protest of the National Guard and police violence at the 1968 Democratic Convention in Chicago, although much less known is that this was orchestrated by geographers already active in anti-racist activities within the discipline. For example, 1964 saw the first attempts to address the segregation of African-American geography students via a proposal to establish a Commission on Geography and Afro-America (COMGA) within the AAG to support the recruitment of African-American geographers and to conduct research into issues facing African-Americans (Deskins and Siebert 1975). While successful in its aims, COMGA was short-lived; the racialized divide was to prove intractable and addressed by few white geographers outside COMGA, notwithstanding serious attempts by Clark Akatiff, Bill Bunge (1965), and Ron Horvath – geographers who were to be active in the formation of the USG.

It was through the work of the Society for Human Exploration and the Detroit Geographical Expedition and Institute (DGEI: 1968–1971) that most white geographers were to become involved in civil right activities.[3] Precipitated by the June 1967 Detroit rebellion and the April 1968 assassination of Martin Luther King, the DGEI was formed in 1968 later (Horvath 2013). At the 1969 AAG meeting in Ann Arbor, three buses filled with DGEI members "brought the presence of the black streets to the walls of Academe" (Akatiff 2016: 7). Based at Wayne State University, the DGEI was formative in shaping the nature of the nascent radical geography, with its focus on race as much as class and an insistence on involvement in people's struggles for civil rights.[4] And yet the work of Bunge's co-organizer, the African-American Gwendolyn Warren, an experienced community organizer, has only recently been documented (Warren *et al.* this volume), speaking both to the erasure of African-Americans even in activities aimed at solidarity and to how unreflexive radical knowledge production (even still) can produce its own theoretical unconscious.

A contentious figure, subsequently estranged from academic geography, Bill Bunge worked in the 1960s with Clark Akatiff, another lifelong socialist, to galvanize other geographers into radicalizing geography. The first (informal) class taught on Marxism and socialism by a North American geographer was most probably by Akatiff in 1961, while a graduate student at UCLA (Akatiff 2016). Akatiff was to move east in 1966, hired as an assistant professor of geography at Michigan State University (MSU). He influenced Ron Horvath to likewise relocate from Santa Barbara for MSU (1967–75), forming "a core of progressive radicalism" (Akatiff 2011: 8) with Ed Vander Velde and Charlie Ipcar that made Michigan 'ground zero' for radical geography.

At the 1968 AAG, Bunge and the DGEI were given a plenary session entitled "Toward Survival Geography: Reports in Human Exploration" (Akatiff 2016). It was to prove a pivotal moment when radical geography jelled, igniting a movement that eventually led to formation of the USG. The nucleus included Bunge at Wayne State University, those from MSU, Jack Eichenbaum from the University of Michigan, and Jim Blaut, Dick Peet, David Stea, and Ben Wisner from Clark University (Akatiff 2011: 8). Despite the numerical dominance of those located in Michigan among this grouping, the combination of temporary appointments and departures by key members was to prevent any permanent base for radical geography forming there, and the institutional base of radical geography in the U.S. was propelled eastwards, to the Graduate School of Geography at Clark University, Worcester, MA, where the recent influx of new faculty members formed a critical mass. With considerable graduate student involvement, *Antipode: A Journal of Radical Geography* was first published at Clark in 1969 (Peet 1969; Huber *et al.* this volume). The importance of the north-east was cemented by the arrival of Marxist geographers, most notably David Harvey in 1970, at Johns Hopkins University, Baltimore, MD (Sheppard and Barnes, this volume).

If the 1960s triggered a critical mass of academic geographers that made possible the emergence of socialist geography, the early 1970s was to ferment the need for an organization to give it form, function and visibility. Yet Bunge's failure to secure another academic post and, later, Ron Horvath's return to Australia (in 1976) weakened the linkages to the civil rights movement prominent in the DGEI. The issue of the whiteness of practicing academic geographers faded into the background, naturalized as beyond investigation with the demise of COMGA. If (virtually) all geographers were white, they were not all male, however, and a number of junior women – Myrna Breitbart, Kathie Gibson, Julie Graham, Cindi Katz, Suzanne Mackenzie, Damaris Rose, and Sue Ruddick – emerged in the early 1970s to play active roles in influencing the direction of socialist and feminist geography in the USG. Waning anti-racism and rising feminism were decisive in shaping the contours of socialist geography, recognizable still in the present.

## Geographical expeditions (GEs) and institutes

While MSU, Clark University, and the DGEI in Detroit were the foci of radical geography in the late 1960s, new GEs were formed across the border in Toronto (1972–1975) and Vancouver (1973–1975), and in Sydney, Australia (1976–1980), that were to solidify and define the emergent form of socialist geography. New Canadian universities in the 1960s, such as York University in Toronto and Simon Fraser University in Vancouver, opened up opportunities for new (white, English-speaking, and largely male) faculty hires and for graduate students from the UK and British colonies where the discipline flourished: Ireland, New Zealand, Australia, and South Africa. And while some arrived already radicalized, others were radicalized by their experience of 1960s North America. Their experiences of politics in other places were to prove extremely influential in shaping radical geography in North America (and beyond) in the 1970s.

Bunge moved to Toronto after being fired from Wayne State in 1968, working for a year at the University of Western Ontario (1970–1971) and at York University (1972–1974) as well as being a taxi driver and union organizer (Bergmann and Morrill 2018).[5] He initiated the *Toronto Geographical Expedition* (TGE) in 1972 with graduate students from York University and the University of Toronto (Stephenson 1974). The TGE had seven full-time academic members, including Ron Bordessa as director, with a number of part-time "explorers" who worked alongside "folk geographers" – community members drawn to the ideals and practice of the TGE. Their base was a house on Brunswick Street in downtown Toronto where the full-time members lived, free classes in geography for college credit were held, and from which 12 expeditions took place (Bunge and Bordessa 1975), including a Canadian-American Geographic Expedition (Bunge 1977). Members saw their legitimacy as based in their ability to conduct research, although the latter was not considered complete until it had been disseminated to both the local community and political decision-makers. Bunge was prescribed as the group's theoretician, defining the scope of all empirical work and proposals, especially those that would improve landscapes to increase the survival chances of children living in the local area. The TGE had folded by 1975, but its members played a seminal role in the formation of the USG.

The *Vancouver Geographical Expedition* (VGE) started a year later in the summer of 1973 after members from SFU (and Laval) visited the TGE. Michael Eliot-Hurst, chair of the SFU Department of Geography from 1971–1975, "almost single-handedly oversaw the creation of a virtual graduate school in radical/Marxist geography at SFU in the early 1970s" (Proinnsias Breathnach, email 2011). The VGE became the largest grouping of activist radical geographers in North America, with more than 40 graduate and undergraduate students involved between 1973–1975 (Blomley and McCann, this volume).

"The Vancouver Expedition was modeled on some of the work [of the TGE], but took on its own personality with a number of neighborhood-based projects and a more disaggregated leadership" (Nathan Edelson, email 2010). At first the VGE focused on direct local action with the intent of nurturing a politics of affinity built on solidarity and support with local residents in the neighborhood where it operated and was housed (like the TGE it had its own Expedition house), but by 1975 its focus had narrowed to a Marxist reading group that met regularly and was visited by Bunge, Akatiff, and Harvey. The students' incorporation into other activities concerning radical geography was made further possible (in these pre-internet days) through the material support of Eliot-Hurst who paid for the use of two vans for students to use to attend conferences across North America. Students drove to CAGs in Regina and Toronto (1974) and to the AAGs in San Francisco and New York (in 1975). This in part is why SFU was able to have a strong contingent at the founding meeting of the USG (Nathan Edelson, email 2010). By spring of 1975 however, the VGE had disbanded. An anonymous commentary on using the VGE to engage with the broader issue of socialist practice in Vancouver, noted that it had outlived its usefulness and was draining more energy than it was generating (Anon 1975a). Like the TGE, its members were to play an important role in forming and sustaining the USG.

The *Sydney Geographical Expedition* (SGE) was the last to form, connected to the US through the Australian geographer, Ron Horvath (this volume). Driven by a desire to demonstrate geography's relevance to a wider social struggle, building bridges to the world beyond the ivory tower, it started work in early 1976, led by post-graduate students such as David Owen. Its members engaged in a range of activities relating to Sydney's homeless population and projects linked to low-income housing. In an attempt to link Marxist theory with expeditionary praxis, it ran a Marxist reading group (1978–1980) and staffed a full-time community information center in downtown Sydney (1978–1979) (*USG Newsletter* 4.4). It sought to be more than just a resource center by acting as a catalyst, with members joining community groups and linking them together in class struggle around common issues in the sphere of reproduction (such as housing, health services, and information provision) (Owen and Baxter 1979).

## 1974: Formation of the USG[6]

It was not only the intellectual and political drive of the GEs and the emancipatory geographies to which they gave rise that galvanized radical geography's thought and practice, setting the stage for further transgression of the boundaries of establishment geography through the formation of the USG. It was also the GEs networked and multiple locations that were adopted by the USG. Catalyzed by the GEs, over 30 North American radical and socialist geographers converged on Toronto in May 26–28, 1974, to discuss issues of mutual interest, agreeing to form an independent organization of revolutionary / radical geographers (USG minutes, 1974) that would "promote a socialist paradigm within geography" (Anon. 1975b). The meeting was held in parallel with the CAG annual meeting and "under the roof" of the TGE (Figure 5.1 visualizes the racial and gender bias) (see Anon. 1975c). While geographical circumstances played a part in that Toronto was an accessible city not only to many Canadians but also those from the US, as Nathan Edelson (2011) remarked: "Under other ... circumstances it could have taken place in Boston or Chicago."

The impetus for the formation of the USG is somewhat unclear. The first *USG Newsletter* (1: 1, 9), edited by SFU students, attributes credit for this to the Toronto GE, but others' recollections differ: "The idea of a Union of Socialist Geographers arose out of the meeting in Toronto

**Figure 5.1**   Members of the first meeting of the Union of Socialist Geographers on the steps of the Toronto Geographical Expedition House, 283 Brunswick Ave, Toronto May 1974. Photo courtesy of Colm Regan. Reproduced with permission.
*Note:* Clark Akatiff remembers that the photo was taken by Colm Regan as the final event of several days of meetings that gave birth to the formation of the USG. Present at that meeting, among others, were: Clark Akatiff, Jim Anderson, Joen Barnbrock, Jim Blaut, Susan Bunge, William Bunge, Bob Colenutt, Sue Cozzens, Bernard Curtin, Nathan Edelson, Tom Edelson, Dean Hanhinh, Richard Hansis, Rob Hare, Kirsten Haring, David Harvey, Ron Horvath, Charlie Ipcar, Jim Lyons, Barry Malmsten, Stephen Mills, Marie Murphy, Steve New, Gayle Olders, Gunnar Olsen, Dick Peet, Tom Phipps, Colm Regan, Paul Rosenberg, Elan Rosenquist, Tom Scanlan, Judy Stamp, Derek Stephenson, Allen Van Newkirk, Ed Vander Velde, Richard Walker, Alan Wallace, John Warnke and Wilbur Zelinski.

itself, and not from a pre-scripted agenda. There was a large contingent from SFU, but the idea came from multiple sources, not the least of which were associated with Blaut, Bunge and Akatiff" (Ron Horvath, 2016). Akatiff (2011) similarly recalls that these geographers had prominent roles in the proceeding.: "Bunge played a much bigger role than did Vancouver. Horvath and me, Blaut, Jim Lemon, a big contingent from Michigan State, David Harvey, Peet from Clark."

Debates ensued over naming the organization, with the USG considered a closer fit with the organization's political direction (compared to the alternative offering of the Union of Radical Geographers). Membership was to be open to all those who agreed with the principles of the organization and to pay the $5 annual dues ($2 for those who "were broke": *USG Newsletter* 1 (1): 17). Although Clark Akatiff was a major figurehead of the USG in its formative year, the idea of electing any formal figurehead or form of centralized organization was rejected at the USG's first AGM in Vancouver in 1975, resulting in a loose regional organization of east coast (Clark), west coast (Vancouver) and Prairies/midwest locals (Toronto and East Lansing, with the Midwest group added later).[7] See Figure 5.2 for the USG's mandate,which emphasized links between theory and praxis.

---

The purpose of our union is to work for the radical restructuring of our societies in accord with the principles of social justice. As geographers and people, we will contribute to this process in two complementary ways:

1. Organizing and working for radical change in our communities.
2. Developing geographic theory to contribute to revolutionary struggle (USG 1974). Thus we subscribe to the principle: from each according to their ability, to each according to their need. We declare that the development of a humane, non-alienating society requires, as its fundamental step, socialization of the ownership of the means of production.
3. [*added by the Vancouver local in September 1974*] To try to promote equality within the university by supporting and agitating for student and staff parity in all decision making (USG 1975).

[*added in 1975*]: To these basic principles were added the following interrelated objectives:
4. To continue the pursuit of presenting critiques of bourgeois geography.
5. To develop an alternative socialist analysis of geography.
6. To organize and maintain a sense of mutual help and support for and between persons within and outside the university system (utilizing, yet outside the confines of the present university structure as much as possible) by:
    collective work format
    seminars
    reading groups
    an effective contact system.

---

**Figure 5.2**  Mandate of the USG (written between May 1974 and 1975). *Source:* Anon 1975c; USG 1974.

## USG Locals

The USG locals grew internationally through cross fertilization via two sets of connections: links with SFU generated locals in Montreal, Sydney, Ireland, and the UK, while the presence of socialist faculty members active in the formation of *Antipode* or with other progressive connections contributed to the formation of locals in the US at East Lansing, the North-East, and the Midwest, as well as in Denmark and the UK (Nathan Edelson, email 2010) (see Figure 5.3). The bulk of USG locals were set up between 1977 and 1978, some based on existing reading groups and others quickly establishing reading groups, with connections being sustained by regional contact persons and extensive letter writing by Susan Williams, the corresponding Secretary (*USG Newsletter* 4.1). As a number of those who responded to requests for accounts of this period attested it was from these USG-based reading groups that a new generation of socialist geographers was to emerge that was to retain its commitment to radicalism and from which new generations of critical geographers have sprung. Table 5.1 lists those individuals associated with the USG locals over their lifetimes; the Johnny (and Jane) Appleseeds of socialist geography.

Building on the VGE, the most active and organized USG local, at least initially, was in *Vancouver* at SFU, formed in late 1974.[8] Facilitated by access to a large supply of volunteer labor, the Vancouver local also agreed to produce and distribute the *USG Newsletter*. The group was fairly tight-knit with a horizontal leadership (Nathan Edelson, email 2011): "To quote Monty Python, [it was] probably more like an anarcho-syndicalist commune than anything else" (Alan Wallace, email 2011). Pluralism flourished as Marxists, Maoists, anarchists, feminists, and environmentalists co-mingled. Alan Wallace (2011) recalls:

> the large groups of people who would travel from Vancouver were very insistent on democratic inclusion and against even the faintest whiff of elitism. …The origins of the Vancouver USG in the Expedition house in East Van…were, I would imagine, very different from the lineages of radicalism emerging at Clark and Johns Hopkins.

Conferences were held and reading groups met frequently, serving a self-education function with undergraduates and graduates working together. The local also engaged in parties and social gatherings, the glue holding people together, with meetings often spilling over into long evenings and early mornings of intense political discussion. Some of the women members established a socialist feminist reading group (in 1975):

> As was true of many such groups at the time, women were involved, but men were most prominent. Geography as a discipline was a tough nut for

women to crack. Their analyses were at that stage Marxist or socialist, never feminist.… It was only when some undergraduate geography students and partners of graduate students formed our own alternative feminist women's reading group that we began to … grapple with the ideas of the slowly growing number of feminist authors. That group included Suzanne Mackenzie, then an undergraduate student and later one of Canada's earliest feminist geographers (Bradbury, 2015: 263–264).

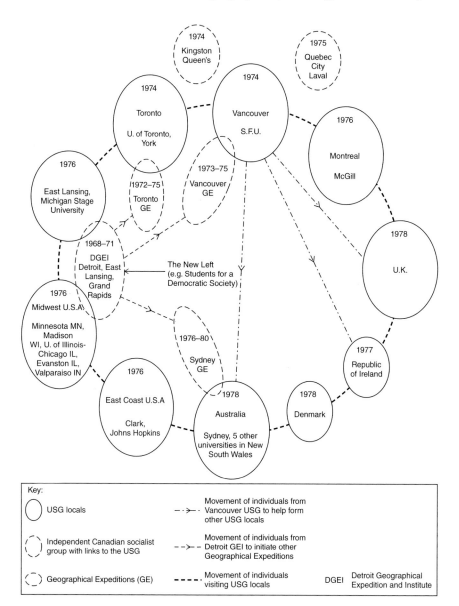

**Figure 5.3**  The USG and its Affiliates. *Source:* Author.

**Table 5.1** The USG locals and their members.*

| USG locals | Year formed | Individual members |
|---|---|---|
| Toronto | 1974 | Francine Beaudin, Jane Bonshek, Garry Crowfoot, Ruth Fincher, Suzanne Geddes, Gord Garland, Richard Harris, Jim Lemon, Suzanne Mackenzie, Barry Malmsten, Marie Murphy, Damaris Rose, Elan Rosenquist, Showkry Roweis, Sue Ruddick, Tom Scanlan, Derek Stephenson, Alan Wallace, John Warnke, Katherine Willson |
| East Lansing, MSU | 1974 | Clark Akatiff, Ron Horvath, Charlie Ipcar, Ed Vander Velde |
| Kingston (autonomous group) | 1974 | Sophie Bissonet, John Campling, Richard Harris, John Holmes, Fran Klodawsky, Greg Levine, Ray Schmidt, Carol Town |
| Vancouver | 1974 | *Faculty*: Michael Eliot-Hurst, Ron Horvath, Jim Overton *Graduate Students*: John Bradbury, Bettina Bradbury, Proinnsias Breathnach, Bernard Curtin, Nathan Edelson, Wendy Eliot-Hurst, Bob Galois, Ace Hollybaugh, Eric Leonard, Alan Mabin, Manfred Malzahn, Paul Raynor, Colm Regan, Lee Seymour, Terry Simmons, Peter Usher, Alan Wallace, Yvonne Zarowny *Undergraduate students*: Lew Jones, Suzanne Mackenzie, Manfred Malzahn, Steve New, Rob Phare, Tom Phipps, Elan Rosenquist, Susan Williams (nee Barry) |
| Québec City (autonomous group) | 1975 | Marc Dion, Jacques Giguère, Loïc Grasland, Rodolphe De Konnick, Juan-Luis Klein, Robert Lavertue, Jean Nadeau, Lyse Pelletier, Monique Piot, Aimé Roy, Paul Villeneuve |

**Table 5.1** (Continued)

| USG locals | Year formed | Individual members |
|---|---|---|
| Montreal | 1976 | Warwick Armstrong, John Bradbury, Bettina Bradbury, Colm Reagan, Sue Ruddick, Frank Tough, Susan Williams (nee Barry) |
| Sydney and New South Wales | 1976 | Members of *University of Sydney, University of New South Wales* and *University of Newcastle*: Steve Baxter, Kathie Gibson, Ron Horvath, Richie Howitt, Vivienne Milligan, Debbie Mitchell, Dave Owen, Peter Rogers, Colleen Shipman, Frank Williamson |
| Midwest USA: Minnesota, MN U of Wisconsin-Madison, WI U of Illinois at Chicago, IL Evanston, IL Valparaiso, IN | 1976 | *U of Minnesota*: Trevor Barnes, Mark Bouman, Michael Curry, Bryan Higgins, Mickey Lauria, Don Olson, Ken Olwig, Bill Pisarra, Eric Sheppard, Rob Warwick. |
| U of Wisconsin-Madison, WI | | *U. of Wisconsin-Madison*: Howard Botts, Mark Garner, Al Gedicks, Nicole Gotthelf, Sean Hartnett, Russel Kirby, Greg Knapp, Frank Magilligan, Kent Mathewson, Ken Olwig, James Penn, Yda Saueressig, Mark Wiljanen, Ben Wisner. |
| U. of Illinois at Chicago, IL Evanston, IL | | *U of Illinois at Chicago*: Jim Bash, Jim Blaut |
| Valparaiso, IN | | *Valparaiso U.*: Dick Hansis Other members included: John Chapell Jr., Gary Edelman, Abid Kureshi, Cornelius Loeser, Jim Lyons, John Rury |
| Ireland | 1977 | Colm Regan, Prionnsias Breathnach |

(*Continued*)

**Table 5.1**   (Continued)

| USG locals | Year formed | Individual members |
|---|---|---|
| East Coast: Clark U. / Johns Hopkins U. | 1978 | *Clark U*: Jim Blaut, Myrna Breitbart, Ann Dennis, Julie Graham, Kathie Gibson, Kirsten Johnson, Cindi Katz, Jim Lyons, Phil O'Keefe, Dick Peet, Chrystine Rodrigue, Paul Susman, David Stea, Bobby Wilson<br>*Johns Hopkins U.*: Joern Barnbrock, Lata Chatterjee, David Harvey, Neil Smith, Richard Walker |
| Denmark | 1978 | Hans Folke, Karen Fog Olwig, Ken Olwig |
| British Isles (Ireland and UK) | 1978 | James Anderson, Martin Brennan, Rod Burgess, Terry Cannon, Bob Colenutt, Ian Cook, Mel Evans, Jo Foord, Malcolm Forbes, Chris Gibson, Tommy McLaughlin, Tom Manion, Doreen Massey, Phil O'Keefe, Colm Regan, Damaris Rose |
| Founding members in other geography departments in North America | | *Aquinas College, MI*: Tom Edison, Dean Haninh; *McMaster U.:* Sue Cozzens, Michael Dear; *Maryland*: Stephen Mills; *Michigan*: Bryan Hennigan, Howard Horowitz, Gunnar Olsen, Judy Stamp Olsen; *Nova Scotia*: Alan Van Newkirk; *Ottawa*: Gaye Olders; *Penn State*: Wilbur Zelinski; *SUNY*: Jack Eichenbaum; *SUNY-Binghampton*: Paul Rosenberg; *Syracuse U*: Joey Helms. |

*Source:* Author
*Members may be listed in more than one local as they moved throughout their career. Members listed include those engaged at any point in the period of existence of the local and not just at its commencement. While due diligence was paid to providing as comprehensive a listing of USG members as possible, it is unlikely that this list is exhaustive.

The fervor of the Vancouver local was also evident in its having its own aims:

1 To develop and disseminate revolutionary theory within geography;
2 To promote community involvement on the part of socialist geographers;
3 To try to promote equality within the university by supporting and agitating for student and staff parity.

These aims were partially met by the overlapping membership of the VGE and the USG Vancouver local, allowing the latter's members to engage in community development.[9]

The *Toronto* local formed at the time of the first meeting of the USG, building on the TGE membership and the involvement of other active socialists, including Jim Lemon and Shoukry Roweis. Figure 5.4 shows members of the USG crammed into a house meeting in Toronto in 1978. *Montreal* became the third active Canadian local in 1976, with the move

**Figure 5.4**   USG members at a meeting May 1978, Toronto. Photo: Courtesy of Damaris Rose. Reproduced with permission.

of John Bradbury (1942–1988) and his partner, the historian, Bettina Bradbury from SFU to McGill. Its participants (also including Damaris Rose and Sue Ruddick) were further catalyzed by the left-leaning Parti Québéçois, notwithstanding little direct engagement in Québec politics.[10]

Two other Canadian groups loosely affiliated with the USG. There had been an identifiable group of geographers at *Queen's University* since 1974 working on socialist issues in development and Canadian political economy who while, never forming a local, produced an issue of the *USG Newsletter* (5.4/6.1) in 1980. A number of members of this department became members of the USG after 1977, but others developed their own Socialist Study Group and departmental links were predominantly with the Conference of Socialist Economists (CSE), the (ongoing) Union of Radical Political Economists, the Kingston Socialists (1975–1980) and a local socialist feminist reading group (Harris 1980).[11] In 1975 a francophone radical research group of students and faculty, led by Rodolphe De Koninck and Paul Villeneuve, also formed in Québec City at *Laval University*, le Groupe de Réserches sur l'Éspace, la Dépendance et les Inégalités (GREDIN;'scoundrel' in French) to study the application of Marxism to geography (Rudolph de Koninck, *USG Newsletter* 4.3: 11; Klein, this volume). Although they had connections with the TGE and saw their work as part of a collective struggle, they remained autonomous from the USG, their focus being on the specific context of the capitalist mode of production within Québec.

In the US, the longest established group of socialist geographers at MSU in *East Lansing* formed a short-lived USG group that did not officially constitute themselves as a local. Although seminal in socialist geography from the mid-1960s onwards, by 1974 the group was close to dying out; Ron Horvath was to leave in 1974 for SFU and Clark Akatiff in 1976 when he was denied tenure. In its place, the Midwest and East Coast locals were to thrive. The former was centred around Minneapolis/St. Paul (MN) but also incorporating Chicago and Evanston (IL), Madison (WI), and Valparaiso (IN); it grew from 1976 onwards when it started to hold its own regional AGMs. The East Coast local was centred in particular at Clark University (*USG Newsletter* 4.3: 3). Despite housing *Antipode*, formal links to the USG were not fully made until 1978 when the interest in socialism expanded at Clark. The Clark local grew in numbers (to 25) after a study group was formed by members in a class taught by the Marxist political economist, Don Shakow, in the Economics Department (*USG Newsletter* 4.3: 4). This group of graduate students (including Julie Graham and Kathie Gibson) and faculty members from geography, economics and sociology, went on to form the Regional Development Unit, with interests in local industrial decline.

Members of the Clark local were also active in campus organizing to save the jobs of socialist professors who were being denied tenure.

USG locals were also to form beyond North America. In Australia, the SGe members formed the USG local, with 10 members based in *Sydney*, New South Wales, and five others scattered through New South Wales. The Sydney local started in 1978, but had been meeting as members of two Marxist reading groups since 1976 (Horvath, this volume). A small local also formed in 1978 in *Denmark* by Hans Folke (Roskilde University) and Ken Olwig, who had moved there to teach at the Danish Pedagogical University in 1979, after having been active in the Midwest local.[12] Other active USG locals beyond North America were formed in the U.K.: in *Ireland* in 1977 and the (self-named) *British Isles* in 1978. Discussion about forming a "British Isles" local occurred at the 1978 Institute of British Geographers (IBG) conference in Hull, England, triggered by a meeting chaired by Phil O'Keefe and attended by some 50 people, including USG members from Canada and the U.S.A. The argument was made that the British Isles local would organize within geography to: act as a bridge to the CSE (the largest grouping of left academics in Britain); aid internationalism; develop collective activity; serve as a support for isolated academics; and defend "comrades in struggle" (*USG Newsletter* 3.3: 55). The local was designed to be autonomous, as its organizing priorities were different from those of the North American locals (*USG Newsletter* 4.1).[13]

Plans were made to form an organizing committee to establish a USG meeting at the 1979 IBG at Manchester, as well as to compile a mailing list, circulate the newsletter, and collect subscriptions.[14] At this meeting, questions arose about the relationship between the North American and British Isles USGs. Given the existence of the CSE and the Social Geography Study Group of the IBG, it was determined that the British Isles section would function best solely as a communication network (*USG Newsletter* 4.3: 9). Yet the enthusiasm generated at the Manchester conference led to the formation of four further groups: A *Capital* reading group; an urban political economy group; a development geography group to read Lenin's *The Development of Capitalism in Russia*; and a general theory group constituted by USG members based in London. Three successful meetings were held, on the topics of nature, ideology, and women and space, but by the end of 1979 all these groups were virtually defunct. Some members of the London local did produce another *USG Newsletter* (5.2 and 3) in 1980, though they struggled with this task due to members' over-commitment to other political activities (*USG Newsletter* 5.1).

By 1978 the *USG Newsletter* (4(3): 33) could claim interest in the USG by geographers from India, Africa and Latin America. AGMs were held independently of but alongside and alternating between the CAG and AAG annual meetings, allowing socialist geographers to organize sessions not just outside but also within the CAG and AAG meetings: "[a]lthough at times relations were antagonistic. The AAG in particular verged on a role of the 'Dark Force'" (Edelson, email 2011). Although coming together at the annual meetings, locals operated independently of each other. As Table 5.2 shows, from the formation of the USG in May 1974, its locals soon were engaged not only in national meetings but also in well-attended regional-based meetings that started to take off in 1977. For example, the Midwest local held its own AGMs between 1976 and 1980 and the 1978 East Regional USG meeting was attended by 80 people from Québec, Worcester, Toronto, Kingston, New York, Baltimore, and Montreal. USG members were also instrumental in organizing the joint meeting with *Antipode* in Dublin in March 1978. By the 1977 AGM, "all the trappings of an academic organization were in place" (Peake and Sheppard 2014: 312). Thus, by the mid-1970s socialist geography was up and running: the USG had its own finance committee, conference organizing committee, posts of Newsletter Editor and Corresponding Secretary, financial statements, meetings, study groups, means of dissemination, key centers, formative texts, and forms of theorizing.

## Newsletters and Other Publications

During its existence the USG engaged in a significant level of publication.[15] Individual members published in the *AAG Newsletter* and *Transition*, while the SFU local edited one issue of *Antipode* at the invitation of Dick Peet (in 1976, 8#3), including articles on anarchism, environmentalism, Ireland and Latin America – an eclecticism contrasting with the Marxist orientation of the other issues of that year. The USG was also to branch out into independent publishing.[16] Products included the volume *Study Papers on Imperialism*, an 87 page document prepared for a workshop on "Marxist Perspectives on Geography" organized for the AAG annual meeting in New York in 1976 (Jim Overton, email 2011), the *Canadian-American Geographical Expedition Papers* by Bill Bunge (comprising *Field Notes 1: The North American Working Class* and *Field Notes 2: Canadian-American Geographical Expedition*), *A Review of Nicos Poulantzas's "Classes in Contemporary Capitalism"* by the SFU Working Group on Social Class, and the 1983 publication *Society and Nature: Socialist Perspectives on the Relationship between Human and Physical Geography* by the USG London Group.[17] Most importantly, the USG began to publish its own *USG Newsletter*, which expanded to become a regular substantive journal (with some 50–80 pages

THE LIFE AND TIMES OF THE USG

**Table 5.2** Timeline of the formation of USG locals, AGMs, national, and regional meetings, 1974–1981.

| Year / month | AGMs and USG locals formed | National meetings | Regional meetings |
|---|---|---|---|
| 1974, May | 1st AGM Toronto local formed | CAG annual meeting, Toronto. USG organized 2 sessions at the CAG | |
| 1974, Sept. | Vancouver local formed | | |
| 1974, Nov. | | | 3 day seminar in Marxist geography, Clark University |
| 1975, April | 2nd AGM | CAG annual meeting, Vancouver. USG organized 3 sessions, plus a fieldtrip in East Vancouver | |
| 1975, April | | USG meeting hosted by Blaut in Chicago, alongside the AAG annual meeting in Milwaukee | |
| 1976 | 3rd AGM Montreal and Midwest locals formed | AAG annual meeting, New York | 1st AGM of Midwest local |
| 1976 | | USG Session held at Ontario CAG annual meeting | |
| 1977 | 4th AGM Ireland local formed | CAG annual meeting, Regina. USG Executive formed. 4 USG sessions held in the CAG plus a separate USG meeting | |

*(Continued)*

**Table 5.2**   (Continued)

| Year / month | AGMs and USG locals formed | National meetings | Regional meetings |
|---|---|---|---|
| 1977 | | | USG West Coast regional meeting, SFU |
| 1977 | | | USG regional meeting, Toronto, held in conjunction with the Political Economy of Urbanization Working Group |
| 1977 | | | Midwest local 2nd AGM, Madison |
| 1977, July | | | Political Economy of Urbanization workshop at Queen's |
| 1977, Oct. | | | USG East Coast regional meeting, Clark University |
| 1978 | East Coast USA, British Isles, Sydney and Denmark locals formed | 1978 Hull IBG conference, the British Isles local was formed | |
| 1978, Jan. | | | USG Midwest/ West Lakes meeting, Valparaiso, IN |
| 1978, Feb. | | | USG West Coast regional meeting, Vancouver |
| 1978, March | | | *Antipode*/USG conference, Dublin |

**Table 5.2**  (Continued)

| Year / month | AGMs and USG locals formed | National meetings | Regional meetings |
|---|---|---|---|
| 1978, April | 5th AGM | AAG annual meeting, New Orleans. Parallel USG meeting | USG Midwest regional meeting, Minnesota |
| 1978, May | | CAG annual meeting, Western University, London, Ontario, with 1 USG session. Preceded by a USG 2 day workshop organized by the Montreal and Toronto locals in Toronto (see Fig. 5.4) | |
| 1978, Oct. | | | East coast USG regional meeting |
| 1979, Jan. | | | British Isles USG held the first session on socialist geography at the IBG, Manchester, UK. Members of Vancouver USG also attended |
| 1979 | 6th AGM | Victoria, Vancouver Island, CAG annual meeting. USG AGM in Vancouver | |
| 1979, May | | AAG annual meeting, Philadelphia | |

(Continued)

**Table 5.2** (Continued)

| Year / month | AGMs and USG locals formed | National meetings | Regional meetings |
|---|---|---|---|
| 1979 | | | Midwest local 3rd AGM, Chicago |
| 1980 | | | British Isles local meetings held parallel to the IBG conference, Lancaster |
| 1980, April | 7th AGM | AAG annual meeting, Louisville, KY. | |
| 1980, Oct. | | | Midwest local 4th AGM, U. of Illinois–Chicago |
| 1980 | | | USG led Canadian Housing Policy Workshop |
| 1981, August | 8th AGM | CAG annual meeting, Memorial U. of Newfoundland-Cornerbrook | |
| 1982, April | 9th AGM | AAG annual meeting, San Antonio, Texas | |

*Source:* Author

per issue) – a third radical geography journal alongside *Antipode* and *Transition*.[18] Eight volumes of the newsletter appeared between 1975 and 1983 with 21 issues (available at https://antipodefoundation.org/supplementary-material/the-union-of-socialist-geographers-news letter-1975–1983/).

Technically, the USG newsletters constitute "gray literature." They had no recognizable "official" library classification coding – such as ISBN or ISSN numbers – no consistent contact for publishers to enable libraries to subscribe, and a changeable format produced on department banda or photocopying machines). As Norcup (2015: 65) states:

Such publications from the 1970s and 1980s were a part of the culture and productivity of not simply being an academic at the time, but being what might now be regarded as a precursor to the work of contemporary activist-scholar or 'public-scholar,' interested in ideas which connected across educational spaces into collaborative and creative works with thinkers, researchers, educators, writers, and community activists independent from academic institutions.

Radical geographic grey literature in North America in this period constituted a common academic form of knowledge dissemination that included *Antipode* (prior to 1980) and *Transition*. These alternative journals were all locally produced efforts, especially by graduate students, allowing for the development not only of a "counter voice but a counter-canonical ambition" (Norcup 2015: 74), educating against establishment geography. The *USG Newsletters* served to disseminate socialist literature to a broad audience of radical geographers as well as document the development of a socialist infrastructure of locals and meetings and other USG activities. Its intent was to keep its widely dispersed membership abreast of developments in various places, as well as serving as a repository for teaching and research resources, including full-length articles, course outlines, annotated bibliographies, book reviews, correspondence, and a Treasurer's report. The space given over to articles increased with each successive volume, making the moniker of "newsletter" less appropriate.

The SFU local volunteered to produce and distribute the newsletter for the first two years (continuing for two more) with initial plans for it to be a bi-monthly production. Starting in 1975, an SFU editorial collective edited five substantial issues of the *USG Newsletter*. Requests already in Issue 1 for contributions to be sent 'ready for publication,' due to the shortage of production staff at SFU, indicated the toll this work was taking. The two issues of Volume 2 (1976–1977) were edited by USG locals at Johns Hopkins and McGill (comprising former SFU students) respectively. In 1977 and 1978, Volume 3, with four issues, was edited successively at SFU, the University of Toronto, Minnesota and McGill. The editing of issues of the newsletter in Volume 4 (1978–1979) alternated between SFU, Clark and two sites outside North America: Sydney and the UK (edited by Damaris Rose and Suzanne Mackenzie). By this time the newsletter had adopted a more formal presentation with an organization based on grouping similar items: announcements of future events, summaries of papers given at conferences, meeting reports, correspondence and articles. With the move to a rotating editorship the *Newsletter* also began to focus more on local editions pertinent to the area in which an issue was produced. By late 1978, many folks had departed SFU with only a few left to produce the newsletter in a department where radical geography was under attack (Blomley and McCann, this volume). Thus, Volumes 5 and 6 (from 1979–1981) settled at Minnesota (with Eric Sheppard as co-coordinator), including issues submitted from Queen's, McGill and London (UK), all of which had an increasing focus on theory. The reduction in the number of articles submitted led to joint issues starting to appear. Joint issue 5.4/6.1 concentrated on the work of geographers at Queen's, with very little USG news. Volume 6.2, produced at McGill, was also devoted entirely to articles by local members.[19] The final newsletter (8.1), which appeared in December 1982, written by Chrystine Rodrigue, who was then at California State University, Northridge, was reduced to a three-page essay outlining reasons for the decline of the USG.

Reading through the newsletters one is struck by the familiarity of the material, its political vantage points, and the trajectories to present day issues of concern to critical geographers. Feminist debates are evident from Volume 1.1. Articles also covered socialist perspectives on anti-racist, development, and environmental geography and on the triumvirate of class, race, and gender, providing the building blocks for current knowledge production among critical geographers with interests in intersectionality, positionality, political ecology, globalization, and critiques of neoliberalism. The newsletters also contained some of the first discussions on sexuality in geography. In Volume 4 Michael Elliot Hurst wrote a report on the Gay Caucus in the AAG that had been formed in 1976, as well as on a proposal to start a Special Interest Group of Gay Geographers, although the relationship between "homosexuality" and socialism depicted (in depressingly negative terms) in an earlier volume (2.2) is indicative of the huge chasms in progressive thought in the 1970s. Also prescient were reviews of non-English language publications, which have only relatively recently been reintroduced into critical geography journals.

The *USG Newsletter* was unquestionably successful in terms of its aim of developing geographical theory to contribute to social change, with work on social justice, race and racism, feminism, the position of women in geography, environmental issues, anarchism, uneven development, public space, underdevelopment, and the social construction of nature evident in its pages. Moreover, the wide-ranging coverage of the *USG Newsletter* disrupts the orthodoxy of the material origins of radical geography being funneled solely through a theoretical Marxist base. In the 1970s both the *USG Newsletter* and *Antipode* were positioned marginally with respect to established geography journals. But the incorporation of *Antipode* into the professional publishing machine of Blackwell in 1980 resulted in it being the only one to feed into the mainstream disciplinary canon. A focus solely on the contents of *Antipode* would appear to vindicate Peet's (1977) view that by 1972 (when Harvey published the first Marxist geography paper in the journal) radical geography had changed emphasis, from socialist to Marxist geography. Yet there also was a recognition of diversity at the time. At the 1978 AGM meeting of the USG in New Orleans, reporting on *Antipode*, Phil O'Keefe stated "there will continue to be a balance between Marxist/radical stances as a wide range of such stances exists within the USG and within the journal's readers" (Anon. 1978). The diverse voices of USG members were loud and strong in the 1970s, publishing in their own journal, and adopting a counter hegemonic stance in a serious and sustained effort to oppose establishment geography.

## The USG as Radical Geographical Praxis

The main impact of the USG was to encourage and facilitate the development of left-leaning analysis in geography at a crucial point in its development and at a time when there was still considerable hostility to "politics" in academic geography. It also provided support for an activist orientation for some who were marginally attached to academia (Jim Overton, email 2011).

If the first half of the 1970s was characterized by "letting a hundred flowers bloom," by the late 1970s radical (including socialist) geography had become less preoccupied with breaking away from the institutional structures of the discipline and more interested in breaking into them (Peake and Sheppard 2014). A number of factors – both internal and external – contributed to depletion of the affects, energies, and politics that had led to the emergence and growth of the USG.

Internal considerations included the difficulty of maintaining the commitment to producing the *USG Newsletter* with dwindling supplies of free labor, not only at SFU but also elsewhere – new members coming into the academy after the mid-1970s were much less likely to have been radicalized and to engage in USG activities – leading to more erratic production schedules for the final volumes. The difficulty was also sounded of securing subscription fees from individual members, serving to undercut what was possible in terms of producing the newsletters and supporting regional meetings. The issue of inadequate funds to produce the newsletter (each costing approximately $100 plus postage costs) was raised as early as the 1977 AGM: 85 of the then 125 newsletter recipients had not paid their fees. In 1978 the situation had hardly improved, with 56 paid up members, 33 not fully paid up, and a further 67 "laggards". The last issue of Volume 4 in 1979 continued to bemoan the failure of members to pay subscriptions; of 180 North American members, 72% were not paid up and there were insufficient monies to cover the costs of printing.

Externally, the 1970s was a retreat from the urgency and movements of the early-1960s, marking a time of general purging of left academics, especially those whose politics were activist. Although radicalism continued to grow within geography, struggles on the ground were less evident (Akatiff 2016; Smith 2000). Many of the USG's seminal members failed to secure academic posts. At SFU in 1975 Michael Eliot-Hurst was forced out of his position as chair of the department, a move he felt reflected not only his politics but also his sexuality (he came out as gay in 1974). Jim Overton's and Ron Horvath's contracts also were not renewed at SFU, contributing to the termination of student engagement and opportunity for progressive work (Edelson, email 2011), and neither Alan Wallace, Bob Galois nor Nathan Edelson secured full-time jobs. In the US, Clark Akatiff was denied tenure in 1976, Bill Bunge had been fired (twice), and Jim Blaut did not secure tenure at Clark. Others managed to head off a similar threat (e.g., Dick Walker at Berkeley) (Peck and Barnes, this volume), in part as a result of campaigns that the USG helped catalyze. Some moved to different disciplinary fields. At SFU Alison Hayford left to teach Sociology at Regina, while Jim Overton moved to the Sociology and Anthropology Department at Acadia University, curtailing his links with the USG.

Those who remained did not necessarily have stable futures, with a preponderance of temporary positions, such as instructorships and post-docs, leading to a precarious economic existence and the increasing inability to commit to USG activities. Moreover, the "Socialist Diaspora" (*USG Newsletter* 8.1) was dissipating and increasingly unable to reproduce itself. If the early-1970s had resulted in groupings of graduate students who had emerged from Ph.D. granting institutions and the mid-1970s had seen these students gain posts in the still expanding major research universities where they started new locals, others were not so fortunate. As university enrolments declined, USG members accepted posts in institutions without Ph.D. programs or had no future in the academy, limiting opportunities to train a new generation of socialist scholars. New Ph.D. students did not necessarily have positive memories of 1960s radicalism, and in an increasingly competitive environment for employment socialist geography became a path less trodden. Although the number of socialist geographers increased – as it became increasingly acceptable for departments to admit a Marxist or socialist scholar to their ranks in the 1970s and 1980s – many found themselves geographically but also ideologically isolated. And while critical geographers today may find themselves in a majority in North American geography departments, for those currently with Ph.Ds in hand it is the echo of precarity that most resonates.

Discussions in the newsletters also indicate self-critique about whether the USG was living up to its objectives. While the USG was successful in constituting a forum for discussion among English-speaking socialist academic geographers it was much less successful in its aims of working for radical change in local communities, even though the GEs preceding it had come some way to achieving this aim (albeit to greater and lesser degrees, with Detroit and Sydney being the most sustained efforts). Schmidt (1980: 9) noted that while the USG had led to an "intellectual rebirth of Marxism [it] remains almost completely divorced from the working-class movement." He argued that the USG's participation in radical political practice was specific and limited with indiscernible long-term effects, not equivalent to participation in a political movement. Others saw scant evidence of the implementation of theory into practice, particularly in relation to teaching and planning, of research reaching non-academic communities, and of maintaining socialist principles in the advent of postmodernism. Difficulties were also noted of personal alienation from socialist theory and the lack of internal agreement on issues beyond class led to rifts, such as that caused by the derogatory discussion of homosexuality by Bill Bunge in the *USG Newsletter* (Volume 2.2: 18–21). Increasingly, ideological fragmentation was cutting off discussion among members, leading both to the exclusion of new-comers and to members moving on to champion other causes. This last was especially the case in relation to feminist geography, which had important roots in socialist geography and the USG (*USG Newsletter* 8.1). Suzanne Mackenzie, for example, seminal in forging the SFU GE and USG local while an undergraduate at SFU, in 1978 went to complete a Ph.D. in the UK; she also was instrumental in forming the Women and Geography Study Group of the IBG, and subsequently the Canadian Women and Geography Study Group of the CAG.

Moreover, although the USG was evidently a cross-national enterprise, the 49th parallel was also to act as a barrier. In a discussion at the 6th USG AGM in Vancouver in 1979, for example, it was felt by some that the principle of alternating the AGM location between Canada and the US should be relaxed on the grounds that not many Americans would travel to relatively isolated locations in Canada; at this AGM there were 35 people, 32 Canadians were joined by only two from the USA plus one from the UK.

The final nail in the USG coffin was the formation of the Socialist Geography Specialty Group (SGSG) within the AAG, draining money, time and energy from USG members. In 1980, at the 7th USG annual meeting alongside the Louisville, KY, conference, a debate was undertaken on whether to disband, with some USG members having already taken the initiative to apply for approval of a SGSG within the AAG's recently created Specialty Groups initiative. The crux of the debate was whether incorporation within the AAG would blunt radical geography's radicalism (Neil Smith, email 2011). There was unease about the cost of joining the AAG, of becoming a more formal organization, and of the international nature of communication among socialist geographers being lost, including that between the United States and Canada. Neil Smith (2011) recollected that:

> I didn't think then and don't think now that it was the SGSG that wrote the epitaph for the USG.... Jim [Blaut]'s position was that participating in a socialist Specialty Group at all would inevitably weaken the USG and suck away its power. He was right, of course, and we knew it at the time but that wasn't the whole calculation. His was the most adamant rejection of the SGSG. My sense was that much as Jim's prognosis was surely correct, and an independent socialist group was obviously ideal, the USG had worked well but wasn't especially building, wasn't expanding quickly, and the political mood was dissipating. We wanted organization for the USG but were resistant in our post-60s way to heavy organization, and this made the USG a bit haphazard. I think many of us felt that for all its wonderful energy, the USG was not likely to become more central, indeed had already served its best role, and, far from ideal as it was, we couldn't afford not to be involved in the SGSG. The SGSG led to a professionalization of the left much as Jim understood it would—indeed I think many of us who argued for working with an SGSG also understood it would bring a professionalization—but in retrospect through the 80s and 90s, I would argue that we would have been in a worse position, especially as regards bringing in grad students, if there had been no SGSG and the USG had fizzled, as already seemed likely in 1980.

Although the first mention of the plan to form the SGSG was made in October 1979 (*USG Newsletter* 5.1), the SGSG was only formally constituted as an AAG Specialty Group in 1981 (its officers coming from the USG). Despite some strident opposing voices and wariness of the cooptive power of the academic profession (Novick 1988), USG members were largely willingly professionalized (Smith 2000) and appeared to accept the SGSG's weakened-down statement of purpose:

> To examine geographical phenomena critically, questioning the implications of geographical research for the well-being of social classes. To investigate the issue of radical change toward a more collective society (Socialist Geography Specialty Group 1980: 1).

Although USG meetings continued for another couple of years, its death knell had been sounded at the 1980 meeting in Louisville, after which it was effectively dissolved as an independent force. There was an ebbing of activities and a much-reduced presence at the AAG and CAG meetings, as members joined the SGSG, with the last recorded meeting of the USG alongside the San Antonio, TX, AAG meeting in April 1982.

The USG excelled in creating a socially and politically engaged discipline, building practice-based solidarities and aiming for a practice of radical geography. But geographical and political fragmentation, the purging of the left in academia, the rise of neoliberalism and the mainstreaming of critical debates, operated as forces of dissipation within the USG – causing it to move from being outside establishment geography to becoming a part of it. The USGs legacy haunts contemporary critical geography, with lessons being taken up from the USG: of organizing across national boundaries; of activism both within and beyond the campus; of operating both within and outside established professional organizations; of working across hierarchies of faculty members and graduate students (and of not forgetting that undergraduates can be allies); and of the insistence still that there is no production without social reproduction. And, of course, critical geographers have gone beyond the lacunae in USG organizing, recognizing: the necessity of an anti-racist analysis that is not erased by class analysis; challenging the hegemony of whiteness in geographic knowledge production; and moving beyond the circulatory pathways of British imperialism.

## Conclusion

The purpose of this chapter has been to give the USG a tangible presence in the history of the discipline. The USG, like COMGA and SERGE, has been largely omitted from or overlooked in the annals of radical geography's historical accounts, which have emphasized the emergence and roles of radical initiatives – the DGEI and *Antipode* – of the 1960s. On one level its absence speaks to the USG's multifarious structure; it is not a neatly contained tale of one organization in one place. On another is a curious lack of interest in going beyond the now hegemonic "1969 story," resulting in an uneven appreciation of the multiple geographical centers – universities, towns, cities, countries – from which radical geography emerged as well as the people involved – women as well as men, undergraduates as well as graduate students and faculty members. This web of connectivity had a wide range of interests and fostered not only Marxist but also anarchist, anti-racist, feminist, and queer geographies.

The USG arose from discontent with the liberal consensus of the 1950s and engagement in struggles in the civil rights, women's, environmental, and labor movements. All of these changed North American society in fundamental ways, making the life and times of the USG possible. The USG was a model of intellectual and political collegiality; a political space for socialist geographers to acknowledge, engage, and challenge establishment geography and a capitalist world system. Its relatively short life does not undermine its collective ambition nor the impact it had on the development of critical forms of geography. Remembering the critical voices of the USG sheds light on radical geography's diversity and contributes to a more comprehensive account of our disciplinary past in which networks of people, crossing multiple geographic locations, shared the ambition to reconstruct geographical knowledge in a way that was to have an ultimately liberatory impact upon the historically conservative discipline of geography.

## Notes

1  Horvath (2013) dates this activist period from 1963 to the mid-1970s.
2  Including Thelma Glass (Kobayashi, this volume).
3  The DGEI had largely wound down by 1971 (Warren *et al*. this volume).
4  Bunge marched with Martin Luther King in Selma in 1965. He was listed as one of 65 members of Students for a Democratic Society in *The New York Times*: a "radical and /or revolutionary campus speaker (alongside Muhammed Ali, Angela Davis, and Stokely Carmichael) according to the House Internal Security Committee (previously the Un-American Activities Committee)" (Rosenbaum 1970: 23).
5  Bunge was to move to Quebec in 1978 with his second wife and child, where he continued to self-publish (Bergmann and Morrill 2018).
6  I draw here on the *USG Newsletter*, minutes of USG meetings and personal communications documenting the recollections of Clark Akatiff, Prionnsias Breathnach, Bettina Bradbury, Nathan Edelson, Jack Eichenbaum, Bob Galois, David Harvey, Alison Hayford, Charlie Ipcar, Audrey Kobayashi, Alan Mabin, Phil O'Keefe, Jim Overton, Colm Regan, Damaris Rose, Sue Ruddick, David Stea, Neil Smith, Alan Wallace, Susan Williams, and Ben Wisner. Communication with the above was by email from 2010 onwards, in response to a set of questions about the formation, membership, and structure of the USG and its linkages to other sites of radical geography. Some of these communications were originally generated for Peake and Sheppard (2011).
7  Although records indicate that the group finished the initial meeting with an agreement to refer to themselves as the Provisional Organizing Committee (of the USG), it appears that the title of USG did not merit any further discussion.

8  The degree of organization varied between places with members at Clark, for example, as late as 1978 seemingly unaware that a membership procedure existed.

9  This latter activity led to divides between the SFU local and Bill Bunge, who disagreed with their approach (Blomley and McCann, this volume).

10  John Bradbury's radicalism was influenced by Keith Buchanan and Terry McGee in New Zealand.

11  The Kingston Socialists were a broad-based group including Trotskyists and Communist Party members. Their organization was part of a network of socialist centers across Canada actively engaged in educational and pro-paganda work with radical working communities, the suggestion for which first appeared in the *Canadian Dimension*, an independent left-wing publication published in Winnipeg.

12  Ken Olwig completed his Ph.D. at Minnesota and spent a year (1977–1978) in a post-doctoral position at UW Madison, before moving to Denmark.

13  Although the Irish local arranged an *Antipode* conference in Dublin in 1978, other activities took place primarily through the British Isles local.

14  Members of this organizing committee included Martin Brennan (Winchester), Damaris Rose (Sussex), Tommy McLaughlin (Northern Ireland), Chris Gibson and Tom Marion (Lancaster) and Colm Regan (Dublin) (*USG Newsletter* 3.3: 56).

15  There were also plans for other publications, including a socialist Monograph Series (the first to be by James Anderson on "The political economy of urbanization"), intended to be self-financing by selling at a small profit. Plans were also made for textbooks, by Mickey Lauria, Dick Hansis, and Eric Sheppard, and in 1978 there was a plan for a textbook on socialist geography to be edited by Neil Smith and Phil O'Keefe and published by Methuen. By mid-1979 the plan appeared to be shelved due to a lack of contributions and a reluctance among some USG members to agree with the framework adopted for the book (*USG Newsletter* 5.1).

16  There were also discussions about whether to formalize the relationship between the USG and *Antipode* (a move that Dick Peet opposed).

17  USG members John Bradbury, Alan Mabin, Manfred Malzahn, Jim Overton, Colm Regan, and Yvonne Zarowny were involved in the *Study Papers on Imperialism*. The *Canadian American Geographical Expedition Papers* were based on a study undertaken by 21 Canadian geography students on a fieldtrip to Detroit and Windsor in November 1976.

18  *Transition* was published by the Socially and Ecologically Responsible Geographers (SERGE), a group formed in 1971 by Larry Wolf and Wilbur Zelinsky (Peake and Sheppard 2014).

19  Volume 7, which was to have been produced by the British Isles local, is missing. All attempts to locate copies have been unsuccessful.

# References

Akatiff, C. (2011). A Great Day in Toronto: The Founding of the Union of Socialist Geographers. Conference paper, *Association of American Geographers Annual Meeting, Seattle* (available from author: cpakatiff@yahoo.com).

Akatiff, C. (2016) The Roots of Radical Geography: AAG Convention, San Francisco 1970, A Personal Account. Conference paper, *Association of American Geographers Annual Meeting, San Francisco* (available from author: cpakatiff@yahoo.com).

Anon. (1975a). Socialist practice: Tenant organizing in Vancouver. *USG Newsletter* 1 (2): 14–16.

Anon. (1975b). Annual General Meeting. *USG Newsletter* 1(1): 16–17.

Anon. (1975c). The First Annual Meeting of the Union of Socialist Geographers. *Antipode* 7 (1): 86.

Anon. (1978). AGM of the USG, New Orleans, April 9–12, 1978 *USG Newsletter* 4 (1): 10–17.

Bergmann, L. and Morrill, R. (2018). William Wheeler Bunge: Radical Geographer (1928–2013). *Annals of the American Association of Geographers*, 108 (1): 291–300. DOI: 10.1080/24694452.2017.1366153.

Bradbury, B. (2015). Twists, turning points and tall shoulders: Studying Canada and feminist family histories, *Canadian Historical Review* 96 (2): 257–285.

Bunge, W. (1965). Racism in geography. *The Crisis* 72 (8): 494–497, 538.

Bunge, W. (1977). *Second Call: The Society for Human Exploration: Field Notes 5* (Canadian-American Geographical Expedition, The Society for Human Exploration).

Bunge, W. and Bordessa, R. J. (1975). The Canadian alternative: Survival, expeditions and urban change. *York University Geographical Monograph No. 2.* (York University, Toronto).

Deskins, D. R. and Siebert, L. (1975), Blacks in American geography: 1974. *The Professional Geographer* 27 (1): 65–72.

Harris, R. (1980). Community action and class struggle in Kingston. *USG Newsletter* 5 (4)/6 (1): 11–21.

Horvath, R. J. (2013). Radical geographical research and pedagogy: The Detroit Geographical Expedition and Institute in 1969. Conference paper, *Association of American Geographers Annual Meeting, Los Angeles*.

Horvath, R. J., Deskins, D. and Larimore, A. (1971). The status of Blacks in American Geography: 1970. *The Professional Geographer* 23 (4): 283–289.

Jöns, H. (2011). Centre of calculation. In J. Agnew and D. N. Livingstone, eds, *The SAGE Handbook of Geographical Knowledge*. Thousand Oaks, CA: Sage Publications.

Latour, B. (1987). *Science in Action: How to Follow Scientists and Engineers Through Society*. Cambridge, MA: Harvard University Press.

Norcup, J. (2015). Geography education, grey literature and the geographical canon. *Journal of Historical Geography* 49: 61–74.

Novick, P. (1988). *That Noble Dream: The "Objectivity Question" and the American Historical Profession*. Cambridge: Cambridge University Press.

Owen, D. and Baxter, S. (1979). Geographers and praxis: The Sydney Geographical Expedition. *The USG Newsletter* 4 (4): 2–6.

Peake, L. and Sheppard, E. (2014). The emergence of radical/critical geography in the USA and Anglo-Canada. *ACME* 13 (2): 305–327.

Peet, R. (1969). A new left geography. *Antipode* 1 (1): 3–5.

Peet, R. (1977). The development of radical geography in the United States. *Progress in Human Geography* 1 (2): 240–263.

Rosenbaum, D. (1970). House panel lists "radical" speakers. *New York Times.* October 15, 1970, p. 23.

Schmidt, R. (1980). Intellectuals and the class struggle. *USG Newsletter* 5 (4): 3–10.

Smith, N. (2000). What happened to class? *Environment and Planning A* 32: 1011–1032.

Socialist Geography Specialty Group. (1980). *Annual Report*. Mimeo.

Stephenson, D. (1974). The Toronto Geographical Expedition. *Antipode* 6 (2): 98–101.

USG Minutes (May 28 1974). Union of Socialist Geographers. *Unpublished document.* 1pp. (Available from Clark Akatiff, cpakatiff@yahoo.com).

## Email exchanges

(conducted by Linda Peake and Eric Sheppard)
Clark Akatiff, November 19th, 2011
Proinnsias Breathnach (aka Fran Walsh), March 24th, 2011
Nathan Edelson, July 13th, 2010
Ron Horvath, May 14th, 2016
Jim Overton, August 1st, 2010
Neil Smith, March 22nd, 2011
Alan Wallace, August 11th, 2010

# 6

# Baltimore as Truth Spot

## *David Harvey, Johns Hopkins, and Urban Activism*

Eric Sheppard and Trevor J. Barnes

Johns Hopkins University, Baltimore, was in many ways an unlikely place for North American radical geography to flourish. Although Isaiah Bowman, "Roosevelt's geographer," was appointed University President in 1935, there were never many geographers on Hopkins' staff (Smith 2003).[1] Owen Lattimore, its most well-known faculty member before David Harvey arrived, had been denounced as communist in the early 1950s by the Departmental Chair, George Carter. Lattimore was added to Joseph McCarthy's list of 205 names for scrutiny by the House Un-American Activities Committee (HUAC), later leaving the US altogether (Harvey 1983).[2] Harvey, who was shortly to become a Marxist himself, joined the Department in 1969. The year before, though, Geography was further diminished at Hopkins, losing its separate departmental identity. It joined with the Department of Environmental Engineering (formerly Sanitary Engineering), the principal research focus of which was waste management.

Despite such unpromising beginnings, we argue in this chapter that David Harvey along with his students made Johns Hopkins University and Baltimore a key geographical node in North American radical geography. The role that geography plays in the production of knowledge has been conceptualized in several forms. In Foucault's (1986) work, place is a heterotopic site, where the clash of difference within it produces new knowledge. In Bruno Latour's (1987) work, place is a center of calculation, where knowledge from elsewhere is brought to the center, sorted,

*Spatial Histories of Radical Geography: North America and Beyond*, First Edition.
Edited by Trevor J. Barnes and Eric Sheppard.
© 2019 John Wiley & Sons Ltd. Published 2019 by John Wiley & Sons Ltd.

combined, and abstracted to generate new knowledge. In Sandra Harding's (2011) work, place is a marginal or peripheral space, where new knowledge is sparked by the tension between those on the margin and those in a dominating metropole. Yet a fourth approach, which we take up in this chapter, is place as "truth spot." Coined by the sociologist of science Thomas Gieryn (2002, 2006), his concern is with how knowledge originating at a particular site is granted a special warrant, a credibility, that makes it appear as the Truth, as "authentic all over" (Gieryn 2002: 118). As Gieryn (2006: 29, fn. 3) writes, a truth spot is defined as "a delimited geographical location that lends credibility to its claims."

For Gieryn, the paradox of a truth spot is that for the knowledge it generates to have credibility, to be authentic all over, that site must be decontextualized, to lose its distinctiveness as a place. As Gieryn (2002: 113) writes, a truth spot must "escape place, ... [to] transcend its suffocating particulars." If it did not, the knowledge generated would apply only to the place in which it was produced, ungeneralizable to other places. Those operating within a truth spot must therefore find a strategy to enable their site to lose its geographical particularity, to become a stand in for all places, to be a "placeless place" (Gieryn 2006: 6)

Gieryn's two exemplars of successful strategies are the Chicago School of urban sociology of the interwar period (Park *et al.* 1925), and the self-declared Los Angeles School of urban theory (Dear 2002) of the 1980s and 1990s. The Chicago School's strategy to escape place, to escape the particularities of Chicago and to speak instead to the universal, was by "shuttling" between two scientific methods, ethnographic and laboratory (Gieryn 2006: 10). Their scientific character, the School claimed, yielded truths about Chicago, but also the Truth about all cities. Chicago as Chicago was erased, becoming a universal city. In contrast, the LA School's strategy to escape place was to assert that Los Angeles was *the* "prototype [city of] the urban future ...." (Gieryn 2006: 26). LA was defined as the universal urban blueprint for what was to come, "predicting what eventually will happen everywhere" (Gieryn 2006: 26). So, while each school began with locally produced knowledge, each adopted a methodological strategy that enabled that knowledge to be separated from its immediate geographical context, given new voice at a universal register. Interestingly, a similar move has been made by some contemporary urban theorists who have argued that particular postcolonial cities have become truth spots, producing a generalized postcolonial urban theory unmoored from the specific local geographical contexts in which it was derived (Sheppard *et al.* 2013; Roy 2016). Subsequent debate, however, has turned on the extent to which local knowledge continues to remain important; that is, rather than postcolonial cities becoming placeless, elements of their localness persist, undermining their role as a truth spot and the apparent universal knowledge created (Sheppard *et al.* 2015).

This is also the issue raised by Gieryn (2002, 2006). He suggests that although various strategies such as the Chicago and LA Schools' are used to expunge place from knowledge, its removal (the aim of a truth spot) is more apparent than real. The specter of place, Gieryn argues, continues to haunt and disturb, undermining universal claims to knowledge.

We make a similar argument about the role played by Baltimore in Harvey's radical urban geographical theory that he began to produce there shortly after he arrived. That theory became one of the pillars of Anglophone radical geography that, as other essays in this book demonstrate, circulated and traveled globally. As with the Chicago and LA Schools, there was at least an implicit claim that Harvey's theory was untethered from the original urban context in which it was produced, 1970s Baltimore. It was now a universal theory, with the truth spot of Baltimore standing in for all places. In this chapter, however, we argue that there was still a lot of Baltimore remaining in the theory. While you can take Harvey's theory out of Baltimore, you cannot take Baltimore out of Harvey's theory. Our task, therefore, is to recoup the native origins of Harvey's theory, to show the continued importance of the local and its impress on the knowledge it generated. Almost as soon as Harvey arrived in Baltimore, the city was integrated into his theoretical architecture. Baltimore did not stand apart, separate from his theory, as if it were a laboratory used to test an already worked out abstract explanatory structure. The process was much messier, organic, even dialectical, with the city and theory mixed up together, muddled and twisted. As Harvey learned about, studied, and became politically active in Baltimore, it shaped how he drew on Marx (1972 [1885]). And as he learned more Marxist theory, the city similarly changed. There was no outside Archimedean viewpoint. Everything was in motion as things were worked out on a moving earth, a churning city, a fluid theory.

In trying to recover the local in Harvey's radical geographical theory we use a broad definition of the local that includes connections a place has with other places (Agnew 1987; Massey 1991). Local knowledge produced in a place is shaped by its geographical links with other nodes. Baltimore as a radical geography truth spot was fashioned by its relationship with events occurring elsewhere, just as it also influenced the trajectories of radical geography happening in other places.

We start with Harvey's long journey, both geographical and intellectual, from the U.K., and the Universities of Cambridge and Bristol, to the U.S.A. and Johns Hopkins University. We discuss his writings during this early period, very different from the radical geography he later took up, but in both cases, following Gieryn, forged from the places where he lived and worked. Second, in the longest part of the chapter we examine how Harvey's engagement with Hopkins and Baltimore shaped the development of his Marxist radical geography. We focus especially on his first decade in Baltimore during which he published his book, *Social Justice and the City* (1983), and experienced a conversion from some form of liberalism to socialism. Finally, we examine how the geographical movements of Marxist radical geography going on outside Hopkins and Baltimore joined with Harvey's own, forming the complex spatio-temporal evolution narrated in this book.

## Slouching toward Baltimore

Harvey's work before arriving at Hopkins was distant from the radical geography he practiced afterwards. Pre-Hopkins, he was concerned with giving a philosophical and methodological rationale for geography's quantitative revolution, and the discipline's shift to a quasi-positivist spatial science that began in the late 1950s. He also provided a set of illustrative empirical and theoretical case studies set within that same spatial scientific tradition. This early work culminated in his tome, *Explanation in Geography* (Harvey 1969). Published the same year he arrived at Hopkins, it quickly became the bible for the quantitative movement. In line with Gieryn's argument, however, *Explanation* was not the view from nowhere, notwithstanding various claims to value-free knowledge made for it by spatial scientists and their fellow travelers. Rather, it very much reflected a view from somewhere, from a densely embodied and geographically material setting steeped in cultural politics: early post-War Britain, and Cambridge in particular, where Harvey conceived the project (Barnes 2006).

Harvey grew up in rural Kent in a working-class family, his father a dockyard ship repairman. Excelling at grammar school, in 1954 he won a scholarship to read Geography at Cambridge. Harvey said he chose geography because "my father was close to the navy and the naval tradition, I always thought from an early age that knowing the world and sailing the world was very important."[3]

In 1957, after gaining a First-class degree, now as a first-year Cambridge doctoral student, he took a job as a "Demonstrator" (teaching assistant) for a first-year class taught by the newly hired "terrible twins," the early modelers and quantifiers of geography, Richard Chorley and Peter Hagget (Barnes in press). Their course covered techniques never taught before in British Geography Departments, "statistical methods, matrices, set theory, trend surface analysis, and network analysis" (Chorley 1995: 361). Another important Cambridge influence was the young lecturer and historical demographer, Tony Wrigley, who introduced Harvey to August Comte's positivism and more generally to nineteenth-century thought, including Marx's (Harvey 2002: 165; 2003). Wrigley's philosophical approach and Haggett and Chorley's emphasis on "scientific methods," as Harvey (2002: 165) later put it, were "enmeshed" in his 1962 doctoral dissertation about the historical dynamics of the Kent landscape that so captivated him when growing up: "Aspects of agricultural and rural change in Kent, 1815–1900."[4]

In 1961, Harvey was appointed Assistant Lecturer in Geography at the University of Bristol. There he continued pursuing both the philosophical justification of the scientific method, in his case through logical empiricism, as well as applying various spatial mathematical techniques and theories to real-world problems. Logical empiricism, he believed, would provide an overarching, integrative rational principle for the fragmented discipline of human geography, allowing it to move from a deadening exceptionalism toward generalization, systematicity, and universal explanation. Further, technical problem-solving from within the perspective of logical positivism would exemplify the practical worth of scientific geography, demonstrating its potential contribution to such state administrative planning goals as improved social efficiency and equity.

Harvey believed that his exposition in *Explanation* of logical empiricism and assorted spatial scientific techniques was him simply being rational, measuring out iron-clad, universal logical propositions one by one on the page. It was seemingly irrelevant where he set them out. In later interviews and autobiographical reflections, however, and effectively channeling Gieryn, he appears now to recognize that where he had been, the places he inhabited, made an enormous difference to that early work. *Explanation* was in large degree a response to a specific time and place: Britain in the immediate post-WW2 period. Harvey was part of the younger generation reacting against the social rigidity, conservatism and traditionalism of 1950s and early 1960s "Little Britain." Along with others, from "angry young men" playwrights (e.g., John Osborne) and novelists (e.g., Kingsley Amis), to TV satirists (e.g., David Frost), and comedy sketch performers (from the casts of Beyond the Fringe to Monty Python), he demanded modernization:

Mine was the generation that spawned the *Footlights Review* that became *That Was The Week That Was* – a television show that mercilessly ridiculed the ruling class as well as almost everything else that might be regarded as "traditional" in British life. Cambridge was populated by an intellectual elite, and if something was seriously wrong with the state of Britain (and many thought there was), then this elite was surely in a position to do something about it. The modernization of Britain was firmly on the agenda, and a new structure of knowledge and power was needed to accomplish that task (Harvey 2002: 164).

*Explanation* was not a single-handed attempt to modernize Britain, but it was a contribution to that larger end, providing a "new structure of knowledge" for the hitherto conservative discipline of geography. As part of a Cambridge intellectual elite, Harvey believed he was "in a position to do something about it." There was "the idea that we could break out of tradition," Harvey said. "There was a modern geography waiting to be constructed and we were the ones who could do it" (Harvey 2003).

The idea of modernization also bore on his politics. It turned on what he called "socialist modernization … backed by technological efficiency" (Harvey 2002: 165). It was the same politics that lay behind the election in 1964 of the British Labour Party under Harold Wilson, and captured by his catchphrase, "the white heat of the technological revolution."[5] Increased technological efficiency, rational planning, and progressive social change would unfold in a new and modern Britain, which broke the old conservative order. While *Explanation* was not exactly a "how-to" book, it drew its spirit in part from wider cultural and political concerns in Britain that emerged in late 1950s, and discussions about the best means to move forward. As Harvey (2002: 166) now says: "For those of us involved in geography [during the 1960s], rational planning (national, regional, environmental, and urban) backed by 'scientific' methods of enquiry seemed to be the path to take." *Explanation* was Harvey's contribution to "'scientific' methods of enquiry." While not overtly political, it was a thoroughly political text. It was Harvey's contribution to modernizing Britain, its politics, and geography.

May, 1968, when Harvey submitted the manuscript of *Explanation* to Edward Arnold publishers, was not exactly propitious for science or modernity, however (Harvey 2002: 167). Things were falling apart, the center no longer holding. Martin Luther King was assassinated in a Memphis motel in April, there was almost a revolution in Paris in May (spreading to other cities around the world), and Robert Kennedy was murdered at the Democratic Party convention in Los Angeles in June (Barnes and Sheppard, this volume). Science's technical problem-solving capacity, the manipulation and application of mathematical formulas and associated calculations, were also bringing untold horror and misery to large numbers of people in Southeast Asia through the Vietnam War. Yet these political events did not infuse the book:

> I was so busy writing a book in 1968 that I never really noticed what was happening, and I confess this. I finished the manuscript almost in May of '68, and I was so preoccupied with it that I put the manuscript on the desk and I went to Australia.... I got to Australia, and everybody in Australia said, "What the hell is happening in Europe in '68?" I said, "I had no idea."[6]

## Some Revelation is at Hand

Before he moved to Baltimore, Harvey already had spent time in America. In 1964, he visited Bill Bunge in Detroit where he went on his Detroit Expedition (Katz *et al.*, this volume). In 1966, he lived for a year in almost splendid rural isolation at Penn State University, although he still managed to experience anti-Vietnam war campus protests.[7] Turning down an offer to return to Happy Valley,[8] in 1969 he moved permanently from Bristol to Baltimore, reiterating a historically long-standing connection between these two port cities that dated back at least to the era of slavery and the triangular trade (Morgan 1993). Harvey (2002: 169) said he was attracted to Baltimore precisely because it was "deeply troubled … A city of crises and uprisings."[9] There, as he put it, "I was determined to grapple with the urban crisis."[10]

Baltimore's crisis was particularly acute. Located in effect between the US North and South – "the largest East-coast city below the Mason Dixon line" (Fee *et al.* 1991: xiv) – from the 1950s Baltimore was a full participant in the nation-wide struggles unfolding over civil rights, racial discrimination and urban impoverishment. Particularly issues of race historically cleaved the city. Harvey thought of Baltimore when he first arrived as "the largest plantation in the South."[11] The year before he arrived, parts of Baltimore's inner city had "gone up in flames" following the assassination of Martin Luther King (Harvey 2002: 169). There was substantial looting and property destruction, along with six deaths, and hundreds injured. Close to six thousand mostly black residents were arrested.[12] The skewering effect of the city's racial divide in 1968 led the Black Panther Party (BPP) to establish a Baltimore branch, headquartered on East Eager Street. It was outside that building in frigid temperatures during his first year in Baltimore that Harvey, along with other faculty and students, slept in rotation for six weeks to protect the BPP from potential violence following the police killing in Chicago of its leader, Fred Hampton, in December 1969 (Harvey 2002: 170).

Harvey came to Baltimore at the invitation of M. Gordon "Reds" Wolman ("Reds" for his carrot-colored hair, not his politics). Wolman arranged for the new Department of Geography and Environmental Engineering (DoGEE) to hire Harvey. The position was available because a year earlier Wolman and John C. Geyer (an engineer) established the DoGEE by merging Geography and Environmental Engineering, two small departments each with 5–6 faculty.[13] Buttressed by a substantial ($858,000) Ford Foundation grant, there was money to appoint three new faculty members bringing the department's faculty complement to 14, including Harvey who was appointed Associate Professor with tenure. Wolman thought him "the most brilliant geographer [he'd] … ever met."[14]

Wolman was a critically important figure: Open-minded, ecumenical, enthusiastic, energetic, polymathic, a man who seemed to know anyone who was anyone in Baltimore. By force of intellect, persuasive power, and administrative skill, Wolman kept a wildly heterogenous faculty and their interests relatively harmoniously housed within a single Department. While no Marxist, Wolman provided openness and elbow-room for Harvey and his students to thrive, for Johns Hopkins to become a radical geography truth spot. In this sense, Wolman's role was comparable to Donald Hudson, the Chair at the University of Washington, Seattle, who while no quantitative revolutionary was nevertheless midwife to geography's 1950s quantitative revolution (Barnes 2001). Like Hudson, Wolman provided radical geography at Hopkins with protection and cover, independence, opportunity, and on occasion material resources.[15]

## Baltimore housing research

Those material resources were made available the day Harvey joined DoGEE. Wolman provided him with in effect a "turn-key" research project on Baltimore's housing market. Prior to Harvey's arrival, Wolman secured a research grant from the Federal Housing Authority (FHA), and even lined up "a great graduate student" research assistant, Lata Chatterjee.[16] Once the project began Wolman functioned as an *eminence grise*, using his extensive rolodex of Baltimore social contacts to link Harvey with the city's principal actors. As Harvey recalled:

> Reds had influence within the city as a whole. [He was] connected to everyone. He was politically connected; he could talk to bankers; he knew landlords who owned inner-city properties; and he had connections to the FHA. He had access to all the players.[17]

Chatterjee's main role was to collect and analyze large amounts of statistical information, particularly financial data about banks, mortgages, house prices and rents. With Harvey, she also jointly conducted interviews with the main actors, providing an understanding of the principal mechanisms operating within Baltimore's housing market. With this involvement, Harvey said he "became invested in the city. I really got to know it."[18]

The research produced two reports. The first was a long and detailed spatial analysis of the Baltimore housing market, taking into account housing conditions, prices, occupants, tenure status, and housing code enforcement (Harvey *et al.* 1972). Its principal conclusion was that housing code enforcement was too heavily concentrated in low-income, black inner-city areas, with the result that housing within that submarket was withdrawn because the return on landlords' investment in those areas was too low to justify refurbishment of buildings marked as below code by city inspectors. City code enforcement in low-income areas therefore only reinforced the already existing disinvestment in that sub-market. Harvey *et al.* (1972) recommended that code enforcement be more evenly applied across the city, and that the state should implement alternative housing policies, particularly for the inner city, including taking over housing provision.[19] The second report (Chatterjee *et al.* 1974) was an assessment of FHA mortgage activities. It broke the Baltimore housing market into geographical submarkets, categorized them into sub-types (e.g., gentrifying), and analyzed the role of the FHA in each sub-type. The report concluded that the FHA should privatize its consumer protection function, abandon its discount points system, reform its arbitrary and time-consuming appraisal procedures, and work more closely with city planning departments.

## Karl Marx in Baltimore

It was precisely during this very early 1970s period, when undertaking his housing research, that Harvey had his Road-to-Baltimore experience. The scales fell from his eyes, and he was converted from a belief in logical empiricism to something very different. His initial methodological approach to the Baltimore housing project in autumn 1969 was logical empiricist. Hence, Chatterjee's statistical analysis. That quickly changed precisely because of Harvey's engagement with Baltimore, as he "really got to know it." Place mattered; Baltimore mattered. As he put it, "studying Baltimore was always part of what I was doing. The abstractions were always in dialog with … my urban experience."[20] Consequently, as Harvey (2002: 170) later wrote, "the travails of Baltimore formed the backdrop to my theorizing."

*Explanation* "was a book about methods. But it didn't say anything about ethics and justice."[21] Yet ethics and justice were paramount in early 1970s Baltimore, staring Harvey in the face: racial prejudice and segregation, urban poverty, debilitating housing conditions, street violence and crime, riots, arson, looting, armed military patrols. "My move to Baltimore was all about these things," Harvey said. "I had done the methodological side. Now I needed to do social justice."[22] It was here, and relating directly to his housing project, that Harvey brought Karl Marx to Baltimore.

Harvey said he knew little about Marx before he arrived.[23] That began to change as he "participated in the housing study … in the autumn of 1969" (Harvey 2016a: 35). Given his concerns for ethics and justice, as he began writing up his Baltimore housing study he became "deeply frustrated … [and] more and more uncomfortable" with the urban theory then on offer and its "market-based analysis."[24] Harvey thought the leading urban theorists, William Alonso and Edwin Mills (also at Hopkins), were "good progressive people." Their market approach dealt neither with ethics nor justice, however, and missed key causal mechanisms in the operation of housing markets in Baltimore as Harvey understood them.[25] As an alternative, he "experimented with the idea, taken from Marx's *Capital*, of analyzing the commodity character of housing provision in terms of the contradictory character of the relation between use value and exchange value" (Harvey 2016a: 35). This was especially clear in chapter three of the first housing report where he explored (albeit without explicit reference to Marx) the distinction between the use value of housing (its value to residents for shelter, social reproduction and access to urban amenities) and its exchange value (the capitalist market price), from the perspective of residents, realtors, landlords, developers and government institutions.

Within the same report, and again without any explicit reference to Engels or Marx, he introduced the idea that capitalism never solved its crises but simply shifted them geographically. As Harvey (2016a: 35) put it retrospectively,

> I noted [in the housing report] how public policies generally failed to solve housing problems but just moved them around. I concealed for obvious tactical reasons the origin of these ideas in the report and I was pleasantly surprised that city officials, landlords, and financiers found such formulations helpful, commonsensical, and intriguing.

He remembered one occasion when he presented his Baltimore housing study at a conference attended by the Vice President of Chase Manhattan Bank. Some in the audience attacked Harvey after his talk, but the vice President defended me. He said one of the biggest problems we have in New York is that once we improve the neighbourhood, the people we are trying to help leave. He said to me that my account was such a good idea. "Where did I get it from?" I said, "I got it from Engels." "Is he at Harvard?" he said. "No, Friedrich Engels." He was a little bit shocked.[26]

Harvey "wasn't a convinced Marxist at the time." Rather, he thought of his analysis as "deep common sense. What Gramsci called 'good science.'"[27] As he put it in *Social Justice and the City*, "I do not turn to [Marx's analysis] out of some a priori sense of its inherent superiority but because I can find no other way of accomplishing what I set out to do or of understanding what has to be understood" (Harvey 1973: 17). Marx gave Harvey an "ah-ha" moment. Rosalyn Franklin, who competed during the early 1950s with Francis Crick and James Watson to understand the structure of the DNA molecule, said when she was presented with the latter's double-helix model, "So, that's how it's done" (quoted in Haggett 1990: 184). It was the same with Harvey and Marx: "So, that's how Baltimore's housing market works." Consequently, Harvey continued reading Marx for fresh insights. He gave his first graduate seminar on Marx's *Capital* in 1972 and has taught the same books annually ever since: at Hopkins (1972–1989; 1993–2001), Oxford (1989–1993), the CUNY Graduate School (1993-present), and now on YouTube.[28] As he reflected:

> Just like you walk the streets of a city, and get to know it better and better, I've been walking through Marx's *Capital* [every year]… Just as you notice more things as you walk through a city, it is the same as I walk through *Capital*. As I wander through it, I see something new. I am not a fast learner, but I get there in the end. It took me a long while to work out what Marx was saying… If *Capital* remained the same as when I first read it, I wouldn't have enjoyed reading it again. But each time I get to know it better. I don't mind teaching it again. It is not, "Oh no," but, "Oh good." "What am I going to find this time?"[29]

## *"Socialist Formulations"*

Harvey's record of his conversion to Marx in Baltimore, and his perambulative relation to both, is best found in the first book he wrote after arriving in Baltimore: *Social Justice and the City* (Harvey 1973). As Harvey recalled, "it all came together in *Social Justice and the City*. It came out of the political atmosphere of ant-war protests, civil rights disputes, environmental ferment, along with the hedonistic side of the early 1970s, sex, drugs, and rock and roll."[30]

Divided into two parts, "Liberal Formulations" and "Socialist Formulations," the book was in effect an account of Harvey's intellectual turn from the former to the latter. In 1992, when the volume was selected as a *Progress in Human Geography* "Classic in Human Geography Revisited," Harvey (1992: 73) recognized that "it was a most unpromising book for publication." It was mainly a collection of previously published articles without, at least initially, a clear structure. One reviewer of the early manuscript at Johns Hopkins University Press even called it "incoherent" (Harvey 1992: 172).

Harvey carried the book manuscript to Edward Arnold Publishers in Merrick Street, Mayfair, London, in the summer of 1972: Arnold had published *Explanation in Geography* three years earlier. John Davey, one of Arnold's editors, was in his tiny office that day. As Davey recollected,

> David just arrived with it in my office. He had come back from the US. He was clutching his handful of typescript. It was not very neat, partly off-prints and partly typescript. Some of it was *Antipode* stuff. In those days the journal wasn't printed. Completely unedited. Dreadful stuff. ... The book had no organization. ... I had modest input by dividing the book manuscript into two: liberal formulations, socialist formations. You know thesis, antithesis, synthesis. ... I feel I was the midwife for the book. It arrived in such a mess. But I unified it. It became a beautiful object. We used Garrod type. It was slightly eccentric, but it was elegant... There was nothing like it around. Nothing... "What shall we call it," I asked? And he immediately said, "*Social Justice in the City*." It sold 25,000 copies in its first year. Kept on selling.[31]

Its success was partly because there was "nothing like it around," but partly also because of a narrative structure in which Harvey was a Pauline figure converted to Marxism on the streets of Baltimore (St. Paul was converted on Straight Street, the road to Damascus). The importance of the concreteness and specificity of Baltimore was clear from Harvey's (1973: 9) Introduction, writing that the book

> could not be pursued by abstractions ... I was therefore determined to pursue it in a context with which I could know at first hand and yet which was broad enough to provide material examples and a fund of experience upon which to draw whenever necessary. Since I had just moved to Baltimore it seemed appropriate to use that city ... as a backdrop to explore questions that arose from projecting social and moral philosophical considerations into the matrix of geographical enquiry.

## Reading Capital

Richard Walker, then a graduate student at DoGEE, recalls reading and discussing chapters from Harvey's *Social Justice and the City*. As he put it, "we were there for his conversion."[32] Harvey's conversion was reinforced in part through an informal reading group that met every two weeks. Initiated by graduate students, they convened to discuss Marx's works and associated literatures. The first reading group focused on Volume One of *Capital*. Early graduate student participants included Walker, and also Jörn Barnbrock and Gene Mumy (all handsomely acknowledged by Harvey 1973: 19 as "contribut[ors] to the writing of this book").[33] Harvey (2000: 80) recalls the first meeting of the reading group as both "a decisive moment" and a "wonderful experience." Recently reflecting on that reading group he said:

> We were all confused and discontented with the social scientific frameworks of knowledge that seemed incapable of helping us to understand the turmoil in the cities, the diverse responses to the civil rights movement, a war in Vietnam that was obviously more than a little bit animated by imperialist ambitions and the stifling approach to knowledge in academia. We turned to Marx to see if there was anything there that could enlighten us, and most of us concluded that there was, even though none of us (that includes me) considered then ourselves "Marxist" (Harvey 2016b).

In terms of those contributors, Dick Walker had arrived at Hopkins in fall, 1971. He had majored in economics at Stanford, entering Hopkins as a "closet positivist."[34] He was no political conservative, "no Goldwater Republican," however. At Stanford he was exposed to the Cambridge (U.K.) radical economist, Joan Robinson. As a visiting professor, she taught her classes "outside on the grass," signaling her left-wing politics by "wearing a Mao suit." Her central argument was that the neoclassical economics that Walker was learning inside his Stanford classrooms was all "bosh." Walker said, "at first I couldn't understand her. What the hell is she talking about?" Later he got it. Walker also was radicalized by the anti-Vietnam War movement, which reached a crescendo on US American college campuses during the late 1960s and early 1970s (Barnes and Sheppard, this volume). After graduating from Stanford and looking for a graduate program focusing on the environment, Hopkins' DoGEE was one of the very few possibilities. The Ford Foundation money that Wolman received to inaugurate DoGEE was to train people for the US Environmental Protection Agency. Arriving in winter 1972 to study with Stephen Hanke, an environmental economist hired at the same time as

Harvey, Walker enrolled in Harvey's course on social and philosophical issues and the environment. Joining the *Capital* reading group that same year, Walker was convinced only at the third reading: "It wasn't an instant conversion. We weren't red diaper babies. But we were all drifting to the Left."

Jörn Barnbrock studied architecture at the Berlin Technical University and spent a year at Wayne State University's Center for Urban Studies on a Ford Foundation grant studying with Wilbur Thompson. In 1972, he joined Johns Hopkins' Center for Metropolitan Studies and Research on a one-year fellowship. Having read *Capital* in German, he enrolled in Harvey's environmental course, and then his winter 1972 seminar on *Capital*. Harvey recruited him into the Ph.D. program later that same year, and he became a full-time member of the reading group.[35] His knowledge of *Capital* was sufficient for him to sub for Harvey in the seminar when he was away.

Gene Mumy arrived in Baltimore in fall, 1970, from the University of Colorado, Boulder, where he took classes with Kenneth Boulding.[36] He first met Harvey in 1971 when DoGEE moved into a new building, Ames Hall. That same year he enrolled in Harvey's course on social and philosophical issues and the environment that involved reading Clarence Glacken's (1967) *Traces on the Rhodian Shore*, and other equally large and dense tomes by Hegel and Marx. This was not geography as Mumy imagined it, as "cartography and making maps." The next quarter he took Harvey's graduate course on *Capital*. Here he was better placed. He knew Marx's work through Boulding's courses and debated Harvey on the labor theory of value and the transformation problem. Harvey did not show Mumy drafts of *Social Justice and the City*, but gave him a copy when it was done saying, "here is the final product. It is going to be pretty big." Mumy thought so too. It was.

The reading group often met at the house of Harvey's then girlfriend, the investigative journalist Barbara Koeppel.[37] She lived in Baltimore's leafy inner suburb in Mount Washington (classified as an "upper-income submarket" in Harvey and Chatterjee's housing research). Koeppel remembers that Harvey never dominated the reading group. Rather, he participated primarily by summarizing the meetings as they came to a close, with always "an original way of looking at things." There were occasional visitors, including Ralph Miliband, and the political scientist Nancy Hartsock. Hartsock told Koeppel after the meeting she joined: "I'm never going back there. Don't you see how the guys interrupt? They don't listen or respond." That was not Koeppel's impression, however.

Other faculty members of DoGEE generally regarded Harvey's turn to Marx with equanimity. John Boland, who came to Johns Hopkins in 1968, completed an environmental economics DoGEE Ph.D.

(1971–1973), and then spent his career there as a faculty member, recalls that under Wolman the department had no cliques, no barriers between faculty and students, and a culture of respect for one another's scholarship. It was a white male liberal department, within a white liberal university located a few blocks from black South Baltimore where Harvey would take students on redlining field trips. It was clearly "fertile ground" for radical geographers.[38] Harvey had cordial relations with DoGEE's environmental engineers, even trying to teach them dialectics (Harvey 1996: ch. 2). According to Boland, the only tension was when Stephen Hanke, Walker's and Mumy's former doctoral supervisor, returned from sabbatical as a "right wing monetary economist." He opposed Harvey's work by writing newspaper op-eds. Hanke was ignored by the other DoGEE faculty, however.[39]

## "Out in the streets"

Harvey (2007: 15) said in an interview,

> I am not a good organizer, I am not a good militant in the sense that I can be out in the streets. But I am good at certain things, and I try to use those skills in a very specific way to be as active as I can.

He continued,

> I am intellectually trained ... [and] given ... that I've read political economy in the depth I have, the fact that I've read Marx in the depth I have, and the fact that I've worked in urban history in the way I have, I have an obligation, given those understandings to tell it how it is (Harvey 2007: 20).

That's what he was doing in those Baltimore housing reports. Telling it how it is. He is too modest, however, because he was also "out in the streets." There was that first winter in Baltimore when he slept on an icy pavement to protect BPP members. This active encounter with the streets of Baltimore intensified his knowledge of the city. As he put it, from that experience, "I learned a great deal about what ... life was like in central Baltimore."[40] In turn, such knowledge became a vital component of his larger social theorizing.

Later Harvey (1996: ch. 1) conceptually explored the relationship between place, activism, and theorization in an essay he wrote on "militant particularism," an idea he took from the Welsh Marxist cultural theorist Raymond Williams. Militant particularism meant engaging actively and critically with the political issues of the place in which you lived. When

Harvey wrote that essay it meant being involved in the activism around contesting the downsizing of the Cowley automobile assembly plant in Oxford, the city he had moved to in 1989 to take up the Mackinder Chair of Geography at Oxford University. Harvey argued that a critical engagement with one's place was politically and theoretically crucial for radical geographers: "Theory is never a matter of pure abstraction... Theoretical practice must be constructed as a continuous dialectic between the militant particularism of lived lives and a struggle to achieve sufficient critical distance and detachment to formulate global ambitions" (Harvey 1996: 44). To realize the global ambition of overthrowing capitalism, Harvey was suggesting, one must start in one's own backyard. He recognized the difficulties in scaling-up politically from the local to the global, or as he also called it, "translating [among] different levels of abstraction" (Harvey 1996: 39). But it was a fundamentally necessary task, exemplified by Harvey's insistent involvement with local activism during his Baltimore years.

At first, local activism meant primarily protesting the Vietnam War. Harvey went on anti-War marches accompanied by his students, Barbara Koeppel, and other faculty, particularly Vincente Navarro, a Professor and MD at Hopkins' School of Public Health, and Ric Pfeffer, an activist lawyer and Professor of Political Science (subsequently denied tenure).[41] By the mid-1970s Harvey had become part of a "strong social group involved in political action," concerned especially with labor issues.[42] He was active in supporting strikes by miners in Appalachia (1974–1975) as well as those by garbage workers, teachers and hospital workers in Baltimore.[43] He also forged connections with the local United Steelworkers union, with some of its members coming to his class on *Capital* taught off campus. On campus, he gave "teach-ins" on the economy to interested students (Figure 6.1).

A highpoint in Harvey's local activism was his participation in the Baltimore Rent Control Campaign that culminated in a successful 1979 city referendum (72,000 votes in favor versus 67,000 against) (D'Adamo 2002). Opposition had come from Baltimore's elite, which reputedly outspent the Campaign of progressive political organizations by fifty-to-one. Harvey contributed an editorial to *The Baltimore Sun* that drew directly on his housing research with Chatterjee (Harvey 1979). It was a short-lived victory, though. Two weeks after the referendum, a Maryland judge ruled the rent control measure unconstitutional, a judgment reaffirmed by Maryland's State Supreme Court in June 1980 – symbolically on the eve of the neoliberal "Reagan revolution."[44]

That revolution increasingly blocked progressive urban reform. In response, Harvey worked with others to create a local grassroots

**Figure 6.1** Teach-in on the Economy at Johns Hopkins (including a lecture by Harvey). *Source:* Drawn by Jörn Barnbrock. Reproduced with permission.

progressive initiative informed by political economy, an approach he believed increasingly germane as neoliberalism took hold. As Harvey subsequently reflected, "When the welfare state was strong, Marx's writings were less relevant. But Marx made more sense as the welfare state was whittled away."[45] He became a founding member of the Progressive Action Center (PAC), which opened on October 22, 1982, in the former Enoch Pratt Library in Roland Park (Figure 6.2).[46] "The initial group included a librarian, a machinist, a truck-driver, a dockworker, two lawyers, a few social workers, and several professors. Their politics were council communist, democratic socialist, Gramscian, independent Marxist, Maoist, and socialist-feminist" (D'Adamo 2002). Bought from the city for $1,000, and renovated with $48,000 from Baltimore's

**Figure 6.2**  The Progressive Action Center, Enoch Pratt Free Library building, Roland Park, Baltimore. *Source:* Research Associates Foundation, Baltimore MD. (http://rafbaltimore.org/about/history/). Reproduced with permission.

Commercial Revitalization Program, the PAC was run by the Research Associates Foundation, a non-profit 501(c)3 organization and initially limited partnership. The PAC hosted various radical organizations including the Alternative Press Index (http://www.altpress.org/mod/pages/display/8/index.php?menu=pubs), sponsored talks by notable left intellectuals (e.g., Rudolf Bahro, Barbara Ehrenreich, Ralph Miliband, Paul Sweezey, and Harold Zinn), was the site for Harvey's outreach teaching on *Capital*, and organized the People's History Tour of Baltimore: "a bus trip to places of important struggles of African-Americans, women, and workers" (D'Adamo 2002).

PAC activities had no direct connection to Johns Hopkins. Its participants also struggled to connect with radical activist groups organized around race such as the Black Panthers and the Nation of Islam. Harvey recalls the PAC as "basically a white radical group" that actively campaigned for black political candidates but was not involved in "collective organization with the black community."[47] The PAC remained involved in progressive Baltimore politics until 2002, tailing off until 2010, when PAC was sold to another organization. "We were getting old; younger folks did not want to live in this kind of way."

## Realizing the Truth Spot: The Circulation of Baltimore-Style Radical Geography

In turning to radical geography during the early 1970s, Harvey of course was not alone. History was at his back. Other social sciences in America were making similar radical moves as dissenters left the street for the lecture hall (Katznelson 1997). Moreover, as discussed in the introductory chapter, there was a long radical tradition in geography rooted in anarchism with which Harvey's own work in effect now joined hands (see Springer 2014, and Harvey 2015). More immediately, there were also the other embryonic nodes of radical geography surfacing such as in Detroit with Bill Bunge and Ron Horvath (Katz *et al.*, this volume), or at Clark University in Worcester, MA, with Jim Blaut, David Stea, and Dick Peet (Huber *et al.*, this volume).

It was especially with Clark and Dick Peet that Harvey formed an early connection. Peet recalls: "Hopkins and us were about the only places around where radical geography was [then] found."[48] At the 1971 annual meeting of the Association of American Geographers (AAG), Peet organized a session on American poverty and invited Harvey. The previous year's AAG meeting in San Francisco had not gone well (Barnes and Sheppard, this volume). The Boston meeting would be different. Harvey presented the crucially important hinge chapter from *Social Justice*, "Social justice and spatial systems". He argued that the critical spatial issue was not the efficient use of resources, and the concomitant assemblage of optimizing mathematical theories and methods, but their social distribution. It was "questions of distribution that ... ha[d] for so long been left in limbo" (Harvey 1973: 118). Harvey's paper resonated powerfully. Afterwards, Peet thought to himself, "God, we've arrived"[49] (Barnes and Sheppard, this volume).

Others thought that too. The British geographers David Smith and Hugh Prince provided "letter-from-America" style reports for a special section of the UK geography journal, *Area* ("America! America?"). Smith

(1971: 153) observed after attending the session that "a new wind of change is beginning to blow, in the form of an emerging 'radical' geography and an embryonic 'revolution of social responsibility.'" While Prince (1971: 153 noted the stark contrast between "the vitality and sense of commitment displayed" in the radical geography sessions compared to what went on elsewhere in the conference, marked by "private debates, [a] preoccupation with trivia and ... a cold lack of moral sensibility and human compassion."

Harvey's AAG paper did not refer explicitly to Marx, but a second paper that he "circulated at the Boston meetings" resolutely did, with about a third of the references citing Marx and Engels (Harvey 2016a: 35). That essay, "Revolutionary and counter-revolutionary theory in geography and problems of ghetto formation", was published shortly afterwards in *Antipode* (Harvey 1972). It was followed by series of commentaries including an especially spirited riposte from Brian J. L. Berry (1972), then-doyen of urban geography and spatial science. "Revolutionary and counter-revolutionary theory" became the chapter immediately following "Social justice and spatial systems" in *Social Justice*, leading off the "Socialist Formations" part of that book.

"Revolutionary and counter-revolutionary theory" represented Harvey's coming out, declaring his love for Marx, and for Marx's political rendering of questions of social distribution, of social justice. For our purposes, its significance was, as Harvey (2016a: 36) later put it, that it "emerged from [the] detailed studies of housing conditions ... in which I participated ... shortly after arriving at Johns Hopkins." In the series of papers that followed he used the Baltimore housing research to exemplify his now radical approach, set within the revolutionary theory of Marxism.

It was also at this point that Harvey's work jumped scale to exert a broader effect on the evolution of radical geography in (and beyond) North America. It was then that Baltimore became a truth spot, sloughing off its particularity and increasingly standing in for everywhere. Thelocal became global. That change was partly realized through the travels of Harvey and his students, like Dick Walker, taking Baltimore to other places. Partly it was through the tactics of radical geographers who, to make their approach dominant, shut down rivals and denied alternatives. Radical geographers from Clark and Hopkins would arrive early at the AAG meetings, and identify conservative, "counter-revolutionary" AAG sessions and participants that should be disrupted. Peet said in an interview:

> You did whatever you could to make your way into the discipline and achieve a position of power.... For example, before the AAG meeting, we'd have a radical meeting.... We would look through the program,

and we'd look for nasty people giving nasty papers, and we say, "All right comrades, which of you is going to go and get that person?" And we'd go and ask really difficult questions, and some of the people were pretty good. I mean, imagine David Harvey sitting there for the 20 minutes that you're making your delivery, and thinking of the most difficult question he could think of, and asking it to you.... I saw people almost faint when they saw this group of people, the comrades sitting there with their arms folded. I mean really nasty stuff![50]

Finally, Baltimore became a truth spot as the Baltimore approach increasingly dominated radical geography's key organ of dissemination, *Antipode* (Huber *et al.*, this volume). Radical geography in the first three volumes of *Antipode* was left-leaning, but eclectic. It became increasingly Marxist from Volume 4 in 1972, however (Philo 1998). Volume 4 (1) was organized by Walker (1972) and included an article by Mumy (1972), while Volume 4 (2) was ordered around Harvey's paper, "Revolutionary and counter-revolutionary theory." It highlighted the argument that "Radical geography must be Marxist" (Folke 1972). It is no exaggeration then, to see radical geography's Marxism as at least in part made in Baltimore.

The truth spot further spread its wings as the DoGEE doctoral students who read *Capital* graduated from Johns Hopkins. Walker was hired at UC Berkeley in 1972 (before completing his thesis, the first draft of which Harvey told him was "terrible"). Having survived a titanic battle for tenure, he collaborated with Michael Watts, Allen Pred, and graduate students to catalyze the Berkeley node of radical geography (Peck and Barnes, this volume). Gene Mumy was hired by Economics at Ohio State University after failing to receive tenure at Virginia Tech. Once at Columbus, Harvey connected him with Kevin Cox, and together they convened a radical discussion group of geography and economics graduate students. Mumy's own scholarship continued to appear in both mainstream and "extreme left" journals.[51] Finally, Neil Smith, who arrived at DoGEE in 1977 to work with Harvey, exerted a major scholarly influence in and beyond radical geography (like Harvey, receiving Distinguished Scholarship Honors from the AAG), shaped radical geography more generally as an active member in the Union of Socialist Geographers (USG) and then in the AAG Socialist Geography Speciality Group (now the Socialist and Critical Geography Speciality Group) that replaced the USG – actively influencing the decision of the USG to self-disband in favor of the SGSG (Peake, this volume).[52] Smith also infiltrated the by now less-staid AAG, serving on its National Council. Smith and Harvey, with Cindi Katz, reunited in the City University of New York's Graduate School in 2001, creating a new radical geography node there.

In sum, through these various means, radical geography became identified more and more with the version developed by Harvey in Baltimore, particularly as *Antipode* took an increasingly Marxist turn during the 1970s. Paradoxically, Baltimore's own presence grew more faint as Harvey's theorizing, initially sparked by the city, became more abstract and footloose.

## Conclusion

Baltimore became a means through which Harvey understood and interpreted Marx, allowing him to determine which theoretical categories were useful, and which were not.[53] At the same time, Marx became his means to understand Baltimore, to appreciate why it became the city it was, and especially to comprehend its 1960s history around racial and class division and its various cycles of growth and destruction. Harvey's (1982) later *Limits to Capital* solidified the position that was more tentatively worked out in his earlier papers around housing markets in Baltimore: radical geography was Marxist geography. In *Limits* he was explicit that Marx's principles applied everywhere, drawing in an ever-expanding audience including many non-geographers in the social sciences and humanities. Our point, though, is that Harvey's geographical Marxism had its own geography. Harvey's was in part a local achievement, with Baltimore serving as an any city, making it a truth spot. Harvey's students did not work in Baltimore (Smith came closest, researching gentrification in Philadelphia), but living in Baltimore undoubtedly fashioned their thinking. As Walker put it, "That's why we have face-to-face contact: Things happen in places; ideas bounce around. Something happens."[54]

If Baltimore was the place that shaped the variant of Marxism that took root in radical geography, its status as truth spot was the result of the circulation of those ideas and their proponents beyond Baltimore. As Gieryn (2002: 113) argued, for a place to become a truth spot the ideas developed there need to circulate, during the course of which they seemingly "escape place." That was so for Harvey's theorization of radical geography and its circulation beyond Baltimore. As his version of radical geography traveled, infusing the 1970s Marxist turn in *Antipode* and the USG, its connection to Baltimore became ever-less apparent. Yet circulation does not elide geography. In the case of radical geography, its circulation was not only inflected by Baltimore but also was reshaped by the engagements of its various protagonists with other places, and with other strands of critical and radical thought. These various geographical traces have been submerged over time but resurrecting them helps us understand how seemingly abstract theoretical

debates emerged from the socio-spatial positionality of their protagonists. Here we have sought to uncover those lines of influence, inking
them more legibly, to reveal an underlying map placing Baltimore at the
center of the Marxist turn in North American radical geography.

## Notes

1   The Department of Geography at Hopkins was founded in 1943 and remained a separate entity until 1968 when it was amalgamated with Environmental Engineering. https://engineering.jhu.edu/dogee/wp-content/
uploads/sites/.../jhu-DoGee-history.pdf, accessed October 23, 2017.

2   Lattimore was subject to 17 months of investigation by HUAC, and in
1952 was charged with perjury. He was acquitted in 1956 and left for
Leeds University in 1963 to become the founding Professor of its
Department of China Studies.

3   http://globetrotter.berkeley.edu/people4/Harvey/harvey-con1.html, accessed January 5, 2018.

4   Harvey (2002: 165) says that throughout his whole life the Kent landscape
has been a "central obsession." A short version of his doctoral dissertation
was published as Harvey (1963).

5   Wilson first used the phrase at the 1963 annual congress of the Labour
Party in Scarborough.

6   http://globetrotter.berkeley.edu/people4/Harvey/harvey-con2.html, accessed January 5, 2018.

7   David Harvey interviewed by Trevor Barnes and Eric Sheppard, Encino,
CA, October, 2013.

8   Peter Gould's (1999) moniker for College Park, PA, where Penn State is
located.

9   Harvey interview, 2013.

10   Harvey interview, 2013.

11   Harvey interview, 2013.

12   Almost 30 years later, in April, 2015, race-based riots broke out again on
Baltimore streets. The death of Freddie Gray while under Baltimore police
custody provoked a series of city protests: 235 people were arrested, some
20 police officers injured, and 380 businesses damaged. Marbella, J. 2015.
The day the Baltimore riots erupted: New details of Baltimore riots after
Freddie Gray's death. *The Baltimore Sun* October 23, 2015.

13   John Boland phone interviewed by Trevor Barnes, Baltimore, June 2017.

14   Barbara Koeppel phone interviewed by Trevor Barnes, Washington, DC,
June 2017.

15   There is an interesting comparison between the early experiences of
David Harvey at Hopkins and Richard Walker at Berkeley. Harvey was
on Walker's Ph.D. thesis committee at Hopkins, in effect his *de facto*
doctoral supervisor (the official supervisor was DoGEE's Steve Hanke).
Once Walker was appointed at Geography at Berkeley, he received

protection and Departmental space from a senior colleague, Allan Pred. However, Walker, as an untenured Assistant Professor, needed much more protection. Walker's tenure battle was epic, with Pred the key to its eventual successful resolution (see Peck and Barnes, this volume)

16  Harvey interview, 2013
17  Harvey interview, 2013.
18  Harvey interview, 2013.
19  The report includes two appendices detailing housing inspection simulation models.
20  Harvey interview, 2013
21  Harvey interview, 2013.
22  Harvey interview, 2013.
23  Harvey interview, 2013.
24  Harvey interview, 2013.
25  Harvey (1973: ch. 4) damned the work of Alonso and Mills as "counter-revolutionary" or "status quo" theory. In Harvey's (1973: 95) terrible pun, "the status was nothing to quo about".
26  Harvey interview, 2013
27  Harvey interview, 2013.
28  https://www.youtube.com/watch?v=gBazR59SZXk, accessed January 21, 2019.
29  Harvey interview, 2013
30  Harvey interview, 2013
31  John Davey interviewed by Trevor Barnes, Oxford, July 2016.
32  Richard Walker interviewed by Trevor Barnes and Jamie Peck, Los Angeles, CA, April 2013.
33  Barnbrock estimated there were about 10 members in the reading group when he joined in 1973, but he could remember only two others by name apart from Walker and Mumy: Carolyn Hawk and Maria Beatriz Nofal. We were not able to contact either. (Barnbrock, phone interview with Eric Sheppard, Munich, August 2018).
34  All quotations in this paragraph are taken from Walker interview, 2013.
35  Having written a thesis critiquing Von Thünen, written up as an early *Antipode* paper (Barnbrock 1974), he could not find an academic position in North America. After working for Abt Associates, he founded a consultancy firm in Munich, JBConsulting, providing advice to the building materials sector.
36  All quotations are from a phone interview with Gene Mumy by Trevor Barnes, Columbus, OH, June 2017.
37  All quotes are from a phone interview with Barbara Koeppel by Trevor Barnes, Washington DC, June, 2017.
38  All quotations in this paragraph are from Boland interview, 2017.
39  When Harvey sought to return from Oxford in 1993, requiring a fresh hire as Full Professor, members of the Department of Economics – a quasi-Chicago School program at the time – were asked for advice. One Economics faculty member, charged to read *Capital* and report back, informed his colleagues that *Capital* represented serious scholarship and was not merely a political tract. Opposition to Harvey's reappointment evaporated (Boland interview, 2017).

40  Harvey interview, 2013
41  On Pfeffer's life and activism see his 2002 obituary in the *Baltimore Sun*, http://baltimorechronicle.com/pfeffer_jun02.shtml, accessed, December 27, 2017.
42  Harvey interview, 2013
43  Harvey interview, 2013.
44  Some 175 US municipalities enacted rent control laws by the early 1980s, some 140 of which are still in force ("Rent Control", http://www.kintera.org/site/c.lkIXLbMNJrE/b.6644939/k.9948/Rent_Control/apps/nl/newsletter2.asp, accessed March 18, 2016)
45  Harvey interview, 2013.
46  Pratt was a business scion of nineteenth-century segregationist Baltimore; Roland Park began as a white garden suburb bulwarked by a racial covenant.
47  All quotations in this paragraph are from Harvey interview, 2013.
48  Interview with Richard Peet by Trevor Barnes, Worcester, MA, May, 2002.
49  Peet interview, 2002.
50  Peet interview, 2002
51  Mumy interview, 2017
52  In 2008, dissatisfied with *Antipode's* turn away from Marxism and the emergence of critical geography as an alternative label to radical geography, Peet created a new Marxist house journal at Clark University: *Human Geography* (https://hugeog.com/).
53  In his *Antipode* paper with Chatterjee that reported his Baltimore housing findings, Harvey made use of Marx's concept of absolute rent, which was also discussed in *Social Justice and the City*. He quickly realised, though, that absolute rent was inappropriate, substituting the term "class monopoly rent." In a later interview, Harvey said, "I now think that Marx's theory of absolute rent is meaningless and irrelevant. Marx does not develop monopoly rent but mentions it on the edges. But it has a clear relevance in urban situations" (Harvey 2013).
54  Walker interview, 2013

# References

Agnew, J. A. (1987). *Place and Politics: The Geographical Mediation of State and Society*. London: Unwin Hyman.
Barnbrock, J. (1974). Prologomenon to a methodological debate on location theory: The case of von Thünen. *Antipode* 6: 59–66.
Barnes, T. J. (2001). Lives lived, and lives told: biographies of geography's quantitative revolution. *Society and Space: Environment and Planning D* 19: 409–429.
Barnes, T. J. (2006). Between deduction and dialectics: David Harvey on knowledge. In N. Castree and D. Gregory, eds, *David Harvey: A Critical Reader*, pp. 26–46. Oxford: Blackwell.
Barnes, T. J. (In press). Game changer: Peter Haggett. In P. Atkinson, S. Delamont, M. Hardy, and M. Williams, eds, *The SAGE Encyclopedia of Research Methods*. London: Sage.

Berry, B. J. L. (1972). Revolutionary and counter-revolutionary theory in urban geography – a ghetto reply. *Antipode* 4 (2), 31–33.

Chatterjee, L., Harvey, D. and Klugman, L. (1974). *FHA Policies and the Baltimore Housing Market*. Baltimore, MD: Center for Metropolitan Planning and Research, Johns Hopkins University.

Chorley, R. J. (1995). Haggett's Cambridge: 1957–66. In A. D. Cliff, P. R. Gould, A. G. Hoare, and N. J. Thrift, eds, *Diffusing Geography: Essays for Peter Haggett*, pp. 355–374. Oxford: Blackwell.

Dear, M. J. (ed.) (2002). *From Chicago to L.A.: Making Sense of Urban Theory*. London: Sage.

D'Adamo, C. (2002). Progressive Action Center celebrates twenty years! Baltimore, MD: Baltimore Independent Media Center.

Fee, E., Shopes, E., and Zeidman, L. (1991). Introduction: Toward a new history of Baltimore. In E. Fee, L. Shopes, L. Zeidman, eds, *The Baltimore Book: New Views of Local History*, pp. vii–xvii. Philadelphia: Temple University Press.

Folke, S. (1972). Why a radical geography must be Marxist. *Antipode* 4: 13–18.

Foucault, M. (1986). Of other spaces. *Diacritics*, Spring: 22–27.

Gieryn, T. F. (2002). Three truth-spots. *Journal of the History of the Behavioral Sciences* 38: 113–132.

Gieryn, T. F. (2006). City as truth spot: Laboratories and field-sites in urban studies. *Social Studies of Science*, 36: 5–38.

Gould, P. J. (1999). *Becoming a Geographer*. Syracuse, NY: Syracuse University Press.

Haggett, P. (1990). *The Geographer's Art*. Oxford: Basil Blackwell.

Harding, S. ed. (2011). *The Postcolonial Science and Technology Studies Reader*. Durham, NC: Duke University Press.

Harvey, D. (1963). Locational change in the Kentish hop industry and the analysis of land use patterns. *Transactions, Institute of British Geographers* 33: 123–40.

Harvey, D. (1969). *Explanation in Geography*. London: Edward Arnold.

Harvey, D. (1972a). Revolutionary and counter revolutionary theory in geography and the problem of ghetto formation. *Antipode* 6: 1–13.

Harvey, D. (1972b). On obfuscation in geography, a comment on Gale's heterodoxy. *Geographical Analysis* 4: 323–30.

Harvey, D. (1973). *Social Justice and the City*. London: Edward Arnold.

Harvey, D. (1974a). Class-monopoly rent, finance capital and the urban revolution. *Regional Studies* 8: 239–255.

Harvey, D. (1979). Rent control and a fair return. *The Baltimore Sun* September 20.

Harvey, D. (1982). *The Limits to Capital*. Oxford: Basil Blackwell.

Harvey, D. (1983). Owen Lattimore: A memoir. *Antipode* 15: 3–11.

Harvey, D. (1992). Author's response. *Progress in Human Geography* 16: 72–73.

Harvey, D. (1996). *Justice, Nature and the Geography of Difference*. Oxford: Basil Blackwell.

Harvey, D. (2000). *Spaces of Hope*. Berkeley: University of California Press.

Harvey, D. (2002). Memories and desires. In P. Gould and F. R. Pitts, eds, *Geographical Voices: Fourteen Autobiographical Essays*, pp. 149–188. Syracuse: Syracuse University Press.

Harvey, D. (2003). E-mail to Trevor Barnes, August 26th.

Harvey, D. (2007). Interview with Steve Pender. *Studies in Social Justice* 1 (1): 14–22. https://brock.scholarsportal.info/journals/SSJ/article/view/978/948, accessed January 21, 2019.

Harvey, D. (2015). "Listen anarchist." A personal response to Simon Springer's "Why a radical geography must be anarchist." Available at: http://davidharvey.org/2015/06/listen-anarchist-by-david-harvey/, accessed December 31, 2017.

Harvey, D. (2016a). Commentary. *The Ways of the World*. Oxford, Oxford University Press.

Harvey, D. (2016b). Books interview: David Harvey. *Times Higher Education Supplement*, February 4th. https://www.timeshighereducation.com/books/interview-david-harvey-city-university-of-new-york-graduate-school, accessed January 7, 2018.

Harvey, D., Chatterjee, L., Wolman, M. G., Klugman, L. and Newman, J. S. (1972). *The Housing Market and Code Enforcement in Baltimore*. Baltimore, MD: Baltimore Urban Observatory.

Katznelson, I. (1997). From the street to the lecture hall: the 1960s. *Daedalus* 126: 311–22.

Latour, B. (1987). *Science in Action: How to Follow Engineers and Scientists Around Society*. Cambridge, MA: Harvard University Press.

Marx, K. (1972 [1885]). *Capital, Volume 3*. Penguin: Harmondsworth.

Massey, D. (1991). A global sense of place. *Marxism Today* June: 24–29.

Morgan, K. (1993). *Bristol and the Atlantic Trade in the Eighteenth Century*. Cambridge, UK: Cambridge University Press.

Park, R., Burgess, E., and MacKenzie, R. (1925). *The City*. Chicago: University of Chicago Press.

Prince, H. (1971). Questions of social relevance. *The Professional Geographer* 3: 150–153.

Philo, C. (1998). Eclectic radical geographies: revisiting the early *Antipodes*. Paper presented at the Annual Meeting of the Association of American Geographers, Boston, MA, April. http//www.eprints.gla.ac/uk/116527/, accessed January 10, 2018.

Roy, A. (2016). Who's afraid of postcolonial theory? *International Journal of Urban and Regional Research* 40: 200–209.

Sheppard, E., Gidwani, V., Goldman, M., Leitner, H., Roy, A. and Maringanti, A. (2015). Introduction: Urban revolutions in the age of global urbanism. *Urban Studies* 52: 1947–1961.

Sheppard, E., Leitner, H. and Maringanti, A. (2013). Provincializing global urbanism: a manifesto. *Urban Geography* 34: 893–900.

Smith, D. M. (1971). Radical geography: The next revolution? *The Professional Geographer* 3: 153–157.

Smith, N. (2003). *American Empire: Roosevelt's Geographer and the Prelude to Globalization*. Berkeley and Los Angeles: University of California Press.

Springer, S. (2014).Why a radical geography must be anarchist. *Dialogues in Human Geography* 4; 249–270.

# 7

# Berkeley In-Between
## *Radicalizing Economic Geography*

### Jamie Peck and Trevor J. Barnes

Economic geography was rather slow off the mark in getting its radical makeover. Until the late 1970s, the discipline was variously in the sway of the classical location theories of Von Thünen, Weber, Christaller, and Lösch, several strands of neoclassical regional science, and a descriptive, often rather dull, industrial geography. In each case, the economic world depicted tended to be ordered, predictable, set in static equilibrium. All that changed from the late 1970s when a cluster of Marxian and marxisant approaches gripped and transformed the discipline. Out went the old lexicon of bid-rent curves, isodapanes, and hierarchically ordered hexagonal markets, and in came a new one with its talk of profit cycles, corporate restructuring, creative destruction, and deindustrialization. Radically different economic worlds came into view. They were jagged and lumpy, often chaotic, sometimes on the brink of careening out of control, punctuated by abrupt dislocations, and skewered by pervasive social conflicts. The promise of equilibrium seemed ever more irrelevant; the new language was of crisis, disruption, instability, and unevenness.

Partly propelling this change was a radically disruptive political-economic climate and geography. It was driven by intensifying conditions of crisis during the 1970s: stagflation, the oil shocks, the appearance of the "rust belt," large-scale job losses, "runaway shops," and the adoption of a new managerial arsenal, including concession and whipsaw bargaining. In this context, the old economic geography was neither able to cope, nor was it fit for purpose. Its portrayal of a steady, cordial, and well-adjusted world was at marked odds with the

*Spatial Histories of Radical Geography: North America and Beyond*, First Edition.
Edited by Trevor J. Barnes and Eric Sheppard.
© 2019 John Wiley & Sons Ltd. Published 2019 by John Wiley & Sons Ltd.

new realities of jarring, conflict-ridden, and path-altering change. Economic geography was ripe for revolution. New tools were needed for what looked and felt like a wider conjunctural realignment.

Our chapter argues that one of the germinal sites for the radical reconstruction of economic geography during this period, and a birthplace of what became known as the "restructuring approach," or the "new industrial geography" (see Pudup 1992; Storper and Scott 1986; Storper 1987a; Walker 2018; Barnes and Gertler 1999), was the University of California-Berkeley. As a place it sparked a series of remarkably consequential and transformative developments in the field – many of which can be traced to roots in *and routes through* mid-1970s to early 1980s Berkeley, but also drawing on longer-distance connections, that resulted in a body of transformatively radical work and ideas that traveled across North America and internationally. Radical economic geography was not "made in Berkeley," but it was a site, a gathering place, an incubator of an alternative project that ultimately ignited a blaze of extra-local effects. As we will show, the Berkeley moment involved the meeting of an unusually talented cohort of researchers, collaborative by disposition and mobilized around a shared purpose, with a set of galvanizing political-economic circumstances. On the cusp of the Reagan era, the U.S. economy was in the throes of a phase of intense restructuring. Within the Bay Area, especially, a new kind of "flexible," high-tech economy was emerging, the moniker "Silicon Valley" only recently coined. It was a turbulent, disorienting, challenging, energizing time – and place. Berkeley between the mid-1970s and the early 1980s was a crucial intersection point, a site of local multiplier effects, and an intellectual forcing house for a radical realignment in the theory and practice of radical economic geography.

Richard Walker (2001: 34), then an untenured assistant professor in Berkeley's Geography Department, believed that the origins of that realignment from the mid-1970s were with "an extraordinary circle of people gathered around the City Planning and Geography Departments [at Berkeley]." "This circle of friends with a leftish bent" would go on to "transform the fields of economic geography and regional development in subsequent years," most of them later moving to other locations (Walker 2001: 34). "Timing was really important," Walker added. "As a group, we were drawn together. Electrifying! We were all of the same age ... We just picked up things. If you are charged like static electricity, things just stick to you ... Things happen in places. Ideas bounce around. Something happens ... [Together] the [economic] crisis, the students, and charismatic people like Bennett Harrison and Doreen Massey," who were visitors during this time, "created a Vienna circle moment. It was terrific."[1]

That moment was partly about a rising generation, carrying new ideas, raising new questions, who were highly politicized, situationally shaped by the anti-war, women's, environmental and civil-rights movements (see Introduction). Ann Markusen – Walker's counterpart in the Department of City & Regional Planning (DCRP) – says theirs was an "uppity" generation, which would form a close-knit group of junior faculty and graduate students who "were all sort of the same mind. We had this feeling that we were different from our elders. And that we had this experience of being really engaged."[2] Engagement meant not only learning and deploying Marxism and radical political economy, although it invariably involved that. It was also political activism and policy advocacy, campaigns and projects. It was about developing leftist *praxis*. It "was *not* a reading group," as Markusen emphasized.

While the departmental affiliations of Walker and Markusen, along with the graduate students with whom they were most closely involved, were readily identifiable,[3] the intellectual and sociopolitical origins of this Berkeley "moment" – understood as a distinctive comingling of currents, connections, and contextual conditions – were not with one department or the other. They were not even with one discipline or another. Rather, they arose in the spaces in-between. Michael Storper said, "there wasn't a sharp line between Geography and Planning ... It was one space that we all circulated in."[4] Something similar can be said also of the extra-disciplinary work and worldviews of the two visiting interlocutors, economist Bennett Harrison and geographer Doreen Massey, who in each case were inspirational, catalytic. Walker believed that "together [Harrison and Massey] triggered the whole Berkeley school."

This was not a "school" in the conventional sense, however, nor did it behave like one. While it is possible to discern a distinctive "Berkeley point of view," it derived not from a single authorized (or institutionalized) core. Nor did it dethrone some *in situ* orthodoxy. Rather, the Berkeley view emerged from interstitial spaces, some of which were notably vacant or otherwise unoccupied: between an old economic geography, losing its traction, and a more radical style of inquiry mixing political economy with political (and policy) engagement; between two different departments and disciplines but at some distance from their respective centers; between a rising generation of scholar-activists, radicalized off campus, and their generally more conservative predecessors; and between public and private spaces, between workplace and homeplace, between seminar rooms and living rooms.

It was also liminal or in-between in a broader, one might say "structural" sense. It took shape between the old and the new economy in California and beyond – between the hollowing out of once-dominant regional-manufacturing complexes and the rise of the "new industrial spaces," at the cusp of the Reagan era. Members of the Berkeley group were involved both in campaigns against manufacturing plant closures and in pioneering efforts to convert abandoned industrial spaces into "postindustrial" uses, harbingers of a green(er) economy. In turn, these struggles occupied a distinctive political space lying between an arch-neoliberal current in California politics, represented by Ronald Reagan's two terms in office as State Governor (1967–1975), and the dynamic jumble of radical, countercultural, socially liberal, and progressive politics found in the Bay area, and Berkeley in particular. UC-Berkeley was the epicenter of the free-speech movement in the 1960s, a flashpoint in the campus culture wars that Reagan had exploited on his path to power. While San Francisco was long a hotbed of liberal, libertarian, and alternative politics. The local culture was countercultural, multiple, and contested. From the mid-1970s, many of the faculty and graduate students who found their way to the Department of Geography and the DCRP did so from involvement in environmental and anti-war activism, from the civil-rights and women's movements, and from a wide range of disciplines.[5] Variably politicized, they found common cause around issues new to most of them – pressing questions of economic dislocation and regional restructuring. It was an "intense [political] intersection," as AnnaLee Saxenian recalled.[6]

In examining this roiling and propagative period at Berkeley from the mid-1970s to the early 1980s, we divide the chapter into two. The first part, "Cohort and club," presents a close-focus analysis of intellectual and social life on the campus and its immediate environs. It is about Berkeley as a place, concerned especially with the material and sociological landscape that the Department of Geography and DCRP shared. Partly it is a story of a cohort – of graduate students, visitors, and new arrivals on campus, often on journeys to and from other places – and partly a story of a close-knit inclusive group, a "club," with informal membership and a collectivist ethos based on overlapping social, political, and intellectual affinities. The second part, "Crossings and conjunctures," ventures beyond the geography of the campus. It describes the Berkeley moment articulating with, and mutually conditioning, a host of off-campus ideas, events, political movements, and connections. Here we situate Berkeley in its regional as well as national context, recognizing that the events in Berkeley (and their downstream effects) reflected a wider confluence of circumstances and conjunctural conditions, in addition to intersecting personal and political trajectories.

## Cohort and Club

We begin with the two departments. The Berkeley moment was made possible, if not enabled, by a series of incremental but cumulatively significant changes in the Department of Geography and the DCRP. Both departments were long characterized by what Amy Glasmeier called "heavy-handed" control, centering on a tight (but at times riven) network of "old boys," in the wake of which space was gradually opened up (or at least quietly appropriated) for more variegated modes of enquiry, social engagement, and cross-disciplinary exchange, and (more slowly) for a more diverse faculty.[7] As Saxenian put it, "both departments had a conservative past, but [by the 1970s] that center wasn't holding, which allowed new stuff to emerge."[8]

Geography was the older of the two Departments. Founded in 1898, it did not come into its own until 1923 when Carl Sauer was appointed as Head, also known as "the chief," or "the Old Man." "Less a Department than an individual" (Parsons 1979: 9), Geography was shaped by Sauer but also by his students who were favored in hiring decisions (Hewes 1984). Believing that they had "caught at least a glimpse of Isaac Newton's great unexplained 'ocean of truth'" (Leighly 1979: 9), Sauer's Berkeley School pioneered a form of cultural history based on the investigation of past human uses of the earth and their effects on shaping the landscape (Parsons 1979: 13). Requiring muddy boots, fluency in multiple foreign languages, immersion in the past, and selective knowledge of physical and biological sciences, Berkeley cultural geography was the antithesis of social science, especially the model-based, theoretical, and quantitative form of social science ascendant in the U.S. after the Second World War (see Introduction). Yet not even Sauer could prevent the Department from hiring just such a social scientist in 1963. Following the injunction of an external Departmental review to recruit a "new economic geographer," Berkeley "swallowed the pill" and appointed the urban modeler and theoretician, Allan Pred, fresh from the University of Chicago and trained by Brian Berry.[9]

Initially spurned by Berkeley's "Sauerian old guard," Pred used a competing offer of a full professorship from the University of Michigan to secure tenure and guarantee an option to teach a regular graduate seminar in methodological and philosophical issues in the social sciences. The worm had turned. By the 1970s, Pred's seminar became a meeting place for geographers and students from across the Berkeley campus, including DCRP. Those students sought not only more systematic social science but its more politically radical forms, as did Pred. Walker observed that Pred was the only person he knew who "became even more radical as he got older … [His] seminar brought in the lefties, and linked geography and planning. Allan was essential."[10] He was also essential in another way as we will later discuss, becoming something of a protector of Walker, who would go farther than Pred both in his radicalism and in his engagement with DCRP. Michael Storper, who found his way from History and Sociology to work with Walker, felt that many in Geography at the time had "no interest in the social sciences … The language that people were using in Geography was weird … I just didn't know what their world was … It was a pretty foreign culture."[11] From the 1970s, however, the hold of Sauerian cultural geography began to loosen as faculty like Pred and then Walker pushed for something different, and as graduate students like Storper, Susan Christopherson, Michael Heiman, Judith Carney, and Susanna Hecht arrived on the scene.

Here it is crucial to note, although it cannot be pursued in any detail, that a second strand of radical geography was germinated at Berkeley's Geography Department during this same late-1970s, early 1980s period. Its focus was not the restructuring regions of the Global North, however, but rural areas across the Global South. The appointment of Michael Watts in 1979 brought to Berkeley the radical geography of both political ecology (where he would make field-shaping interventions) and Southern agrarian studies (quickly forging a link with Alain de Janvry at Berkeley's Department of Agricultural and Resource Economics). His ground-breaking book, *Silent Violence* (Watts 1983), a historical analysis of poverty and famine in rural Nigeria, was informed by a close reading of Marxist studies of nineteenth-century agrarianism, emphasizing kinship, gender, and ethnicity. The radical geography of Watts and his many students was consequently quite different from the Walker-Markusen variant, although in its own way no less significant (see Chari *et al.* 2017). It was internationalist, drawing on theory from the Global South such as Franz Fanon's and Samir Amin's, primarily ethnographic, often with a historical component, stressing the importance of familial, racial, gender and indigenous social relations, and aspiring to the dialectical and non-linear. Like the Walker–Markusen version, though, it pursued both praxis (to the end of third-world liberation), matched with an insistent logic of class relations and transformation. With Bernard Nietschmann, who arrived at Berkeley two years before Watts, and later Gillian Hart, this brand of radical geography was shaped and developed by feminism; indeed, was foundationally feminist, and later further enriched by the encounter with post-colonialism and critical race theory. This trajectory of radical geography, while by no means hermetically sealed from the economic-restructuring project led by Walker and Markusen, nevertheless has its own story to tell and its own diverse legacies (see Chari *et al.*

2017). The radical economic geographers, meanwhile, were constructing their project at the interface between Geography and a rising generation of planners at DCRP.

DCRP had been established in 1948. After more than a decade of deferred urban construction due to the Great Depression and World War II, its mission was to "bring planning to California's cities" (Kent 1984; Webber and Collignon 1998: 1). None of the original faculty possessed a Ph.D., however. The program's early orientation was less on social-scientific research than instructing hands-on practices of physical planning and urban design. That changed in 1953 with the hiring of Donald Foley, a Ph.D.-bearing, research-focused sociologist. By the early 1960s, student drafting tables were increasingly being replaced by desks, as DCRP's staff and students engaged with the social-sciences and public-policy evaluation (Webber and Collignon 1998: 3; Teitz 2016: 51; Weiss and Schoenberger 2015: 64). Initially, the social science was quantitative and theoretically formal. Several new faculty members were full-blown mathematical modelers such as the Chicago-trained economist John W. Dyckman, as well as Michael Teitz and William Alonso, both of whom received doctorates from Walter Isard's Department of Regional Science at Penn. The 1970s brought further changes, with a shift toward social theory, including political economy, not least as a response to the political turmoil of the 1960s (see Introduction): civil-rights protests, anti-War demonstrations, Johnson's war on poverty, and an all-out urban crisis, as inner-city areas from Los Angeles to Detroit and Baltimore witnessed riots and mass protests, triggering both repressive actions and alternative social mobilizations. Faculty and students were radicalized, especially at Berkeley, which became "a national magnet" for oppositional activism and research (Weiss and Schoenberger 2015: 64).[12] With the agreement of Dean Wheaton, Wurster Hall, home of DCRP, became a temporary center for the anti-war movement, including a facility for producing posters.

These moves in the direction of political engagement and social theory anticipated the subsequent turn to regionally engaged political economy. Intensifying internal and external pressures caused DCRP's culture to fracture. As in the case of Geography, the old center failed to hold. Teitz recalled that when he first arrived at DCRP in 1963, "the Department was very tight, very close-knit."[13] This was followed by as he character-ized it "the great breaking apart," when DCRP hived off its different specialties into three separate graduate programs. After that, Teitz continued, "the Department had at best modest coherence. I guess it was about as coherent as the Geography Department [at the time], i.e. not at all! ... There [was] no whole department after that," as different segments of DCRP began to do their own thing.[14] When Ann Markusen arrived as an assistant professor in 1977, she found the place "riven with conflict." Faculty meetings, she said, could be "disastrous." The sensible strategy for junior faculty was to stay out of the firing line. The intellectual climate was also sometimes tough. Markusen remembered it was not unusual for seminars held in faculty homes to involve "shouting matches."[15]

A feature that DCRP shared with Geography, and with so many programs in Berkeley and beyond, was a largely unreconstructed, complacent, and in some respects reactionary gender culture. Neither department had much, if any, experience of recruiting, mentoring, and promoting women faculty. Their cultures were heteronormative, often intransigently masculinist, and reliant on old-boy styles of decision-making. As of the late 1970s, Geography had yet to tenure any women faculty. It was "was still an all-male bastion," Markusen said.[16] At DCRP, the experiences of Janice Perlman, Judith deNeufville, and Ann Markusen were often challenging. When Perlman was hired in 1973, there were no women faculty members at DCRP. Yet women were entering the graduate program in significant numbers. For some years, DCRP was reportedly "under the gun" to hire a woman, including pressure from higher up in the administration. But the issue was treated flippantly by many male faculty members. Catherine Bauer Wurster (1905–1964) had been a key figure in the department's history, but in corridor chatter many faculty were reported to have said, according to Perlman, "We will just have to put up another bust of Catherine Bauer Wurster, and that will be fine for our women's quota!"[17] While Perlman was subsequently tenured, she would later leave, never to occupy a tenured position again. Perlman recalled a scarring experience in which her own role was ultimately that of a "sacrificial lamb." Struggles to diversify the department were often met with indifference and intransigence, sometimes disrespect and resistance.[18] Markusen (2015) would later reflect on Perlman's vital contribution to her own mentoring, along with sympathetic colleagues like Teitz, which at times came close to a form of protection. These bruising and damaging experiences, with Perlman and others paving the way often at considerable personal and professional cost, were instrumental in the development of the Faculty Women's Interest Group (FWIG) in planning programs across the country. Locally, they opened a door for Markusen, who would face different battles, but some of the same ones too.

Discriminatory attitudes, behaviors, and sexist talk were rife at the time, in faculty meetings, in hiring committees, and certainly in less-regulated settings. As Perlman put it, "doing good work, working hard, showing commitment to teaching and to community service, I thought that ought to be enough to be accepted [among male colleagues], but it wasn't the case."[19] This led some, including the female graduate students, to second-guess themselves, to wonder if their work "was worthless ... if it had mojo," as Erica Schoenberger said.[20] Echoing the recollections of others, Amy Glasmeier recalled that, "there were still situations where women students had male faculty hit on them [although perhaps] we weren't different to any other place in that way."[21] As many of the other chapters in this volume attest, Glasmeier was right.

The group of graduate students and young faculty that began meeting from the mid-1970s, did so in part out of an interest in political economy and political engagement, but also as a means to develop a different culture of collaboration and exchange. More often than not, they would do so in meetings, discussions, parties, and writing projects located off campus. Ann Markusen remembered that, "We met often [but] for the first two or three years, we met [in] someone's living room. We didn't meet at the University. ... We didn't try to give ourselves a University presence."[22] Similarly, Meric Gertler recalled a convivial, off-campus atmosphere: "our critiques were born out of many discussions [over] ... pot-luck dinners," comparing the development of their shared project to a "moveable feast."[23] This was not a drinking culture, although Alice Toklas-style brownies reportedly did play a role in gatherings that were social as well as political and intellectual. Marc Weiss explained that "different people would host [the meetings] and we were very casual, it was almost always in someone's home," either a faculty member or one of the graduate students.[24] "At that time," Saxenian noted, "everyone lived close enough to campus, so you were able to bump into each another both during the program and afterward. Today, the place is so expensive nobody lives nearby, so the networks are much harder to maintain."[25]

Then, people lived close enough to campus that their homes effectively became extensions of the university itself (see Figure 7.1). Living rooms were where new ideas were hatched and worked, where visitors socialized and presented their work, where grant proposals were developed, and manifestos conceived. Saxenian's place was especially attractive because it "had a hot tub in the back, so that made it easy to invite people over," she recalled; "Hot tubs were a big thing then ... the culture of the time." While the discussions and debates were often serious, this was clearly a group of close friends as much it was an academic circle, with constant circulation between nearby homes and hangouts like Café Roma, La Val's, and Moe's bookstore. As Marshall Feldman remembered,

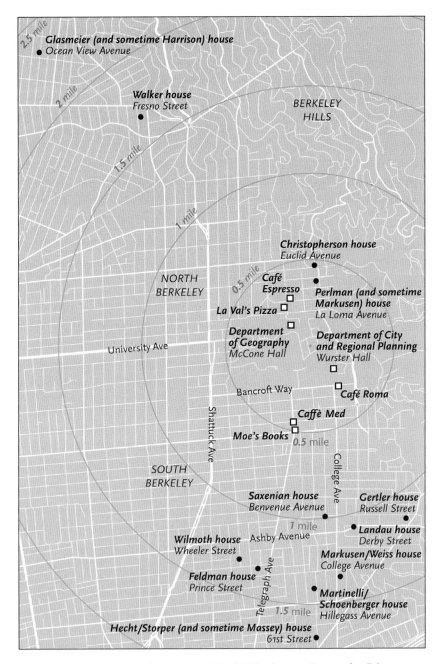

**Figure 7.1** Berkeley circles, circa 1978–1983. *Source:* Drawn by Eric Leinberger (UBC cartography) based on interviews by authors.

We would have dinner at each other's places. We would go to parties together. We knew people's significant others. We knew what people were working on. It was rich intellectually and it was rich socially.[26]

Feldman joked that when "I was invited to get into the hot tub at Anno Saxenian's house … I knew I had become a member of the club!" In such ways, social and intellectual activities and their spaces were conjoined and comingled. But if there was a clubbish aspect to the Berkeley moment, it was not one built upon exclusions. The group was open at least to anyone with compatible intellectual or political interests. There were some ironies, though, as Erica Schoenberger now recognizes. While the Berkeley group was committed to radical political economy, to activism, "to being engaged on the street, we were doing all this from the cushiest possible places!"[27]

Pred's graduate seminar also met at his home, "in his living room," Gertler recalled, "overlook[ing] the Golden Gate Bridge, [where you could] watch the sun go down." This setting

> allowed participants to be more honest and open and have perhaps more fun, find more stimulus.… Allan would make references to Sauer, often without naming him. It was as if he was a ghost walking the halls. I think that's why he liked to meet at his house. He felt doing it in the Department would not be conducive to free-flowing discussion. [That] was an oppressive environment, and to do something creative it had to be done in a different space.[28]

The Berkeley group was also trying to find a different space theoretically. Theirs was a radicalism unmistakably informed by Marx, although it was not classically or indeed exclusively Marxist. Schoenberger thought "marxisant" was more apt.[29] Some of the group, like Walker, Markusen, and Schoenberger, were steeped in Marx (although Walker said he had to read the first volume of *Capital* three times before he was converted). But it was neither feasible nor advisable simply to transfer Marx's ideas *en bloc* to North America or the Bay Area of the late 1970s. Those ideas first needed to be tailored, amended, supplemented, sometimes even corrected, which meant that "we made a lot of it up as we went on," as Schoenberger said.[30] It follows that while there were recurring motifs and themes in the work of the Berkeley group – including concerns with corporate power, with labor struggles, and with the disruptive force of technological change – there was no single Berkeley line. "There was no central dogma," Glasmeier said,[31] no iron-clad theoretical loyalties or commitments, but instead a shared recognition of what Storper termed "the amazing cauldron of learning," the value of which was realized through conversations in those living rooms, hot tubs, pizzerias like La Val's, and (even) seminar rooms. This fostered a climate of openness, tolerance, and creative experimentation that accommodated a range of positions and projects. Research projects, while frequently echoing the group's less-than-programmatic interests and concerns, were individual. "No one signed on to any orthodoxy," Schoenberger maintained, "for our dissertation work we all did our own thing and did not trample on one another's space."[32] Glasmeier added, "It was ecumenical. You didn't have to be a hardcore Marxist to ultimately be considered a friend and a participant in the experience."[33] (This said, some of those on the fringes of, or outside, the club were more likely to see it as "clubbish," especially if their own approach was not (so) marxisant, or if their substantive interests were less focused on questions of first-world industrial transformation.)

There were times when the group operated as a collective, both in name and practice. It never produced formal communally published work, however. Early on, there was a plan to write a full-scale joint critique of the Carter Administration's urban policy, but like its object the effort had a limited life (see Figure 7.2; Weiss and Schoenberger 1979; Markusen and Wilmoth 1982; see Kingsley and Fortuny 2010).[34] More ambitiously, when Ann Markusen spearheaded an initiative to raise more than $2 million in funding for a large-scale project on Western regional development, the bid ultimately failed (see Markusen and Schoenberger 1983). She later realized "that there wasn't really a constituency that would fund us [to do a] project together. So, we just did our own stuff."[35]

Unfinished projects, unfunded grant applications, and people doing their "own stuff" never eroded the group's collective spirit, however. "There *was* a there there" Schoenberger maintained.[36] Maybe it was as a "thought collective," but even that signifies a singularity of purpose that likely was not always present. Perhaps for want of a better word, but also to signal its informalized and social character, some likened it to a club. Even in this loose form, Schoenberger said, "It felt real ... It was most exhilarating ... the most fun and exciting thing ever."[37] Making it exciting, Markusen believed, was the "ongoing conversation. It generated research topics for people ... It informed all of our work ... But it didn't gel in a more intense way than that."[38] So, while in spirit, and formally on some occasions, "the club" operated as a collective, it never became an upper-case C collective. After all, most of the group's graduate-student members were only passing through.

Rather than a failure, it was likely the *absence* of formal institutionalization that facilitated the ongoing conversation Markusen emphasized, along with its generative and creative diffuse outcomes. The circle of

Planners Network, OCCUR and Berkeley Citizens Action
present:

# Carter's Urban Policy: Broken Promises?

Eugene Mihaesco

a one-day workshop at
**EPIC WEST**
2640 College Avenue Berkeley
buses 51 and 58

10 am - 3 pm
Saturday April 29

10-12: keynote speaker
   HUD presentation
   panel discussion
12-1 lunch available
1-3 small workshops on:
   ·jobs and economic development
   .neighborhood 'revitalization' and housing
   .government reorganization
    .energy, environmen ,&resources
**FREE!**
child care
wheelchair access
further information: 8435506

**Figure 7.2** Carter's broken promises, 1978. *Source:* collection of David Wilmoth.

friends, and the aggregation of projects, was more than the sum of its parts. But how did the "club" operate? At least four distinct sets of actors played a role: tenure-track activists, behind-the-scenes supporters and protectors, graduate students, and visitors.

The tenure-track activists, Dick Walker and Ann Markusen, were focal points for the group, providing intellectual leadership, setting the tenor. Walker had come to Berkeley in 1975 directly from completing his doctoral work at Johns Hopkins University (see Chapter 6). Prior to joining DCRP in 1977, Markusen was an assistant professor in Economics at the University of Colorado and before that a research economist at the State of Michigan's House of Representatives. Her substantive interests were government policymaking and regional economics, while Walker's were with environmentalism and urbanism. Theoretically, there were affinities, however. Both practiced political economy, although of different stripes. Markusen tended to theory that was strongly empirically rooted and found, say, in works by Bennett Harrison and Doreen Massey, while Walker was early on more inclined to the abstract and classical, from Karl Marx to Karl Kautsky.[39] Beyond this, Markusen and Walker shared an intellectual and social outlook, including commitments to collaborative work and social activism.

The roles that Markusen and Walker respectively played in the group, as Marc Weiss reflected, were as lodestones, helping to attract and maintain students:

> They were young enough and radical enough that they could be magnets as faculty for the [graduate] students … [They] were close enough in cultural experience to students that we [could] form a bond of friendship. You would forget that Ann and Dick were professors. In the work we did, we were all on a more equal basis.[40]

The two created a space outside the two departments, erasing the distinction between Geography and Planning. "The Geography/Planning axis was between Dick Walker and Ann Markusen," Feldman said, a "linkage … between human geography, policy, and a political-economy centered group."[41] Students from each of the programs would take courses with Markusen and Walker, who would in turn serve on many overlapping supervisory committees. Before long, the group was a "unified thing."[42]

It was not plain sailing, however. Both Markusen and Walker were to face significant tenure battles. Walker's was the first, and the more epic, public, and brutal. Even before he was formally hired, Walker thought there might be trouble given his relationship to David Harvey at Hopkins.[43] He was right. For the old guard at Berkeley, Walker was tainted not only because Harvey was a Marxist but a positivist before that.[44] Harvey was doubly suspect. It only got worse. Walker was further accused of "being not just a commie, but anti-Sauer and overly popular with students; [even] of sleeping his way to popularity." One colleague reportedly said, if he was given tenure, "they should just close the Department of Geography." In the end, the department narrowly voted in favor of promotion and tenure, but that vote was overturned twice by the University's Budget Committee (in 1981 and 1982). Walker believed that the "key guy" on the Budget Committee had been lobbied by an unsympathetic Geography faculty member. A last-ditch appeal to the Chancellor of Berkeley (and a former DCRP faculty member), Mike Heyman, led to the tenure denial being rescinded. Walker kept his job, but it was touch-and-go, requiring student and faculty petitions, an international letter-writing campaign, and in-person representations, not to mention back-room interventions and maneuverings. Off campus, David Harvey's role was pivotal. On campus, vital support came from Allan Pred in Geography, and Peter Hall, Manuel Castells, and Michael Teitz at DCRP.

Markusen's tenure denial was less public and dramatic, but in other ways just as confounding. With a strong publishing and teaching record, and with an accumulated seven years of service as an assistant professor, she went up for tenure "early," with the prospect of a second chance if denied. She was denied, being told (after the fact) that she needed a book on her resume. Markusen attributed the denial partly to DCRP's internal divisions, partly to inadequate mentorship, and partly to plain political prejudice on the part of higher-level University committees and administrators against a young, left-wing professor. Despite Berkeley's ostensibly radical reputation, the University had not tenured an actual Marxist prior to this time.[45] As Markusen recalled, we "faced considerable resistance, because we were uppity. We were writing in a different way for different outlets, and we had an enormous amount of confidence … We were part of that generation who said, 'Don't trust anyone over 30!'"[46] In 1983, at her second attempt at tenure and promotion, and with a book manuscript in hand, written in less than a year (Markusen 1985), she was successful.

While Markusen and Walker may have had enemies at Berkeley, they also had friends. Notable among these for present purposes were behind-the-scenes allies and protectors. In both departments, there were senior professors who took it upon themselves to provide protection, sometimes to run interference, for their more radical junior colleagues and for activist graduate students. Allan Pred occupied this important role in Geography, and in DCRP it was Michael Teitz and Peter Hall. All three combined left-of-center politics with strong ethical commitments. Pred had been "radicalized by the Vietnam War," said Walker.[47] Teitz grew up in a working-class family in the UK, and while skeptical of European-style communism, he was "very progressive, open-minded, a cultured person," in the words of his former advisee, Flavia Martinelli.[48] Weiss characterized both as "liberal patrons,"[49] something that could also have been said of Hall. They believed not just in the right to express dissenting views, but in an ethical imperative to remove obstacles to that expression, for both junior faculty and graduate students.

In Geography, Pred provided protection specifically for Walker. As Gertler said, "Allan created a lot of the space for Dick to arrive and feel comfortable in the place. He was his patron saint." More generally "Pred made it safe for the left. He was *very* important as far as nurturing, and creating a receptive environment for people working on the left … He was genuinely interested in all we did."[50] Pred was a tireless advocate for Walker during his tenure battle, even in what seemed to be the jaws of defeat. Meanwhile, over at DCRP, it was Teitz who "provided cover for Ann."[51] She needed it, not being temperamentally "risk averse" (Teitz 2016: 53). As Glasmeier said, "Ann came [to DCRP] raw [but] it was not a hospitable environment … It was a very, very difficult time."[52] Within a divided department that contained few "shrinking violets,"[53] Teitz was a calming influence, a source of systematicity, wise counsel, humor, and moral principles. He was "very supportive of the graduate students," recalled Glasmeier, "wanting to be on our side, [both] respectful and respected, with a subtle way of dispersing negativity."[54] In these as in other ways, Teitz "played an important role in creating the conditions for the emergence of the Berkeley 'school'."[55]

If Walker and Markusen were key animateurs (with Pred and Teitz at their backs), it was a remarkable cohort of graduate students that energized the effort. Most came to Geography and to DCRP from outside the Bay Area, in several cases outside of California. The place where they met, exchanged ideas, and worked pulled them together, but in both a literal and a metaphorical sense, they were "assembled in other places" too.[56] Many had lived previous lives before going on to graduate school, often in political activism of one form or another, Markusen recalled:

> A lot of these students … were activists already. They had come from campuses where there were strong anti-war movements … Some of them had important ties … to the environmental movement. People were already out there. We didn't have this bias that it was compromising to be active in wherever the struggle was. We saw we had an intellectual job to help and support it and pull it in certain directions.[57]

The forms and sites of activism varied. Storper was involved in the environmental movement, especially around water issues, having been a lobbyist for Friends of the Earth in Sacramento; Schoenberger was part of the protest movement against U.S. imperial involvement in El Salvador's civil war; Shapira, Glasmeier, Weiss, and Storper, along with a number of other students, had joined campaigns to halt Bay Area deindustrialization, one of the flashpoints of which was around a conversion plan for the Colgate-Palmolive plant in Berkeley (in which Teitz was also a significant player); and Glasmeier, Saxenian, and Schoenberger would go on to expose the connections between U.S. military funding and high-tech development in Silicon Valley.

To suggest that the graduate students' scholarly radicalism echoed or paralleled their off-campus activism is to make a false binary, however. These were not separate but overlapping lives. They were scholar-activists, even when activism involved mundane labor. For example, for a time the group took over publishing and distributing a newsletter for the Planners Network, a national organization for progressive planners founded in 1975 in San Francisco by Chester Hartman. It was an important form of activism, helping to facilitate progressive scholarship within the academy. But it was also busy work that included, "besides collecting the information and writing up, ... a lot of maintaining and updating index card files, typing, printing out mailing labels, folding and stapling" (Weiss and Schoenberger 2015: 67). As Schoenberger said, "many people had to be willing to pitch in and do what needed to be done. Younger generations of planners have no idea! [But] I remember it as one of the high points of my life."[58]

The social and collaborative nature of much of these activities worked against and around the individualism (and isolation) built into the doctorial research process. While the epistemological break that the Berkeley geographers and planners helped to engineer, and then consolidate, was undeniably important, the glue that bound the group together was political and practical as much as intellectual. There was a shared commitment to change the world, even if the steps were incremental and the work sometimes humdrum.

Finally, there were those who helped make the Berkeley moment it was, but who were never permanently affiliated, the visitors. As Markusen recalled, during especially the late 1970s and early 1980s "there was a lot of porosity ... There were a lot of people coming in, coming out." Without doubt the two most consequential visitors, who characteristically left more behind than they took away, were Doreen Massey and Bennett Harrison. Massey visited Geography in the Spring quarter 1981 at the invitation of then-chair, Allan Pred; Harrison was a frequent visitor to DCRP.[59] That their combined and separate impact was transformative speaks to the fact that this is not simply a story of endogenous intellectual growth. It is to these geographical worlds farther afield that we now turn.

## Crossings and Conjunctures

There was a significant cohort effect at work in the Berkeley moment. The loosely constituted and relatively porous "club" capitalized upon dense and frequent internal social and intellectual interactions, making for emergent, more than the sum-of-the-parts conditions. These locally assembled and internally generated capacities, the combination of cohort and club, while essential constituents, were on their own insufficient to explain the Berkeley achievement. They required various forms of supplementation. That included a loose network of left faculty across campus, such as Michael Burawoy in Sociology, and Claire Brown and Michael Reich in Economics. There were also longer-distance connections. As Richard Walker's static electricity metaphor implied, the Berkeley moment could attract participants from afar. Cases in point were energizing visits from Doreen Massey and Bennett Harrison. Walker credits both with decisive roles in his own "big turn … away from … early stabs at environmental and (sub)urban geography [and] into industrial and economic geography."[60] Likewise, AnnaLee Saxenian said they were "both amazing," not least for their "tremendous passion and enthusiasm,"[61] and as role models for the rigorous, politically engaged researcher.

Massey stayed in a spare room at the rented house of Susanna Hecht and Michael Storper during her visit. She was not the person who was expected, though. Before she arrived, "her name was in the air," Gertler recalled, "everyone had read her work."[62] Hecht had in mind "some mousey-haired, lady British academic in brogues [and] tweeds [but] then this punk showed up!"[63] Storper said it was as if Massey arrived from another world, with her "cropped blond hair – short, loud, [and] punk," with talk of miners' strikes and combustible class politics. She was "marked as radical from the beginning."[64]

> [Massey] was very important, very important; she showed us what a punk radical could look like. She was great. She was so full of energy … The years we're talking about, punk was the dominant music form … She had a punk attitude and [was] real working class.[65]
>
> Doreen shows up. She has tight black jeans on … really short, bright blonde hair. And the energy of a whirling dervish. She is very different and much more animated than anyone else. She stayed there for a very short period of time, but she had a huge impact on all of us.[66]

And this, at a time when there were no tenured women in the Geography department, and female graduate students like Susan Christopherson and Susanna Hecht were a small minority. Glasmeier recently recalled that, prior to Massey's visit, "virtually no discussion of feminism existed, and politics were of a specific variety" (Gertler *et al.* 2018: 218). Further, if there was any truth to the stereotype of Berkeley as the center of its own universe, Massey gave many pause to think again. These otherwise worldly graduate students "felt a little bit naïve" in her presence, Saxenian remembers. Or as Storper said, they suddenly became much more self-conscious of their identity as "young Americans."[67] Massey also embodied a less conventional (career) path, having herself never obtained a doctorate and at the time lacking anything approaching tenured professorial security.

Bennett Harrison, as a tenured faculty member at MIT, lived the dualized identity of the scholar-activist in a different way (see Christopherson 2001; Markusen 2001a, 2001b), even though he was also remembered for his infectious personality and personal warmth. As Weiss said, just as Massey "had an immense impact on ... all [of us], in showing a different path forward as a scholar," Harrison was a "really big deal" for those uncertain about the career and life trajectory that lay in front of them:

> [Harrison was] inspiring and empowering ... He was telling us that the elite back in Cambridge thought Berkeley was really cool! We were not as established as them ... I would say that Ben, he *changed us*. It was a giant "attaboy!" By the time he left, we all felt much more energized to keep going forward, on the intellectual side, on the critique of Carter [urban] policy, on the plant closures movement in California, on many levels.[68]

Characteristically, Harrison demonstrated his capacity for "pulling everyone into his big embrace and making [us] all feel important" (Walker 2001: 35). And as Michael Teitz would emphasize, Harrison "was a huge influence in Planning ... Bennett was a real mensch, a real human being, a lovely man," not least as a provider of "support for Ann [Markusen] who was a bit beleaguered" at DCRP.[69] An economist who mixed with planners, geographers, community-development and union activists, Harrison was at the forefront of a diverse group of "freshly minted economists, political scientists, sociologists, and geographers, influenced by the on-campus intellectual ferment of the Vietnam War period, [who in the 1970s had begun] to question the narrow and politically disconnected analyses of urban economics and regional science" (Markusen 2001a: 291), never reifying the region nor endowing it with special powers but recognizing it as a significant site of conflict, contestation, and change.

Although their paths were different, Harrison and Massey would both end up wrestling with the problematic of (industrial) restructuring, and the associated nexus of corporate strategy, labor relations, and regional development (Lovering 1989; Peck 2002). More than a new take on the "regional problem," their work placed the shifting geography of class relations at the heart of the political economy of capitalist restructuring. Uneven spatial development was integral to the functioning of capitalist dynamics, not just as a source of contingent variations or confounding complexity. Storper said that Massey had

> a language about regionalism, identity, class struggle ... Doreen's explanation of the spatial differentiation of class relations in Britain ... had a big influence on [my] thinking about ... the geographical puzzle of class struggle. ... Harvey would call it a spatial fix later on ... It is really key to understanding American capitalism. The differentiation of social conditions, the rivalry between the states, the card that these less progressive areas play, and how that reverberates over the structure of American capitalism. This is absolutely key. You know, Doreen had that in her head very early on, when people weren't talking about it here.[70]

Taking regional configurations seriously also involved certain commitments to (primary) research methods and to exploratory modes of theorization, marking out some differences with David Harvey's more foundational and abstract treatment of capitalism a la *Limits to capital* (Harvey 1982). For his part, Weiss found Harvey a "distant figure," and his orthodox and formal renderings of capitalist dynamics "inaccessible in some ways."[71] Rather than some antagonism to Marxism, however, this attitude reflected a more general sense of the insufficiency of any orthodoxy in a time of turbulence, uncertainty, and new forms of social engagement. Along with that went a desire for more open, flexible, experimental, and even eclectic approaches to methodology and theory building. As Flavia Martinelli emphasized, the Berkeley group were "more into analyzing what was going on" in a concrete way, which called for a more grounded approach, closer to the action and to the actors.[72]

Consequently, explanations, working concepts, provisional frameworks had to be more plastic, more adaptable – and more connected to the shifting ground under foot. As Hecht put it, "we invented the theory as we went along," or, as Gertler said of the time, "everything was changing, and there was a sense that we needed new conceptual tools to make sense of what was happening around us."[73] If there was a common methodological conviction across the Berkeley group it was that capitalist dynamics (and the less-than-mechanically determined politics of region and class) could not be "read-off" from even sophisticated understandings of system-wide tendencies, patterns, or laws of motion. This derived from Walker's reading of critical realism, originating in a bond struck with Andrew Sayer in the late 1970s, but there was also a line to structuration theory, via Pred's seminars. The role of critical realism later became a bone of contention with Harvey,[74] but such doctrinal differences were not a source of friction in Berkeley. This was a group with "lots of fire power," as Storper put it, increasingly convinced that "something epochal was happening in the world."[75] And there is no doubt that Harvey was continuously and closely read, a point that Flavia Martinelli underlined: "Harvey was on every reading list, let's not forget that, on *every* reading list!"[76] But in terms of research practice, there was a greater emphasis on exploratory, case-study investigations, bolstered by a flexible interpretation of critical realism along with more open interpretations of Marx(ism), even as this iteration of "the project" had yet seriously to question the prioritization of class relations – in their regionalized, industrial form – over those of race and gender, or to explore wider registers of economic diversity (see Christopherson 1989; Gibson-Graham 1996).

Instead, the moment was captured by two books published in the same year, Massey and Meegan's *Anatomy of Job Loss* and Bluestone and Harrison's *Deindustrialization of America*, the subtitle of which told the story: *Plant Closings, Community Abandonment, and the Dismantling of Basic Industry* (Bluestone and Harrison 1982; Massey and Meegan 1982). They both stayed close to labor, both analytically and politically. Harrison, Markusen, and others worked closely with collaborators in the labor movement for some years, and clearly knew "which side they were on." But it had dawned on Harrison, and then Markusen and several of the Berkeley group, that what the unions *really* needed, in this moment of accelerated restructuring and managerial unilateralism, was a strategic and contextually grounded understanding of corporate behavior. This amounted to a knowing and indeed calculated turn to what would later be known as "capital-centrism," in the service of a labor geography *avant la lettre* (Peck 2002, 2013). As Schoenberger plainly put it, "Labor unions didn't need me to tell them how they experienced capitalism … What they didn't know [and wanted to know] was how *corporations* thought."[77] Ann Markusen remembers one of Harrison's talks at Berkeley in the late 1970s when he made the case for shifting the focus of attention from labor – the "victims" of restructuring processes – to the "perpetrators," corporate capital: "I have been studying labor for all these years [he was reported to have said] but now I think I am studying the wrong thing … I should be studying capital, because that's where the power is. And I was sitting there and thinking '*Yes! That's right!*'," Markusen said, "We should be studying *capital*."[78] Doreen Massey would make a parallel move – problematizing capital, but not in a one-sided or capital-logic fashion – in her formative work on the spatial division of labor (Massey 1984, 1995).[79]

This called for a different approach, different methods. In Harrison's case, it meant "he had to go beyond analyses of published statistics (his forte) and do primary research, including interviewing firms and reading the trade press in specific industrial sectors" (Markusen 2001b: 40). Initially, this took the form of an influential report prepared for a coalition of labor-union and community groups (Bluestone and Harrison 1980). Figuring out, by way of this more granular, face-to-face method, what corporations were doing and why, became a hallmark of this more-than-local Berkeley-style approach. It represented a methodological watershed for an economic geography that had previously relied on an arm's-length analysis of secondary data, premised on a set of assumptions about the supposedly rational decision-making behavior of firms (Schoenberger 1991; Markusen 1994; Barnes *et al.* 2007). It turned out that the new approach of talking to firms, indeed "studying regions by studying firms," generated other surprises. Saxenian recalled that her Master's research, prompted by a newspaper report on rapid employment growth in Santa Clara County, quickly became an exploratory investigation of the electronics industry. Manuel Castells, her advisor, said:

> "Don't do any background research, just go out and interview people" …
> He was the one that sent me out into the field … Erica [Schoenberger] and
> I both did these interviews. We went out found executives, interviewed
> them. [We] told stories, and fitted them into our framework … We were
> all into case studies.[80]

Many of these would soon be numbered among the critical cases of a new industrial geography, one that would seek to account for (new forms of) growth as well as counting the costs of decline.[81] Having "stumbled into" the Silicon Valley phenomenon, courtesy of Saxenian's Master's research, the group was soon theorizing industrial growth and decline as alternate sides of the same coin of capitalist restructuring.[82] As Weiss put it, "in California, we had deindustrialization and the new economy … cheek by jowl … Both were happening at the same time, so you could see them almost side by side."[83] Similarly, Glasmeier compared her experience to "living inside a multispectral case study … in an accelerated state." While it was no rustbelt, Northern California was displaying signs of deindustrialization, but was also on the way to becoming the "mecca" for a new growth paradigm: "We had both the phoenix and the ashes … Timing is everything, and the environment around [us] was so electric, so intense [and] it was moving fast."[84]

This intensity was experienced in political and policymaking circles as well. In 1981, Weiss put on hold his dissertation work on the real-estate industry to take up the role of Deputy Director of the California Commission on Industrial Innovation in Sacramento under Governor Jerry Brown. This involved commissioning research into various branches of the state's high-tech economy, such as biotechnology and software, that later was converted into academic writings, including in Peter Hall and Ann Markusen's (1985) influential collection, *Silicon Landscapes* (see also Markusen *et al.* 1986). According to Feldman, those studies "made us question the received wisdom, the tacit model of capitalism, as a factory grinding out textiles," while also demonstrating that the dynamics of the so-called high-technology "industry" were anything but singular.[85] Although this research in Northern California on industrial restructuring later spawned all manner of transition models, some of which portrayed a sequential movement from old-style to new-style industrialization, the original research raised critical questions about that transition from the start. The future was not going to be simply the mirror of the past; nor was it going to be any less complex and contested in terms of its sectoral dynamics or employment outcomes. Industrial transformation – in its up cycle as well as its down cycle – remained an intrinsically open, political process. Interestingly, Markusen, Weiss, and others worked on a project during the early 1980s that explicitly spanned the old and new economies of the region: "We tried to merge the two, in Berkeley, when Colgate was closing. We tried to turn that into what would have been the first alternative energy research and industrial park in America" (see Markusen and Weiss 1984; Weiss and Schoenberger 2015).[86] The implications of this work would run deep into the *practice* of regional planning and local economic development policymaking. As Weiss explained,

> Because most [in the Berkeley group] became professors, the general thrust [and] overall impact ... was a paradigm shift *in academia*, both in geography and planning ... But an equally important side of the story [concerned] a new to way to really do policy and planning on the ground ... Without a doubt, the whole intellectual root of [a series of shifts in economic-development policy and practice] was the work we did at UC-Berkeley ... This is a side of the story that you wouldn't see as clearly, but it was equally important.[87]

The recognition of what Storper called "this California thing" was positional in several senses. While by no means exclusively focused on Northern California, being there and seeing from there was a source of distinctive insights into the "flipside of deindustrialization, which was these new growth centers," including the "discovery of [new] forms of industrialization."[88] This recognition of the creative as well as the destructive side of processes of creative destruction would contribute to a far-reaching realignment in the rubric and remit of Anglo-American economic geography, which would soon be theorizing regional growth in addition to (or instead of) regional decline. As Storper said, at first

> we didn't exactly know what it was all about. But it was something [to do with] innovation, growth, and agglomeration *going all together* … [It was about] trying to grapple with California, which was a growth economy … We were seeing these different processes, bust in the North East, boom out here, the South in another vein. And we were trying to conceptualize it all, realizing that capitalism was spatially dynamic – it seems banal to say it now – capitalism was spatially dynamic, but differentiated.[89]

As Storper and Walker (1989: 3–4) later recognized, while Massey and Meegan's (1982) formulation of industrial restructuring theory was enormously insightful, it "amounted to no more than comparative statics unless it is encompassed within a suitable theory of capitalist growth." It was the latter that the Berkeley group began to explore. At the same time, they became increasingly conscious of the fact that theirs too was view from somewhere.

As the Berkeley group moved on (and out), perspectives changed. For example, Gertler and Saxenian both moved East to do their doctoral work, to Harvard and MIT, respectively. Saxenian (1996) went on to draw a sharp analytical distinction between the innovation cultures of the East and West coasts, contrasting the buttoned-down, hierarchical culture of Route 128 with the flatter networks and freewheeling style of Silicon Valley, in the process establishing some of the most enduring stylized facts in economic geography.[90] In one of the signature exchanges of the 1980s, Gertler debated Erica Schoenberger over the meaning of "flexibility" in labor regimes, technology, and production systems (Gertler 1988; Schoenberger 1988). Gertler argued the Berkeley view ended up "giving too much credit to capital: this is how we screw the workers," developing a critique that benefited both from the closeness of continuing friendships and the perspective born of a certain kind of separation:

> I could make these criticisms of flexibility because I had moved myself from that milieu, and given myself analytical distance … It was a reality check, separating out the rhetoric from what is real … As much as I enjoyed being at Berkeley … I kept thinking to myself that there is a certain inward-looking character to this place. They convinced themselves

they were at the center of the universe! The only way that they would engage with the world outside would be to have others come and visit. Coming from the East, I knew there was another world out there, and I didn't want to be trapped by that insularity [which] is a characteristic of Berkeley and California even today ... All the smart people are here, so why go anywhere else?[91]

Located at some distance from the actually existing rustbelt, it is understandable that the group's situated understanding of restructuring would be especially attuned to the dynamics of growth, corporate strategy, and innovation. The Berkeley approach stitched together, in new and productive ways, elements of what would become a strand of critical orthodoxy in the "new" economic geographies that followed, including a somewhat eclectic reading of political economy, a focus on organizational-technological drivers and distribution-employment effects, and a meso-analytic concern with regional formations and dynamics (see, Markusen 1987; Storper and Walker 1989). Subsequently, this approach was variously blended, combined, and cut with the field's subsequent cultural, institutional, and relational turns, but its imprint remained indelible (see Scott 2000; Sheppard 2011).

Some of this influence was the result of an unusually productive agglomeration of talented, progressive, and cooperatively-minded researchers, many of whom went on – in a range of other locations – profoundly to shape the theory and practice of economic geography and urban and regional planning, the boundaries between which (partly as a result) became increasingly porous. But the fact that this was a turbulent time and place also made a material difference. Northern California was positioned at the threshold of the new economy, beginning to cut a new path to innovation-rich growth. The wider political climate, moreover, was deeply uncertain, a mixture of late-Keynesian ennui and the opening salvos in the neoliberal offensive, engaged locally by a raft of left, radical, and progressive movements. The moment was conjunctural in a more specifically political sense, too, in that these were formative years for a generation of activist graduate students. Vietnam, the urban crisis, the assassinations of Robert Kennedy and the Reverend Martin Luther King, the ascendancy of Ronald Reagan, the impeachment of Richard Nixon, it was "one damn thing after another" for this generation, Richard Walker emphasized.[92] "We all came out of Vietnam," was how Schoenberger put it. And the Bay Area was a cauldron of creative radicalism since the 1950s – the Beats, queer politics, hippies, draft protesters, third-world strikes, anti-gentrification, free-speech, political countercultures of all kinds. "We were marinating in that context," said Walker.[93]

All of this contributed to what was a radical and activist predisposition in the Berkeley group: "We were *already* politicized," Schoenberger went on, "Most of us were experienced [organizers and campaigners]. We were not naïve waifs ... That made it possible for us to connect instantly and made it possible to do the work."[94] While these backgrounds were diverse, rather than monolithic, they were all located somewhere on the spectrum of progressive causes and struggles, shaping a "cohort of like-minded individuals doing like-minded things."[95] In this context, Reagan's election to the presidency in 1980 came as an especially severe jolt. This was about more than an antipathy to what became known as "Reaganomics," since there was a visceral awareness that Reagan had launched his political career in California by targeting student protesters, "radical" faculty, and even liberal administrators, having earlier promised the state's voters that he would make it his business to "clean up the mess at Berkeley" (Kahn 2004: 1). As Markusen explained of this confluence of political circumstances:

> We were still charged from that moment, the anti-war movement, the discovery by young radicals of the labor movement and the civil-rights movement, the women's movement. It still had staying power. So, the election of Ronald Reagan was a nightmare for us really. It took us a while to absorb it. When you are young and you think you know exactly where the world is going, and everybody that you surrounded yourself with was thinking the same way, it's a little hard to take it in right away that the world is a changed place.[96]

It was into this changed world of the 1980s that the graduate students from the Berkeley circle entered. By then the academic labor market was "the pits," as Glasmeier recalled.[97] Almost all of members of the group eventually followed paths to tenured security, even though for some it was circuitous, with many making a virtue of their identities as disciplinary "chameleons."[98] Others made careers in the world of progressive policy advocacy, underlining that all along this was more than merely an intellectual pursuit.

## Conclusion: Passing Moments?

Our goal in this chapter was not simply to celebrate the Berkeley moment of the late 1970s and early 1980s across the interacting fields of economic geography and urban planning, although it would be churlish not to allow for some measure of this. We also tried to explain it. In this vein, we attempted to trace some of the connections between the widespread embrace of radical political economy during the 1980s and one of its more important spaces of incubation. As such, we have sought to illustrate the larger argument of this book that the making of radical geography was always a grounded, geographical process, the complex outcome of a series of localized struggles, projects, and movements, often connected and energized by longer-distance networks, alliances, and affinities. Radical geography consequently had its own historical geography; it was assembled in places, often quite organically from the ground up, but at the same time intermixed with fragments, experiences, connections imported from other places. It was certainly not handed down, all of a piece, like sacred tablets from on high.

In the Berkeley case, these geographies played out in different ways and across a range of scales. The first half of the chapter stayed close to the scale of the campus. Here the focus was on some of the underlying micro-geographies that shaped (both positively and negatively) these especially fecund years at Berkeley, including a partial estrangement from the formal institutional sites that were its two "home" departments, Geography and the DCRP, and the generative uses made of more informal and non-hierarchical sites like graduate-student and faculty residences, their living rooms, kitchens, and hot tubs. In these places, disciplinary codes and boundaries had little relevance, less still the influence of departments that were each only beginning to shake off deeply entrenched conservative traditions, with their sectarian and masculinist cultures, their narrow definitions of acceptable scholarship, and their skepticism of critical social theory and engaged political activism.

The sparks that were generated by a rising cohort of radical geographers and planners in Berkeley at this time were partly a result of the less-than-formal organization of the group as an intellectual and indeed social and political community – loose, porous, collaborative, non-hierarchical, non-competitive. But the space that the group made and occupied also benefited from some measure of benevolent "protection" from a small number of more senior faculty, who while they may not have always been in agreement, even more strongly disagreed with those attempting to silence them. This is not to give credit to a paternalist class of overseers, but goes to the wider point that the Berkeley moment was enabled by a heterogeneous confluence of actors and circumstances, including young "uppity" faculty, activist graduate students, inspirational visitors, and backstage supporters and enablers, in addition to the legacy of battles fought by their immediate predecessors earlier in the 1970s. In turn, all of this was connected to geographies of political activism of the women's, civil-rights, anti-war, anti-imperialism and environmental movements, which had been galvanized in distinctive ways by struggles around plant closures, around first-wave neoliberalism, and around the emerging shape of the new economy.

The radicalization of economic geography during the 1980s cannot be reduced to only the events at Berkeley from the mid-1970s to the early 1980s. Our argument is that this was an especially important inflection point *and place*, however. As significant as some of the localized conditions were, the enduring relationships and innovative ideas forged in Berkeley at the time would surely not have resonated were it not for a host of wider conjunctural circumstances. To strike a Marxian tone, this was a synthesis of multiple determinations, a story of intersecting biographies, and a slice of moving histories, as well as a by-product of the collision of different political projects and movements. Correspondingly, the intellectual and political consequences of the Berkeley moment were dispersed, as the subsequent lives and careers of its key protagonists illustrate.[99]

That the Bay Area region lay on their doorstep was crucial. It functioned as a kind of laboratory for understanding the emergent spatial dynamics of capitalism, itself in the throes of an intense period of creative destruction. The Berkeley moment spanned both the plant-closure movement and rise of Silicon Valley, and the compressed interregnum between the dislocations of deindustrialization and the rise of a new, more flexible and more unequal, economy (see Walker 2018). These were circumstances, crucially, in which many of the old explanations, routines, and responses were manifestly falling short, when there was not just an opening but an urgent need for new theories, methodologies, and practices that engaged and grappled with the political economy of real-time change, all in the service of more progressive and sustainable futures. The headwinds were formidable, of course, at this breaking moment for all manner of transformations – flexibilization, financialization, precarity, neoliberalization – but as they continued to argue about both the causes and consequences of these developments, the Berkeley group were among the first to see them coming. And in their efforts to make sense of this new economic world in the making, they helped shape a new and more radical style of economic geography. There was nothing rote or mechanical in any of this; no inevitability. Yet if this was something like a Vienna circle moment for economic geography, it was also one when the stars aligned, when connections fell into place, and the pulse of truly transformative change was felt.

## Acknowledgment

This chapter is dedicated to the memory of Susan Christopherson (1947–2016), Bennett Harrison (1942–1999), and Doreen Massey (1944–2016). We are most grateful to all those from Berkeley's past and present who found the time to talk to us, so engagingly and candidly, while recognizing that none of them bears any responsibility for the interpretations we have offered here. And we thank Eric Leinberger for his creative cartography.

## Notes

1 Richard Walker, interview with Trevor Barnes and Jamie Peck, Los Angeles, CA, April 2013. Hereafter Walker interview, 2013. (Subsequent references to interviews follow this format.)

2 Ann Markusen, Skype interview with Trevor Barnes and Jamie Peck, Minneapolis, MN, June 2016.

3 Those graduate students included: from Geography, Judith Carney, Susan Christopherson, Susanna Hecht, Michael Heiman, Kristin Nelson, and Michael Storper; from Planning, Meric Gertler, Amy Glasmeier, Flavia Martinelli, AnnaLee Saxenian, Erica Schoenberger, Philip Shapira, Marc Weiss, David Wilmoth, and Marshall Feldman (a visitor from UCLA's planning program), and Madeline Landau from Anthropology (see Weiss and Schoenberger 2015).

4 Michael Storper, interview with Trevor Barnes and Jamie Peck, Vancouver, BC, March 2016.

5 Michael Heiman recalls that only a minority of graduate students in Geography at the time had previous degrees in the field, while many brought with them "real-world" experiences of many kinds; another reason for the weak hold of any "center" or incumbent orthodoxy. Heiman personal communication, February 2018.

6 AnnaLee Saxenian, Skype interview with Trevor Barnes and Jamie Peck, Berkeley, CA, June 2016.

7 Amy Glasmeier, Skype interview with Trevor Barnes and Jamie Peck, Cambridge, MA, June 2016.

8 Saxenian interview, 2016.

9 Allan Pred, interview with Trevor Barnes, Berkeley, CA, March 1998.

10 Walker interview, 2013.

11 Michael Storper, interview with Trevor Barnes, Helga Leitner, and Eric Sheppard, Encino, CA, June, 2014.

12 It was Berkeley students that had launched the free-speech movement, initiating a wave of campus protests across the country. In December 1964, students had gathered at Berkeley's Sproul Hall to protest the University's ban on on-campus political activities. After more than 800 student activists were arrested, the protests intensified, culminating in the shut-down of the University and, a few weeks later, in the resignation of Chancellor Edward Strong. His successor, Martin Myerson, a planner and formerly the Dean with responsibility for DCRP, negotiated an end to the protests. His spatial planning solution was to designate the steps of Sproul Hall as an open site for student free speech. During the protests, DCRP's Michael Teitz quietly bailed several students out of jail.

13 Michael Teitz, Skype interview with Trevor Barnes and Jamie Peck, San Francisco, CA, June 2016.

14  Teitz interview, 2016.
15  Markusen interview, 2016; Markusen, personal communication, February 2018.
16  Markusen, personal communication, February 2018.
17  The following quotations are from Janice Perlman, Skype interview with Trevor Barnes and Jamie Peck, New York, NY, June, 2016; telephone interview with Jamie Peck, February 2018.
18  While diversifying gender was a key dimension, there were also struggles over intellectual diversification. Perlman made the case for recruiting Manuel Castells, for example, who arrived in 1979.
19  Perlman interview, 2018.
20  Erica Schoenberger, Skype interview with Trevor Barnes and Jamie Peck, Baltimore, MD, June, 2016.
21  Glasmeier interview, 2016.
22  Gertler interview, 2016.
23  Meric Gertler, Skype interview with Trevor Barnes and Jamie Peck, Toronto, ON, June 2016.
24  Marc Weiss, Skype interview with Trevor Barnes and Jamie Peck, Porto Allegro, Brazil, June 2016.
25  Saxenian interview, 2016.
26  Marshall Feldman, Skype interview with Trevor Barnes and Jamie Peck, Providence, RI, June 2016.
27  Schoenberger interview, 2016.
28  Gertler interview, 2016. See also Gertler *et al.* (2018).
29  Schoenberger interview, 2016.
30  Schoenberger interview, 2016.
31  Glasmeier interview, 2016.
32  Schoenberger interview, 2016.
33  Glasmeier interview, 2016.
34  Weiss and Schoenberger (2015: 65) said that a "lengthy academic paper" was developed out of this critique of the new urban-policy orthodoxy, although it was never formally published, and but for a handful of citational traces – which intriguingly included Doreen Massey and Bennett Harrison, as well as most of the Berkeley group – the document itself can no longer be found.
35  Markusen interview, 2016.
36  Schoenberger interview, 2016.
37  Erica Schoenberger, Skype interview with Jamie Peck, Baltimore, MD, April 2013.
38  Markusen interview, 2016.
39  This difference is reflected in among things class reading lists by Markusen and Walker for the early 1980s and given to us by Flavia Martinelli who at different times was both a graduate student attendee and a teaching assistant.
40  Weiss interview, 2016.
41  Feldman interview, 2016.
42  Weiss interview, 2016.

43  David Harvey was a member of Walker's supervisory committee, but the environmental economist Steve Hanke was his immediate doctoral supervisor. Walker, though, was part of the circle of students around Harvey, and participated in the *Capital* reading group held at Harvey's Baltimore home (see Chapter 6, this volume).

44  The next three quotations are from Walker's retirement speech given to the Department of Geography at Berkeley, "From the age of dino-Sauers to the anthropo-scene: Reminiscences of life in Berkeley Geography, 1975–2012." April 12, 2012, www.youtube.com/watch?v=-m1KvOneKjA.

45  Walker believed Berkeley's supposed radicalism was never real, referring to its "faux-radical reputation." Personal communication, February 2018.

46  Markusen interview, 2016.

47  Walker interview, 2013.

48  Flavia Martinelli, Skype interview with Jamie Peck, Reggio Calabria, Italy, July 2017.

49  Weiss interview, 2016.

50  Gertler interview, 2016.

51  Gertler interview, 2016.

52  Glasmeier interview, 2016.

53  Gertler interview, 2016.

54  Glasmeier interview, 2016.

55  Schoenberger interview, 2013.

56  Markusen, personal communication, February 2018.

57  Markusen interview, 2016.

58  Schoenberger interview, 2013.

59  In Spring 1981, Massey had no permanent job in the UK, having been made redundant when the incoming Conservative government of Margaret Thatcher defunded the research institute in which she worked, the Centre for Environmental Studies. While Massey was at Berkeley, she was offered the Professorship of Geography at Britain's Open University, which she accepted (see Peck *et al.* 2018).

60  Walker interview, 2013.

61  Saxenian interview, 2016.

62  Gertler interview, 2016.

63  Susanna Hecht, interview with Trevor Barnes, Helga Leitner, and Eric Sheppard, Topeka Canyon, CA, June 2014.

64  Storper interview, 2016.

65  Michael Heiman, Skype interview with Trevor Barnes and Jamie Peck, Carlisle, PA, July 2017.

66  Glasmeier interview, 2016.

67  AnnaLee Saxenian, 2016; Storper interview, 2016. Storper continued, "Doreen is this … gardener's daughter who goes to Oxford, and she *lets you know it*, right, from the beginning! This is something that on a whole bunch of levels [didn't] compute for a young American."

68  Weiss interview, 2016.

69  Teitz interview, 2016.

70  Storper interview, 2016.

71  Weiss interview, 2016.
72  Martinelli interview, 2017.
73  Hecht interview, 2014; Gertler interview, 2016.
74  Contra Massey and Sayer, Harvey declared his impatience with those approaches that he saw challenging "the direct relevance or power of general theory," contending that the turn to midlevel exploration and empirical specificity in restructuring studies – far from (re)animating processual understandings of capitalism – had incapacitated them: "so laden down with a rhetoric of contingency, place, and the specificity of history, that the whole guiding thread of Marxian argument is reduced to a set of echoes and reverberations of inert Marxian categories" (Harvey 1987: 373; see Sayer 1985, 1987; Storper 1987b; Massey 1995; Peck *et al.* 2018).
75  Storper interview, 2016.
76  Martinelli interview, 2017.
77  Schoenberger interview, 2016.
78  Markusen interview, 2016.
79  Walker, personal communication, February 2018.
80  Saxenian interview, 2016.
81  Interestingly, Saxenian's (1980) Master of City Planning thesis, "Silicon chips and spatial structure," made the argument that the "urban contradictions" of Silicon Valley would drive a dynamic of decentralization. Subsequently it became clear that growing (quite literally) *through* contradictions was a defining feature of postFordist agglomeration.
82  Saxenian said, "the Silicon Valley stuff, I kind of stumbled into it, and the research kind of took off after that. It wasn't part of [the project] initially, but as we started to recognize how it was growing, we channelled that into the worldview we had about deindustrialization." Saxenian interview, 2016.
83  Weiss interview, 2016.
84  Glasmeier interview, 2016.
85  "Biotechnology is not like software," Marshall Feldman said in defense of this pioneering sector study, on which he also worked. Feldman interview, 2016.
86  Weiss interview, 2016.
87  Weiss interview, 2016.
88  Storper interview, 2016; Weiss interview, 2016.
89  Storper interview, 2016.
90  In an echo, perhaps, of the innovation climate more generally, Saxenian also found the intellectual culture of Cambridge and MIT to be "very competitive, hard edged" in comparison to the more collaborative atmosphere of Berkeley. Flavia Martinelli, who took classes with Bennett Harrison both at Berkeley and at MIT, joked that Cambridge students would sometimes "cover their notes" so that their classmates could not see them – symptomatic she believed of an "uptight, competitive" institutional *and regional*

culture quite different to that found in Berkeley. Saxenian interview, 2016; Martinelli interview, 2017.

91　Gertler interview, 2016.

92　Walker interview, 2013.

93　Walker, personal communication, February 2018.

94　Schoenberger interview, 2016.

95　Janice Perlman, Skype interview with Trevor Barnes and Jamie Peck, New York, NY, June 2016.

96　Markusen interview, 2016. For her part, Markusen would ask herself, "What's the target now?" The answer came in the form of a new line of work on the "gunbelt" and the defense economy (see Markusen *et al.* 1991).

97　Glasmeier interview, 2016.

98　Glasmeier interview, 2016. On the subsequent, discipline-crossing careers of the Berkeley group, see note 99.

99　AnnaLee Saxenian and Richard Walker, who made their careers at Berkeley, are the exceptions that prove this more general rule. From the perspective of economic geography, notable in all kinds of ways would be the academic careers of Susan Christopherson (planning, Cornell), Meric Gertler (planning and geography, Toronto), Amy Glasmeier (geography and planning, Penn State and MIT), Ann Markusen (planning, Rutgers and Minnesota), Erica Schoenberger (geography, Johns Hopkins), and Michael Storper (planning and geography, UCLA, Sciences Po, and LSE).

# References

Barnes, T. J. and Gertler, M. S. eds (1999). *The New Industrial Geography: Regions, Regulations and Institutions.* London: Routledge.

Barnes, T. J., Peck, J., Sheppard, E. and Tickell, A. (2007). Methods matter: transformations in economic geography. In A. Tickell, E. Sheppard, J. Peck and T. J. Barnes, eds, *Politics and Practice in Economic Geography.* pp. 1–24. London: Sage.

Bluestone, B. and Harrison, B. (1980). *Capital and Communities: The Causes and Consequences of Private Disinvestment.* Washington, DC: Progressive Alliance.

Bluestone, B. and Harrison, B. (1982). *The Deindustrialization of America: Plant Closings, Community Abandonment, and the Dismantling of Basic Industry.* New York: Basic Books.

Chari, S., Friedberg, S., Gidwani, V., Ribot, J. and Wolford, W. (2017). *Other Geographies: the Influences of Michael Watts.* Chichester, UK: Wiley-Blackwell.

Christopherson, S. (1989). On being outside "the project." *Antipode* 21 (2): 83–89.

Christopherson, S. (2001). Bennett Harrison's gift: collaborative approaches to regional development. *Antipode* 33 (1): 29–33.

Gertler, M. S. (1988). The limits of flexibility: comments on the post-Fordist vision of production and its geography. *Transactions of the Institute of British Geographers* 13 (4): 419–432.

Gertler, M. S., Thomas, M. and Glasmeier, A. (2018). In Memoriam: Susan Christopherson (1947–2016). *Cambridge Journal of Regions, Economy and Society* 11 (1): 211–219.

Gibson-Graham, J. K. (1996). *The End of Capitalism (As We Knew It)*. Oxford: Blackwell.

Hall, P. and Markusen, A. (1985). Preface. In P. Hall and A. Markusen, eds, *Silicon Landscapes*. pp. vii–ix. Boston: Allen & Unwin.

Harvey, D. (1982). *The Limits to Capital*. Oxford: Basil Blackwell.

Harvey, D. (1987). Three myths in search of a reality in urban studies. *Society and Space* 5 (4): 367–376.

Hewes, L. (1984). Carl Sauer: a personal view. *Journal of Geography* 82 (4): 140–147.

Kahn, J. (2004). Ronald Reagan launched political career using the Berkeley campus as a target. *UC-Berkeley News* June 8, accessed at https://www.berkeley.edu/news/media/releases/2004/06/08_reagan.shtml

Kingsley, G. T. and Fortuny, K. (2010). *Urban policy in the Carter Administration*. Washington, DC: Urban Institute

Kent Jnr, T. J. (1984). A history of the Department of City and Regional Planning (1948–79) part I. *Berkeley Planning Journal* 1 (1): 143–149.

Leighly, J. (1979). Drifting into geography in the thirties. *Annals of the Association of American Geographers* 69 (1): 4–9.

Lovering, J. (1989). The restructuring debate. In R. Peet and N. J. Thrift, eds, *New Models in Geography, Volume 1*. pp. 213–242. London: Unwin Hyman.

Markusen, A. (1985). *Profit Cycles, Oligopoly and Regional Development*. Cambridge, MA: MIT Press.

Markusen, A. R. (1987). *Regions: The Economics and Politics of Territory*. Lanham, MD: Rowman & Littlefield.

Markusen, A. R. (1994). Studying regions by studying firms. *Professional Geographer* 46 (4): 477–490.

Markusen, A. R. (2001a). Regions as loci of conflict and change: the contributions of Ben Harrison to regional economic development. *Economic Development Quarterly* 15 (4): 291–298.

Markusen, A. R. (2001b). The activist intellectual. *Antipode* 33 (1): 39–48.

Markusen, A. R. (2015). How real-world work, advocacy, and political economy strengthen planning research and practice. *Journal of the American Planning Association* 81 (2): 143–152.

Markusen, A. R., Hall, P. and Glasmeier, A. (1986). *High Tech America: The What, How, Where, and Why of the Sunrise Industries*. Boston: Allen & Unwin.

Markusen, A. R., Hall, P., Campbell, S. and Deitrick, S. (1991). *The Rise of the Gunbelt: The Military Remapping of Industrial America*. New York: Oxford University Press.

Markusen, A. R. and Schoenberger, E. (1983). The political economy of western regional development. In J. Kim, ed., *An Inquiry into Critical Perspectives in Planning*. pp. 78–128. Tempe, AZ: Arizona State University.

Markusen, A. R. and Weiss, M. (1984). Beyond shopping malls: planning for jobs and for people in Berkeley, California. *Berkeley Planning Journal* 1 (2): 5–23.

Markusen, A. R. and Wilmoth, D. (1982). The political economy of national urban policy in the USA: 1976–81. *Canadian Journal of Regional Science* 5 (1): 145–163.

Massey, D. (1984). *Spatial Divisions of Labour: Social Structures and the Geography of Production*. Basingstoke: Macmillan.

Massey, D. (1995). *Spatial Divisions of Labor: Social Structures and the Geography of Production*, 2nd edition. New York: Routledge

Massey, D. and Meegan, R. (1982). *The Anatomy of Job Loss: The How, Why and Where of Employment Decline*. London: Methuen.

Parsons, J. T. (1979). The later Sauer years. *Annals of the Association of American Geographers* 69 (1): 9–15.

Peck, J. (2002). Labor, zapped/growth, restored? Three moments of neoliberal restructuring in the American labor market. *Journal of Economic Geography* 2 (2): 179–220.

Peck, J. (2013). Making space for labour. In D. Featherstone and J. Painter, eds, *Spatial politics: Essays for Doreen Massey*. pp. 99–114. Oxford: Wiley-Blackwell.

Peck, J., Werner, M., Lave, R. and Christophers, B. (2018). Out of place: Doreen Massey, radical geographer. In M. Werner, J. Peck, R. Lave and B. Christophers, eds, *Doreen Massey: Critical Dialogues*. pp. 1–38. Newcastle upon Tyne: Agenda Publishing.

Pudup, M. B. (1992). Industrialization after (de)industrialization: a review essay. *Urban Geography* 13 (2): 187–200.

Saxenian, A. (1980). *Silicon chips and spatial structure: the industrial basis of urbanization in Santa Clara County, California*. Masters in City Planning thesis, University of California, Berkeley.

Saxenian, A. (1996). *Regional Advantage: Culture and Competition in Silicon Valley and Route 128*. Cambridge, MA: Harvard University Press.

Sayer, A. (1985). Industry and space: a sympathetic critique of radical research. *Society and Space* 3 (1): 3–29.

Sayer, A. (1987). Hard work and the alternatives. *Society and Space* 5 (4): 395–399.

Schoenberger, E. (1988). From Fordism to flexible accumulation: technology, competitive strategies, and international location. *Society and Space* 6 (3): 245–262.

Schoenberger, E. (1991). The corporate interview as a research method in economic geography. *Professional Geographer* 43 (2): 180–189.

Scott, A. J. (2000). Economic geography: the great half-century. *Cambridge Journal of Economics* 24 (4): 483–504.

Sheppard, E. (2011). Geographical political economy. *Journal of Economic Geography* 11 (2): 319–331.

Storper, M. (1987a). The new industrial geography, 1985–1986. *Urban Geography* 8 (6): 585–598.

Storper, M. (1987b). The post-Enlightenment challenge to Marxist urban studies. *Society and Space* 5 (4): 418–426.

Storper, M. and Scott, A. J. (1986). Overview: production, work, territory. In A. J. Scott and M. Storper, eds, *Production, Work, Territory: The Geographical Anatomy of Industrial Capitalism*. pp. 3–15. Boston: Allen and Unwin.

Storper, M. and Walker, R. (1989). *The Capitalist Imperative*. Oxford: Blackwell.

Teitz, M. B. (2016). The way it was. *Journal of the American Planning Association* 82 (1): 50–55.

Walker, R. A. (2018). Foreword. In A. Herod, ed., *Organizing The Landscape: Geographical Perspectives on Labor Unionism*. pp. xi–xvii. Minneapolis: University of Minnesota Press.

Walker, R. A. (2001). Bennett Harrison: a life worth living. *Antipode* 33 (1): 34–38.

Walker, R. A. (2018). *Pictures of a Gone City*. Oakland, CA: PM Press.

Watts, M. J. (1983). *Silent Violence: Food, Famine, and Peasantry in Northern Nigeria*. Los Angeles: University of California Press.

Webber, M. J. and Collignon, F. C. (1998). Ideas that drove DCRP. *Berkeley Planning Journal* 12 (1): 1–19.

Weiss, M. and Schoenberger, E. (1979). Carter throws peanuts to cities. *In These Times* February 28–March 6: 17.

Weiss, M. A. and Schoenberger, E. (2015). Peter Hall and the Western Urban and Regional Collective at the University of California, Berkeley. *Built Environment* 41 (1): 63–77.

# 8

# Radical Geography in the Midwest

Mickey Lauria, Bryan Higgins, Mark J. Bouman, Kent Mathewson, Trevor J. Barnes, and Eric Sheppard

The events that precipitated the formation of the Union of Socialist Geographers (USG) local branches within the U.S. Upper Midwest varied, ranging in geographical scale from the personal to the neighborhood, to the urban and regional, to the national and international. At the international, and especially the national scale, were the Anti-War, Civil Rights, Black Power, American Indian, and Women's movements (Barnes and Sheppard, this volume). The anti-Vietnam War movement was especially potent including such organizations as the Vietnam Veterans Against the War, Students for a Democratic Society, and the Student Nonviolent Coordinating Committee. Protests began almost as soon as the War began, further propelled by the introduction of the draft, especially when college deferments were ended, with a lottery-based draft instituted that affected more and more middle-class young men and their families. Of course, it wasn't only anti-War protests. Discontent was wide spread, cutting a wide swath across social issues and identities, and often conjoined with varied counter-cultural artistic expressions. Protest music was especially important, performed by artists such as Arlo Guthrie, Country Joe and the Fish, Crosby, Stills, and Nash, Pete Seeger, Tom Paxton, Joan Baez, and Joni Mitchell, among many others.

These movements, separately and through their intersections, radicalized a sizable portion of US youth. Their subsequent repression, whether

*Spatial Histories of Radical Geography: North America and Beyond*, First Edition.
Edited by Trevor J. Barnes and Eric Sheppard.
© 2019 John Wiley & Sons Ltd. Published 2019 by John Wiley & Sons Ltd.

internationally in places like Paris, or nationally in cities like Chicago and Washington D.C., produced visceral local experiences (whether it was the Kent State shootings, other brutal clearings of demonstrators on home campuses, or downtown marches) that drove many of us to search for a radical geography that would help change the world. Geography did not so much radicalize geographers; rather, radicalized individuals interested in geography worked to transform geography. We started in the streets, or behind barricades, but were ecstatic to discover the potential contribution of academic research by such radical geographers and social scientists as Jim Blaut, Bill Bunge, David Harvey, Dick Peet, Lisa Peattie, Janice Perlman, Peter Marcuse, Chester Hartman, Manuel Castells, Susan Fainstein, Ed Soja, Ann Markusen, Anthony Giddens, Richard Child Hill, and Erik Olin Wright, to name but a few. The combination of being on the street and in the university lecture hall produced praxis, leading to the creation of local branches of the USG in Midwestern universities, the focus of our chapter.

As the fullest expression of the USG was perhaps realized at Minnesota – including the hosting of conferences and the publication of newsletters – we begin the chapter there, considering the regional and departmental context before detailing the brief but intense engagement with the USG. We write from our perspective as participants in these events. We then discuss two other centers of radical geographic innovation in the Midwest, the Universities of Wisconsin and Michigan.

## Radical Geography at Minnesota: Origins and Development

Minnesota, and the Upper Midwest more generally, was known for its Scandinavian collectivist political culture, a result of historical patterns of Northern European immigration and settlement. Recognized for political activism, the region was the birthplace of the National Grange, the Populist Party, the Farmers' Alliance, and farm cooperatives (many of which later became corporate). These Progressive movements originated at the turn of the nineteenth century and sustained strong membership through the Great Depression. For example, in the early twentieth century, the City of Minneapolis elected a Socialist Mayor (Thomas Van Lear) who, along with his coalition of reformers (socialists and trade unionists), wielded political power in the mill city for over a decade (Nord 1976). Later, the Farmer Labor party, which supported farmer and labor union protection, social security legislation, and even government control of certain industries, dominated local and state politics until Hubert Humphrey merged it with the

Democratic party in 1944 to form the Democratic Farmer Labor party (DFL). Minneapolis was also the origin of the American Indian Movement (AIM), in 1968. By 1970 Minneapolis had the fifth largest population of urban Indians within the United States, becoming an organizational center and a site for radical action by American Indians (Higgins 1982).

During the early 1970s, the University of Minnesota Geography Department expanded in size but remained politically conservative. The department was strongly identified with cultural and historical geography, with little interest in a critical perspective. Some members of the Department were tied to local, regional, and statewide planning initiatives but formulated within a rationalist and technocratic paradigm. In the wake of the 1968 urban crisis, the University created the Center for Urban and Regional Affairs (CURA). CURA functioned as a research and technical support center, underwriting selected projects that engaged local communities.

Some individual faculty members were open to alternative, even radical perspectives. Fred Lukermann, for example, was well-known for his sharp criticism of the ahistorical, apolitical character of rationalist quantitative geography (Barnes 1996). As Vice President of the University, he supported student anti-War demonstrations, and later, as Dean of the College of Liberal Arts, advocated for alternative community development action research in Minneapolis. Phil Porter's anti-colonial development work in Africa, and Yi-Fu Tuan's alternative "humanistic geography" also ran against the prevailing spatial scientific current. In the mid-1970s, the addition of junior faculty produced stronger support for alternative/radical approaches in geography, for example, Bonnie Barton's polyvocal approach to regional economic geography, Phil Gersmehl's environmentalism, Eric Sheppard's and Tony Lea's alternative form of spatial science, and Helga Leitner's political and social concerns for European migrant labor. Faculty in other social sciences and humanities at the University added reinforcement, including: Luther Gerlach, Stephen Gudeman, and Mischa Penn in Anthropology, Joe Galaskiewicz in Sociology, Gerald Vizenor in American Indian Studies, David Noble and Mary Jo Maynes in History, and Martin Krieger in Philosophy.

The specific experiences, backgrounds, interests, activist pasts, and research practices of the individuals who eventually founded the Minnesota USG local varied greatly. Some were from military backgrounds, others from a disrupted working class, yet others from the environmental movement; all had identity-molding, coming-of-age experiences in the 1960s and early 1970s. The crystalizing point for the Minnesota USG local was in 1976, at the national meeting of the USG, a pre-/concurrent conference alongside the 1976 Association of American Geographers (AAG) meetings in New York City.[1] The Minnesotan geographers not only met established radical geographers but also interested graduate students from all over the U.S. and Canada, becoming part of the community of radical geography scholars. They slept on the floors of New York City graduate student and community activists' apartments. They toured squatted buildings, artist lofts, and gentrifying neighborhoods. They drank, talked, and shared experiences and ideas throughout the night. It was a heady experience. This was the first of a number of national USG meetings that Midwestern geographers attended in the U.S. and Canada (Vancouver, Toronto, and Montreal), meetings that always included similarly intellectually invigorating and locally relevant real-world experiences. The community rapport and support also encouraged the organization of regional Midwest USG meetings (Minneapolis, Madison, Chicago, Valparaiso, and Iowa City).

The Geography graduate curriculum at Minnesota was relatively flexible. It allowed us to read the works of radical scholars in other disciplines as well as to create a formal reading group. The first such group was the Anarchist Seminar (discussed in the *USG Newsletter* 3 (3)). Lauria had been active in a Minneapolis social anarchist collective that published its own magazine (loosely speaking), *The Soil of Liberty,* closely aligned with the extensive cooperative movement established in Minneapolis. Lauria became disenchanted after researching the impact of local cooperatives on low-income neighborhoods in Minneapolis as part of his Master's degree. This led to reading group seminars that focused on alternative theories of social change stretching from anarchism to Marxism.

Research at the local scale, a central anarchist priority, resonated with the sense-of-place literature then popular within cultural geography, and pursued by Yi-Fu Tuan (1974, 1979) in *Topophilia* and *Landscapes of Fear* (1979) ("topophobia"). It also resonated with the University of Minnesota Geography counter-culture formalized as The Phillips Neighborhood Geographical Society (PNGS), along with its pink flamingo symbol (Figure 8.1). The PNGS was a place-based venture in several ways: several members (although not all) lived and were active in pursuing community development in Phillips (a low-income American Indian and African-American community in south Minneapolis); two of its members carried out dissertation research on the neighborhood; and the PNGS held neighborhood meetings (often becoming parties!) and field excursions (geographical expeditions). The Viking Bar became PNGS's unofficial headquarters, serving also as an alternative venue to the Department of Geography's official Brown Day celebrations (held annually in May but boycotted by the radical graduate students).

**Figure 8.1**  Phillips Neighborhood Geographical Society softball team, the Pink Flamingos, outstanding on one leg (flamingo style). Back row from left to right: Paul Meartz, Mike McAllister Janeen McAllister, Jack Flynn, Lorinda Anderson, Bryan Higgins, Marge Rasmussen, Jim Hathaway, Barb Shipka, Mike Mueller, and Rob Warwick. Front row from left to right: Katy Kvale and unknown. University of Minnesota East Bank Athletic Fields, Minneapolis, summer, 1975. Photo: © Mike Albert. Reproduced with permission.

The local community activist concern with Phillips was eventually taken up by CURA. Bryan Higgins and Fred Lukermann, then Dean of the College of Liberal Arts, requested that CURA establish a research position for community development and neighborhood planning in Philips Neighborhood. It obliged. This supported two members of the Minnesota USG local, in positions that continue today.

The Minnesota Geography Department also had an active guest speaker colloquium (Coffee Hour) on Friday afternoons that often triggered further discussions in nearby bars. The department brought in many national and international scholars as guest speakers and long-term guests, like Torsten Hägerstrand and Ted Relph who alerted the group to new intellectual currents in geography. Minnesota faculty also were responsive to USG local member requests to bring in scholars with alternative perspectives: Gunnar Olsson (1980) when his *Birds in Egg: Eggs in Bird* was published, Jim Blaut (1993) during the development of his seminal work, *The Colonizer's Model of the World: Geographical Diffusionism and Eurocentric History*); or Bill Bunge (1971) and his accounts of his revolutionary geographical expeditions in *Fitzgerald: Geography of a Revolution* and *The Canadian Alternative: Survival, Expeditions, and Urban Change* (Bunge and Bordessa 1975).

## USG Praxis: Phillips Neighborhood

Three members of the Minnesota USG focused on radical geography as participant observation and action research within Phillips Neighborhood. In the field, the word "socialist" was rarely used, but the conceptual framework was structuralist, critical of both conservative and liberal approaches. One of the formulations was termed the "compost theory of development" to explain how and why private and state interests allowed low-income neighborhoods to continue to deteriorate in order to secure more federal funding that was then used to improve middle-class areas in the city, not the so-called "poverty districts." This work focused on building community-based organizations, and offering planning that served current residents rather than the "back-to-the-city" urban pioneers and subsequent gentrifiers. One of the early victories was to transform a city-recognized home-owner association into a more inclusive neighborhood improvement association that recognized participation of renters, giving them voting rights.

Existing city agencies were viewed as problematic, manipulating the system to serve middle-class interests. There was criticism of how the City Housing Authority and Planning Office policies and programs supported the middle-class, resulting in displacement of existing residents. Both agencies were viewed as instruments for extracting wealth from low-income neighborhoods like Phillips Neighborhood. In response, Phillips community members argued for independent planning in support of current residents, *not* outside gentrifiers. Time was spent writing grants, raising money, and working to successfully organize an alternative community-based organization within Phillips. A neighborhood comprehensive plan was written, very different from the city-prescribed one (Higgins 1980a). Initially the Minneapolis City Council offered to fund a neighborhood plan for Phillips staffed by the City Planning Department. The Phillips' Neighborhood Association opposed this effort, viewing City Planning as part of the problem. Instead, the neighborhood developed a planning process that placed committees of neighborhood residents and the community-based organization in charge. That combination prevented the city from implementing their Phillips' land-use plan, based on demolishing existing housing and allowing the construction of high-density condominiums around the new Minnesota Twins baseball stadium in nearby Elliot Park. Rather, the current low and medium density in the official land-use plan was preserved: no small political undertaking. The housing authority also was persuaded to shift from focusing on primarily single-family home-ownership (serving mostly whites with middle-class backgrounds) to cooperative housing and other forms of social housing, such as Section 8, that better served the needs of current residents. There was a battle with the school district, that siphoned off federal money received for "poverty education" to enhance middle-class schools., forcing it to spend more of this money in Phillips and to establish a "parent advocate" position for Phillips' parents (Higgins 1978b).

The city's citizen participation framework was forced to reorganize. That system had embedded Phillips within a citizen participation region dominated by middle-class interests. The reorganized city-wide partici-pation system designated Phillips as its own district, giving Phillips an improved voice with a more direct say in Minneapolis city policies and plans. After this change, city agencies were frequently called out and opposed when they presented their plans to the Phillips' council, which often involved promoting gentrification. Indeed, Phillips gained a repu-tation among city staff as "Little Vietnam," where grandiose plans and liberal policies would be attacked and resisted (Laurie Louder, personal communication).

We also used an "Indian self-determination" framework to leverage support for American Indian organizations, seeking to derail the "white savior" complex whereby whites told Indians what was good for them. Rather, Indian organizations needed to serve the social needs of Indian people. This included electing many American Indian members to the Phillips Neighborhood council, but also supporting Little Earth of the United Tribes, a large public housing complex in Phillips managed by Indian residents (Higgins 1982). It was also crucial to integrate Indian social projects within the broader Phillips Neighborhood community (Higgins 1978a; Higgins and Lauria 1986).

The Minnesota local did not use the term "geographical expedition" in their activist research or community process. While they appreci-ated Bill Bunge's geographical expeditions, that term carried much academic and political baggage, harking back to the role of geogra-phers within the European colonization of the Americas. The large American Indian population in Phillips, victims of white settler colonialism, were very aware of the effects of European geographical expeditions. There also was an awareness of the consequences of white settler colonialism in other regions, such as the geographical revolutions and revolutionary geographies of Nicaragua (Higgins 1990a), and more broadly a realization of the need to decolonize geographical knowledge in general (Radcliff 2017). David Hugill (2016) recently provided a searing analysis of the depth and ongoing power of settler colonialism, highlighting its contemporary urban role within the "machinery of enforcement."

Another part of the Phillips resistance was building alternative community organizations. Creating new organizations and assuming leadership positions almost became a way of life. Members of the USG local served on: the neighborhood improvement association's board of directors, chairing its committees (planning, housing and land use, and social services); the area community development corporation board of directors; the community federal credit union; and as community development land trust staff. These activities, along with literatures we were reading, led Lauria to propose an alternative community development model to the classical economic urban redevelopment version. This was used to examine community transformations brought about by exogenous influences (Lauria 1982); by land rent, examined through a Marxist theoretical lens (Lauria 1984); and by endogenous development conceived in terms of organization theory (Lauria 1986). At the same time, Higgins (1982) proposed a culturally based model of community development drawing on his analysis of the experiences of Phillips' American Indians.

## The Reading Group

The Minnesota USG local praxis was a combination of direct action but also serious theoretical exploration. As mentioned, the reading group started with Anarchist theory but quickly moved on to Marxian and other alternative social theoretical perspectives. It began with the three volumes of *Capital,* later turning to literature in neo-Marxist philosophy, the humanities, the social sciences, and human geography. While firmly rooted in the "critical" tradition, the group's tastes and concerns favored political economy. But there was also an avid interest in exploring the "cultural turn," along with an openness to alternate modes of geographical thinking and writing. Likely by blind luck, the group explored several issues that later became prominent in geographical discussions such as the "new cultural geography," the scale debate, and the critique of *homo economicus.*

After reading *Capital,* the group worked through Martin Jay's *The Dialectical Imagination* (1973) and the Frankfurt School extracts from Adorno, Horkheimer, Offe, Habermas, and others gathered in Paul Connerton's *Critical Sociology* (1976). The next work was Alex Callinicos's *Althusser's Marxism* (1978). In Fall, 1979, the group took up Derek Gregory's recently published *Ideology, Science, and Human Geography* (1978). There followed other explorations both in literature and about modernism that bore on critical cultural geography, including George Steiner's *In Bluebeard's Castle* (1971) and Walter Benjamin's (1986) essays on Moscow, Berlin, and Paris in his collection, *Illuminations.* The essay form worked well and the next year the reading included Raymond Williams's *The Country and the City* (1973), and Thomas Pynchon's *The Crying of Lot 49* (1965) and *Gravity's Rainbow* (1972). By summer 1981 and into the following year, the group was back to Horkheimer (1968) and Adorno, and then Gunnar Olsson's (1980) *Birds in Egg/Eggs in Bird* and Peter Gould and Olsson's (1980) *A Search for Common Ground.* The reading group finally dissolved in the mid-1990s.

Weekly reading group meetings were first held in Eric Sheppard's office, but soon migrated on a rotating basis to members' homes, in Phillips, Seward, Nordeast, West Seventh (St. Paul), and the Midway. Each week a different person would prepare a 1–2-page write-up that group members would scramble to read prior to the meeting (along with the reading itself). Consistent themes in the discussions were around volunteerism versus coercion by the state, economic value theory, false consciousness, structure and agency, institutional change, the local state, the viability of local socialism within a capitalist world economy, theoretically precise language versus common vernacular, linguistic hegemony, liberal versus radical reforms, non-reformist reforms, decommodification, and cultural practice, identity, and representation.

While the members of the reading group ebbed and flowed with individuals participating sporadically (initially also including Bill Pisarra, Rob Britton, and Jim Hathaway), the early (1977–1980) core/consistent participants were Trevor Barnes (UBC), Mark Bouman (first Chicago State University, now at the Chicago Field Museum), Philippe Cohen (Stanford, Jasper Ridge Biological Preserve), Michael Curry (Emeritus, UCLA), Mickey Lauria (Clemson), Helga Leitner (Vienna, Minnesota, now at UCLA), and Eric Sheppard (Minnesota, now at UCLA). In 1980–1982, the group added Patrick McGreevy (American University of Beirut), Roger Miller (Minnesota, deceased), April Veness (Delaware), and Diane Whalley (formerly at North Texas State). John Pickles (now at UNC) joined the group while he was a visitor in 1983, when he substituted for Yi-Fu Tuan who had departed for the University of Wisconsin, Madison.

Many of the group felt that a lot of our graduate education derived from discussions within the reading group. It opened new fields that many never knew existed, as well as novel vocabularies and theoretical frameworks. At first those vocabularies and frameworks drew directly from political economy, but over time they became much more eclectic and pluralist. This was also one strand of the wider Minnesota's Department of Geography's intellectual tradition, represented during our period by the rich, diverse, opened-minded scholarship of professors like Fred Lukermann, Phil Porter, and Yi-Fu Tuan, who variously supervised, or sat on students' committees, and from whom courses were taken (Barnes 1996).

The readings always seemed difficult, even when, especially when, the reading was novels like Pynchon's *Gravity's Rainbow*. It often seemed as if the group was thrown into the deep end, madly splashing around, trying to keep heads above water. Sometimes you sunk. Other times you were able to grab on to something, a concept, an idea, a word that enabled you at least to keep floating, despite the unnamed flotsam and jetsam all around you. On the rare occasion individuals found themselves in open clear water, swimming along with ease, wondering why others were not doing the same.

There was undoubtedly some competitiveness among members, especially among the men. Discussion was vigorous, occasionally with raised voices, or cutting irony. For the most part it was a free-for-all. People jumped in when they had something to say, with sometimes the loudest voice winning out. There were sometimes hurt feelings, occasional resentment and jealousy, but very few stopped attending. It was too exciting, and you never knew what you might miss. It was also often fun. It was a community, your friends, and there was as much social gossip as academic talk. And because meetings happened in peoples' homes you got to know a different side of a person. It was what graduate student life should be like.

## Regional USG Meetings

Participants from the Midwest attended various international and national meetings of the USG, the Canadian Association of Geographers (CAG), and AAG. In addition, regional USG meetings were held in Minneapolis (Fall, 1977), Madison (Spring, 1979), Valparaiso (Fall, 1980), and Iowa City (Fall, 1981). The regional meetings allowed local collectives to bring forward local instances of critical praxis and to introduce locally based actors to the broader region. They also provided a means to connect directly with geographers outside the major research universities, working at places such as University of Wisconsin-Eau Claire, Cornell College, and Valparaiso University. And they provided opportunities for some in geography's old guard to check out what was happening in the USG world: for example, Fred Lukermann attended the Minneapolis meetings, and Richard Hartshorne the meeting in Madison.

At the first Midwest regional USG meeting in Minneapolis, it was decided that the next Midwest regional USG meeting would be in Madison. Kent Mathewson got things started but then left Madison for most of the next year while he was in Mexico and Berkeley. Historical geography graduate students Mark Garner and James Penn took over planning the meeting. Garner's invitation letter neatly captured the meeting's context and value: "With the growing interest in radical perspectives in Geography, and, not coincidentally, the deepening global crises that confront humanity, we know that there will be both new questions to address and new personal contacts to establish."

The two-day meeting, held in the Wisconsin Union, drew a good number of participants, including non-USG members such as Hartshorne, David Ward, and Bob Sack. The headliner was David Harvey, who, as noted by James Penn in his account of the talk in Volume 5 of the *USG Newsletter*, was wearing a puce velour shirt as he walked through the three circuits of capital in a packed Department of Geography colloquium in Science Hall. For some, it was the first time that they had met Harvey: both exciting and intimidating. It was also the first time that some presented their work at a professional meeting. To receive an encouraging nod afterwards from David Harvey left the presenter in an intoxicated state.

Volume 5 of the *USG Newsletter* provides an excellent sense of those meetings. It contained the first published instance of Barnes's working through his arguments about central place theory after Sraffa, Bouman's efforts to grapple with Scottish nationalism and the model of internal colonialism, Curry's analysis of human rights and public goods, Higgins' contrast between Western and Native American philosophies of land, Lauria's report on a workshop centered on O'Connor's *The Fiscal Crisis of the State,* and Sheppard's critical summary of a workshop on "rethinking space." In addition to these Minnesotans, other places represented during the Madison USG local meeting included Wisconsin (Penn, Garner, Mark Wiljanen, Yda Saueressig-Schreuder, Mathewson), Valparaiso (Hansis), and Minneapolis (powerline activist Don Olson). Garner's introduction to the newsletter also noted the contributions of Bill Bunge, John Chappell, and Al Gedicks.

The third Midwestern meeting, held in Valparaiso in October, 1980, was hosted by Dick Hansis of Valparaiso (who had taught and supervised Mark Bouman as an undergraduate), and organized by James Penn and Ben Wisner (then at Wisconsin). The meeting had five paper sessions: Theory and Method in Political Economy (Wisner and Sheppard); Culture and Class Struggle in the Nineteenth Century (Penn and Garner); Politics and Class in a Changing World (Blaut, Danny Weiner, and Wisner); Origins of Capitalism in the US (Penn and Garner); and Urban Geography (Lauria, then at Northern Illinois University, John Campbell, and Hansis). The program noted that "due to the generosity of local residents and friends, some sleeping space is available." It concluded with a "Meeting of the Society of Red Bird Watchers" organized by Blaut.

The final Midwest USG meeting was held in Iowa City in 1981. Lauria, then a professor at the University of Iowa, hosted with members of the Geography (Dave Reynolds, Fred Shelley, Rex Honey, Mike McNulty, and Diane Whalley) and Planning programs (Jim Harris, Peter Fisher, and Michael Sheehan), especially those deeply involved in studying the deindustrialization of Iowa's meatpacking and farm implement industries. Attendees also included Blaut, Jim Lyons (from Clark, but then at Valparaiso), Penn and Garner from Wisconsin, Bouman, Barnes, and Rob Warwick from Minnesota (driving Sheppard's car), Anthony De Souza from Eau Claire, Fred Blum from Chicago State University, and a surprise visit by Pat Burnett. Participants Lauria and Bouman recall a "good meeting," which included hearing Greg Brown play at the Mill, a local watering hole and café. A pleasant haze has settled over more specific memories.

## Producing the *USG Newsletter*

By 1979, it was clear that Vancouver local's capacity to produce the *USG Newsletter* had run its course. It was decided to rotate the editorship of each issue, while keeping newsletter production in the Midwest. Eric Sheppard was able to secure limited departmental funds for assistance from the University Computer Center: The USG membership list was digitized, enabling computer-generated mailing labels to attach to the Xeroxed newsletter. Bryan Higgins was the contact person for membership renewals; Bouman and Barnes managed the mailing list, copied and stapled the newsletter, and licked the labels and envelopes before being franked and posted.

There was enough editorial material to fill up a newsletter, and the will to get the task done. The first issue of Volume 5 (1979) was edited by Garner, one of the key organizers of their regional meeting. In addition to reports from that meeting, the newsletter contained business items, such as the plea from Sheppard for 100 signatures to create a Socialist

Geography Specialty Group – the ultimate success of which ironically played into the demise of the USG.

In the next issue, as the Minnesota local noted in a "reproductive comment," after waiting for a long time for content from England, the materials that arrived were twice the amount expected. They blessed it as two issues and moved on. Noting that the intention was not to overburden individual locals with editorial responsibilities, the Minnesota local also put out a call for general newsletter submissions. Pointedly they noted that only 35% of USG members had paid the $6 dues.

There was also a lighter-side to the Minnesota local. The brainchild of Bryan Higgins, it was proposed to devote a special issue of the USG newsletter on "praxis and humor." Higgins had earlier contributed satirical cartoons to the newsletters that leavened their tone of high seriousness. He envisioned topics including "descriptions of contemporary approaches to community development, issues of union organizing, multi-media approaches and the range of perspectives, role playing and practical classroom exercises, and people's history and herstory." He also hoped to map "the more unusual contours of humor." As Higgins (1980b) wrote:

> For this spatial analysis we also ask that you forward any cartoons, slips of the tongue, jokes, riddles, satire, practical jokes, stories, lies, or imagined encounters which have never before been offered as possible geographical explanations. For the more exclusively serious scholars we note that all humorous items will be coded and retained for the long-hoped-for-future international data bank of humor.

In the next issue, the Minnesota Collective (1980) noted,

> so far the response [to Higgins's call] has been nil. Does this mean that there are no USG members out there in the real world who want to start a discussion of issues that interest them? Or none that have a sense of humor?

The group is still waiting...

That was also a double issue (with two different colors to cleverly indicate the end of Volume 5 and the beginning of Volume 6). Again, the Minnesota Collective wrote an urgent plea for funds, noting that still only 28% of the USG membership had paid up through the prior year, and just 8% in the current year. It was suggested that the Newsletter might not be able to be distributed to the 72% who had not paid. This was the last issue produced by the Minnesota local. As mentioned by Peake (this volume), this lack of a stable source of funding ultimately contributed to the dissolution of USG.

## The Madison Local

While Minnesota was a key center of USG activities in the Midwest, other places and individuals were also involved. One strong hub of activity was in Madison. Madison was one of the "cultural hearths" of the New Left in the U.S. Though not unique in serving as a college town incubator for progressive and left radical politics, it had a distinctive history that set it apart from the coastal centers like Berkeley or New York City. With its prairie populist and progressivist legacies and Milwaukee with its German trade union/democratic socialist traditions, Wisconsin helped foster UW-Madison's emergence as a university with a tolerance for left-wing academics in the post-war period (despite the fact that Republican Senator Joseph McCarthy represented the state between 1947–1957, spearheading a witch-hunt against the American left during the mid-1950s). Madison's geographic site and situation played into its development as a center of both academic and activist leftism. It was situated in the penumbra of East Coast culture and conceit, just far enough off away to be slightly exotic, but close enough to remain within its influence. With bucolic surroundings buffering this state-capital city, one of its attractions was the lake-front campus with its replica German Rathskeller in the Student Union, and a dozen film societies showing "foreign" and avant-garde movies nightly.

Leftists could be found in a number of University of Wisconsin departments from the 1950s, although its History Department was the mainspring. Progressive Era historiography took hold there during the 1920s and was never uprooted. In the early 1960s, William Appleman Williams turned US diplomatic history on its head, showing the imperialist roots of US foreign policy. He had a large student following. Students in the department founded the journal *Studies on the Left*. Department alumni include a long list of prominent New Left historians such as Gabriel Kolko, Herbert Gutman, and Martin Sklar. Another departmental alum, Harvey Goldberg, a Marxist specialist of French socialism, arrived in the mid-1960s and became a renowned lecturer, radicalizing undergrads who enrolled in his "must take" introductory history courses. Several events in the late 1960 and early 1970s made national headline news. In the fall of 1967 hundreds of radical students blocked recruiters for Dow Chemical, manufacturer of napalm. Police were called, a "riot" ensued, and dozens were arrested. The Dow demonstrations served as a praxis tutorial for student protesting the Vietnam War around the world. From this time on, until the summer of 1970, the campus was scene of frequent protests, campus closures, and general resistance to this war. In August 1970, the protests culminated in the bombing of Sterling Hall, site of an army-funded mathematics project. A post-doctoral researcher was killed, and the building destroyed. It also contributed to undermining the radical movement and its momentum in Madison (and to a lesser extent elsewhere). The group behind the bombing, an affinity faction of four, were only loosely affiliated with established anti-War organizations on campus. In some ways, this typified the scene in Madison; the usual sectarian groups were in evidence, but the hard-core competition among them that characterized some other centers was less in evidence.

By 1973, when Kent Mathewson arrived to study with Latin Americanist/cultural ecologist Bill Denevan, much of the radical intensity of the previous decade had dissipated. There was little, or no radical, activity in the Geography Department at that time. Mathewson had been active at Antioch College since 1967, in 1968 in Students for a Democratic Society (SDS) (he held a co-op job in New York City with SDS, was in NACLA, and helped start *The Rat*, an underground newspaper[2]), and in 1969 was a founding member of the Radical Studies Institute at Antioch College. After SDS exploded/imploded he joined the IWW, Educational IU 620 at Antioch and continued some Wobbly activity once in Madison. (His suggestions that both the geography TAs and the faculty were workers, who should join the IWW, were met with incredulity, at least by the faculty).

The radical contingent in geography at Madison developed because of a visit by Bill Bunge in 1976. Mathewson had known him from earlier connections at Antioch, and he showed up in Madison demanding to give a talk in the department. Bob Sack was chair of the speakers committee and was concerned about the department hosting his talk. Kent suggested an administratively distancing compromise by resurrecting the moribund Geography Club (dissolved in the late 1960s, as rumor had it, by tear gas on the outside of the building, and cannabis smoke on the inside), and having it host Bunge's talk. David Ward and Bob Sack thought this was a good plan but it was suggested that the Club needed a new name, after a suitably radical geographer. Ward proposed Kropotkin, but Mathewson suggested that, following the "greater obscurity principle," Elisée Reclus should receive the honor. The "Elisée Reclus Geography Club" was launched, and Bunge gave, as always, a memorable performance. Greg Knapp and Mathewson began to publish a mimeographed newsletter of the club, somewhat modeled after Yi-Fu Tuan's "Dear Colleague" letters. They called it *"The Pedestrian"* after Reclus' famous advocacy of militant pedestrianism. The reactivated and repurposed Elisée Reclus Geography Club, Bunge's talk, and *The Pedestrian* newsletter rekindled an interest in a radical tradition in geography in Madison.

Ken Olwig came the next year 1977–1978 (a newly minted Minnesota Geographer) as a visiting faculty member. Olwig was aware of the USG meeting to be held in Minneapolis in the fall of 1977. He encouraged several of the newly energized radical geographers to attend. As was the tradition, the Wisconsin radical geographers were provided accommodations with many of the Minnesota local students. This connection between the Minnesota and Wisconsin USG contingents was further solidified at the AAG West Lakes meetings. By 1978–1979 there was a core of Wisconsin geography students who identified with radical geography to greater or lesser degrees, and central in organizing the Madison Midwest USG meeting. From 1979 through the early 1980s there was a lot of student pressure to hire a radical geographer. In 1980 Neil Smith was interviewed, but he took a job instead at Columbia University, NYC.

The Madison USG local's activist activities circulated around anti-war protests, environmental issues and graduate assistant labor issues. They were instrumental in the 1980 graduate assistant strike at UW Madison. In the early 1980s, new students and faculty interested in radical geography arrived in the geography department. In particular, Michael Johns from the University of Massachusetts (later a David Harvey student at Hopkins), and Michael Solot from UCLA. Richard Mahon from Goddard College also arrived with well-developed political interests, as did Tom Klak from Augustana College, developing them further in Madison. Ben Wisner was also hired (largely through Bill Denevan's efforts), but stayed only about a year. The sole geography reading group at Madison was one that Mickey Lauria tried to get going, driving up from Northern Illinois University and later the University of Iowa to initiate it. Ben Wisner and Yi-Fu Tuan were the faculty involved. The group began by reading André Gunder Frank and the dependency theorists, but it fizzled after a few meetings. As newer epistemological currents drifted into the department – postmodernism, poststructuralism, postcolonialism – 1960s and 1970s geographical radicalism faded. In retrospect, the high-points of the Madison Geography Department's engagements with radical theory and practice coincided with the hinge years of the USG local – 1978 through 1980 or 1981.

## The University of Michigan

Clark Atakiff says that "Michigan was ground zero" for radical geography.[3] Bunge and Warren did their pioneering work in Detroit during the 1960s (Warren, Katz and Heynen, this volume), and it was at the

1969 annual meeting of the AAG in Ann Arbor, MI, that as Akatiff put it, North American radical geography first "jelled ... Wisner from Clark, Eichenbaum from Michigan, Bunge from Detroit, the Michigan State Crowd [that included Ron Horvath and Akatiff], Blaut, and Stea"[4] (Barnes and Sheppard, this volume). It was also at the University of Michigan that a different kind of radical geography was forged around the work of Gunnar Olsson and his students (Olsson was a founding member of the USG, and by 1975 had published four papers in *Antipode* including a special issue that comprised papers by him and his students; Barnes and Sheppard, this volume; Peake, this volume).

Olsson arrived at Michigan in 1966, having been a student at the University of Uppsala, Sweden, where early on he was introduced to the regression equation. This became for him a life-long obsession: with the regression equation "a new world was opened up, a world which in a sense I have never left; whenever I don't understand what is going on, I automatically reformulate the problem into a regression model" (Olsson in Barnes 2012: 251). That obsession was further reinforced by sharing an office in 1960–1961 with David Harvey (still a positivist; Sheppard and Barnes, this volume), who was a visiting postgraduate student, and then taking a seminar the following year, 1961–1962, with the regional science visiting professor, Julian Wolpert (Olsson in Barnes 2012: 252). Wolpert helped Olsson secure a fellowship to visit the Department of Regional Science at the University of Pennsylvania the following academic year, where he wrote a much-cited short monograph on distance and human interaction, featuring both the regression equation and the gravity model (Olsson 1965). In 1965, completing his Ph.D. at Uppsala, Olsson went on the job market, receiving multiple offers. He chose the University of Michigan.

The university was a condensation point for anti-Vietnam War protests from the beginning. It was the birthplace of Students for a Democratic Society (1960), of the 1962 origin of the Port Huron Statement (the manifesto of that society), and where the idea of "teach-ins" about the War was first hatched – the anthropologist Marshall Sahlins gave the first in 1965. In this sense, as Watts (2012: 143) puts it, "Ann Arbor had established its credentials as a centre of dissent." It was also where the worm turned for Olsson. As Watts (2012: 143), one of Olsson's doctoral supervisees, continues:

> Olsson's [subsequent] decade-long sojourn in the US became the setting – and the Michigan campus the ether – for what one might call his "epistemological break." He came as a distinguished member of a foundational group who ushered in the so-called quantitative revolution ... [but] a decade later,

the person who scrambled aboard the *Queen Elizabeth II* in New York …
[to return] to Sweden … was hardly the same geographer who arrived on
the Michigan campus armed with the gravity model and a slide rule.

What had changed, driving this epistemological break, was Olsson's
recognition of what he called the problem of geographical inference. As
he put it in an interview, the problem stemmed from determining "what
a description of spatial form can reveal about the social processes that
have generated it. In my mind, it is this problem … that eventually blew
the discipline apart, a torpedo hitting the ship of Geography below the
water line" (Olsson in Barnes 2012: 253–254). Even if you have the
most perfect description of geographical form, a spatial pattern, Olsson
believed it was of no help in understanding what might have caused it.
For Olsson, the realization that geographical form is overdetermined
produced "fireworks exploding, lightning bolts out of control" (Olsson
in Barnes 2012: 253–254).

Those pyrotechnics sometimes happened at his house, which functioned as a kind of intellectual salon for professors and graduate students, also ignited by good food and wine. These largely male events
were not always confined to Ann Arbor: On occasion the Michigan
group met with Reg Golledge and students from Ohio State, and with
Bernard Marchand and students from the University of Toronto (Watts,
Sheppard, personal communication).

From 1968, Olsson's annual graduate seminar in the Geography
Department, Geography 810, Thought and Action, was also a venue for
radical thinking.[5] All the doctoral students he supervised took that seminar, from Stephen Gale (his first) to Michael Watts (his last). Adrian
Pollock, one of those students (1973–1978), wrote that Geography 810
"stayed with me as a guiding light in my life, although I didn't realize it
until I was older and wiser…. [I]t was the most exciting 5 years of my
intellectual life."[6] It was also in the seminar that Olsson introduced and
worked out many of the ideas that later went into his University of
Michigan monograph, *Birds in Egg* (Olsson 1975), from fuzzy set theory
to opaque Wittgensteinian aphorisms ("If a lion could talk, we could not
understand him") to Beckett-like words of existential bleakness.[7]

There were no Marxists as such on the Michigan Geography faculty.
But Bernard Nietschmann came close, and Marx and Hegel were staples
of Olsson's seminars, key sources in *Birds in Egg*. Through Olsson's
graduate students, there also were strong links to Anthropology that
included several left-wing faculty: Sahlins, but also Michael Taussig and
Eric Wolf. Most of Olsson's graduate students were not formally
Marxists, but considered themselves radical. They were engaged in a
critical project, to denaturalize and to demystify, to uncover structures of

social power, and sometimes to advocate, sometimes to participate in, progressive social change, "to see a better society" as Eichenbaum (1972: vii), expressed it in the Preface to his doctoral thesis about the failure of urban renewal in Detroit's inner-city neighborhood of Corktown. It was a departmental requirement that PhD topics should always be empirical. Some of the theses were contemporary and obviously socially relevant, like Eichenbaum's on Detroit or Watts' on food and famine in Nigeria, whereas others were historical and seemingly less directly socially relevant, such as Peter Hoag's on the post-War state sponsorship of Eskimo arts in Alaska and Canada, Bonnie Barton's on early New England settlement, or Adrian Pollock's on witchcraft in seventeenth century England. In each case, there was an attempt to shift the course of geographical research, to make it radical. That goal is illustrated in Eichenbaum's (1972: x;) disciplinary map drawn at the end of his dissertation Preface (Figure 8.2). He urges a movement from the heartland of the "ESTABLISHMENT," contiguous with the geographical territories of "POLICE STATE," "CENTRAL INTELLIGENCE," "GREATER SELL," "Banal Bay," and "Trivia Point," to the off-shore island idyll of "ANTIPODE."

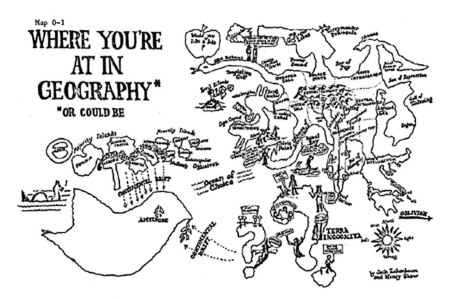

**Figure 8.2** Where you're at in geography*.
*Or could be. Drawn by Jack Eichenbaum (1972). Reproduced with permission.

## Conclusion: The After-Party

Although those forming the Midwest cluster of radical geography had largely graduated and moved on by the mid-1980s, the activities they pursued resonated down the years. Two aspects are worthy of note: How individuals recreated the close mix of community activism and engagement that accompanied their radical scholarship in the places where they moved, and their influence on the USG and *Antipode*.

Examples from key members of continuing activism in new local communities were Higgins who, living in Vermont investigated the praxis of radical change, working for two different government agencies when the Sandinistas were in power (Higgins 1990a, 1990b, 1994) and for Bernie Sanders' mayoral administration in the People's Republic of Burlington, Vermont (Higgins 1986). He also developed community-based plans for the Akwesasne Mohawk Tribe located near his academic institution (SUNY Plattsburgh) to guide the regional growth of that native community (Higgins 1991a, 1991b), and served as the Director of International Education developing new study abroad programs with a radical and political ecological focus. Lauria provided legislative reports to support communities facing the musical chairs of meat packing plant closures in Iowa (part of a business union busting strategy) (Lauria and Fisher 1983; Lauria 1983a, 1983b), later working with low-income minority areas in New Orleans that were struggling with institutional racism and racialization. He helped bring community-controlled planning into the lower 9th ward and other neighborhoods (Lauria *et al.* 1995, 1996; Cook and Lauria 1995; Foley and Lauria 2000, 2003).

As faculty member at the predominantly African-American Chicago State University in the deindustrializing Calumet region, Bouman deployed his Minnesota USG experience in building a community development program for the area. With Fred Blum, a longtime advocacy planner and activist who attended several USG and AAG Socialist Geography Specialty Group meetings, he founded what is now called the Fredrick Blum Neighborhood Assistance Center and Calumet Environmental Resource Center (Block and Bouman 2006). He also led an effort to create a Calumet National Heritage Area based on the region's national significance with respect to environmental degradation and deindustrialization, and its social and environmental justice struggles (Bouman 2001, 2006, 2015). While still in Phillips Neighborhood, Rob Warwick staffed the Phillips Community Development Land Trust.

From Michigan, Jack Eichenbaum was a pioneer in advancing radical and gay geography who, after leaving the academy, remains active as a real-world geographer leading progressive tours of New York City (https://geognyc.com). Adrian Pollock brought his critical thinking to his career in secondary and primary education.

With respect to the USG and *Antipode*, Sheppard (who joined the USG local after moving to Minnesota in 1976) was at the center of debates about whether to institutionalize radical geography by creating a Socialist Geography Specialty Group within the AAG, arguing (with others) in favor of such a move (Peake, this volume). While the USG already was struggling to raise funds from its members and to curate new issues of the *Newsletter*, the inauguration of the SGSG (Sheppard 1979) undoubtedly accelerated its demise as an independent node for radical geography. After accepting Dick Peet's offer to take over *Antipode* in 1986 as co-editor with Joe Doherty (Neil Smith's undergraduate advisor), he supervised the journal's transition to commercial publishing. Influenced by the Midwest radical geographers and their reading group, Sheppard helped open the journal to a more eclectic vision for radical geography (Sheppard and Doherty 1987).

## Notes

1  There was no moment when a request was made to create a Minnesota local of the USG. Rather, after returning from the spring 1976 USG meeting in New York City, the Minnesotan geographers started thinking of themselves as a local. In the fall of 1976, the Minnesota local proposed to the *USG Newsletter* editors that they create the content of a *Newsletter* Volume 3(3). This was the first official collective act of the Minnesota local of the USG.
2  See http://afka.net/Mags/Rat_Subterranean_News.htm for an early article by Mathewson and https://www.ratundergroundnews.com/and www.revelinnewyork.com/blog/09/14/2009/rat-subterranean-news-underground-newspaper for some further contextualization.
3  Email from Clark Akatiff to Linda Peake, February 11, 2011.
4  Email from Clark Akatiff to Linda Peake. That the 1969 annual meeting of the AAG was in Ann Arbor was not the original plan. It was supposed to be in Chicago. But that decision was protested by some AAG members because of the riots, demonstrations and Mayor Daley-directed police brutality associated with the August 1968 Chicago Democratic National Convention (DNC). The suggestion was made that because more student protestors at the Chicago DNC demonstrations came from the University of Michigan than any other university, and that Ann Arbor was in the same Midwest region as Chicago, the AAG should be held at Ann Arbor. Jack Eichenbaum had begun his graduate program at Michigan in 1967, supervised by Olsson. He remembers that

the University of Michigan had the most signatures on the petition to move the 1969 AAG meetings … to Ann Arbor. Students and radical geographers then played significant roles at these meetings …, which had been dominated previously by conservative old WASP professors. I introduced a plenary student-run session which culminated in representatives of the Detroit ghetto brought on stage. (Email from Jack Eichenbaum to Eric Sheppard, July 21, 2018)

5  These two spaces sometimes overlapped. Jonathan Mayer remembered Olsson

> asked a few of us if we would be interested in extending … [the Geography 810] seminar [by] meet[ing] informally each week – frequently in Gunnar's home, in the evening, with some good wine. The topics were a combination of Gunnar's evolving interests – Wittgenstein, Marx, Hegel – and our own interests. We read, discussed, socialized, and got to know one another. (Email from Jonathan Mayer to Eric Sheppard, July 26, 2018)

6  Email from Adrian Pollock to Eric Sheppard, August, 2018
7  Jonathan Mayer who was part of the Olsson group in the end found it too much like "group think," with "condescension to the issue of wanting to find employment after graduate school." "The camaraderie was superb," Mayer says, "but … there was also that dark side that began to emerge. … However, I felt enriched by the experience." (Email from Jonathan Mayer to Eric Sheppard, July 26, 2018)

# References

Barnes, T. J. ed. (1996). Lukermann on location. In *Logics of Dislocation: Models, Metaphors and Meanings*, pp. 299–247. New York: Guilford.

Barnes, T. J. (2012). Gunnar Olsson and me. In C. Abrahamsson and M. Gren, eds, *GO: On the Geographies of Gunnar Olsson*, pp. 245–258. Farnham: Ashgate.

Benjamin, W. (1986). *Illuminations*. New York: Random House.

Blaut, J. (1993). *The Colonizer's Model of the World: Geographical Diffusionism and Eurocentric History*. New York: Guilford.

Block, D. and Bouman, M. J. (2006). Paradoxes of Praxis: Community-Based Learning at a Community-Based University. In H. Rosing and, N. G. Hofman, eds, *Pedagogies of Practice: Interdisciplinary Perspectives on Course-Based Action Research*. Boston: Anker Publishing.

Bouman, M. J. (2001). A mirror cracked: ten keys to the landscape of the Calumet region, *Journal of Geography* 100 (4): 104–110.

Bouman, M. J. (2006). Nurture's metropolis: Chicago and the rediscovery of nature. In R. P. Greene, M. J. Bouman and D. Grammenos, eds, *Chicago's Geographies: Metropolis for the 21st Century*. Washington, DC: Association of American Geographers.

Bouman, M. J. (2015). The slings and arrows of defining the Calumet region. In V. B. Price, D. A. Spatz, and B. D. Hunt, eds, *Out of the Loop: Vernacular Architecture Forum Chicago*. Evanston: Agate Publishing.

Bunge, W. (1971). *Fitzgerald: Geography of a Revolution*. Cambridge, MA: Schenkman Publishing Co.

Bunge, W. and Bordessa, R. (1975). *The Canadian Alternative: Survival, Expeditions and Urban Change. Geographical Monographs Number 2*, Department of Geography, York University.

Callinicos, A. (1978). *Althusser's Marxism*. London: Pluto Press.

Connerton, P. ed. (1976). *Critical Sociology: Selected Readings*. New York: Penguin Books.

Cook, C. C. and Lauria, M. (1995). Urban regeneration and public housing in New Orleans, *Urban Affairs Review* 30 (4): 538–557.

Eichenbaum, J. (1972). *Magic, Minorities and Mobilities in the Urban Drama*. Unpublished Ph.D. thesis, Department of Geography, University of Michigan, Ann Arbor, MI.

Foley, J. and Lauria, M. (2000). Plans, planning and tragic choices, *Planning Theory and Practice* 1 (2): 219–233.

Foley, J. and Lauria, M. (2003). Historic preservation in New Orleans French Quarter: unresolved racial tensions. In H. Thomas and F. Lo Piccolo, eds, *Knights and Castles: Minorities and Urban Regeneration*, pp. 67–89. Burlington, VT: Ashgate Publishing Company.

Gould, P. and Olsson, G. (1980). *A Search for Common Ground*. London: Pion.

Gregory, D. (1978). *Ideology, Science and Human Geography*. London: Hutchinson.

Higgins, B. (1978a). *Technical Assistance to Local Authorities for Community Development*, grant from the US Department of Housing and Urban Development.

Higgins, B. (1978b). *Andersen School Parent Advocate*, grant from the Minneapolis Foundation.

Higgins, B. (1980a). A Special Issue on Praxis and Humor. *USG Newsletter* 5 (2&3): 1.

Higgins, B. (1980b). *Phillips Neighborhood Comprehensive Plan*. Minneapolis, MN.

Higgins, B. (1982). Urban Indians: patterns and transformations, *Journal of Cultural Geography* 2 (2): 110–118.

Higgins, B. (1986). The Sanderistas and a metamorphosis of Burlington Vermont, *Places* 3 (2): 32–39.

Higgins, B. (1990a). Geographical revolutions and revolutionary geographies: nature, space and place in the urban development of Nicaragua, *TRIALOG: A Journal for Planning and Building in the Third World (West Germany)*, 26: 13–20.

Higgins, B. (1990b). The place of housing programs and class relations in Latin American cities: the development of Managua, Nicaragua before 1980, *Economic Geography* 66 (4): 378–388.

Higgins, B. (1991a). *Community Economic Development Action Plan for the Saint Regis Mohawk Tribe*. Hogansburg, New York: Saint Regis Mohawk Tribe,

Higgins, B. (1991b). *Labor Market Analysis of the St. Regis Mohawk Reservation*. Hogansburg, New York: Saint Regis Mohawk Tribe.

Higgins, B. (1994). Urbanization in Nicaragua. In G. Greenfield, ed., *Handbook of Latin American Urbanization: Historical Profiles of Major Cities*. Westport, CT: Greenwood Press.

Higgins, B. and Lauria, M. (1986). Inner-city redevelopment: urban Indians and community-based organizations. In *Minneapolis Field Trip Guide*. Washington, DC: Association of American Geographers.

Horkheimer, M. (1968). *Critical Theory: Selected Essays*. New York: Continuum Pub.

Hugill, D. (2016). Metropolitan transformation and the colonial relation: the making of an "Indian Neighborhood" in postwar Minneapolis, *Middle West Review* 2 (2): 169–199.

Jay, M. (1973). *The Dialectical Imagination: A History of the Frankfurt School and the Institute of Social Research, 1923–1950*. Boston: Little and Brown.

Lauria, M. (1982). Selective urban redevelopment: a political economic perspective, *Urban Geography* 3 (3): 224–239.

Lauria, M. (1983a). *Children of ADC Families Eliminated from the Program in Response to the 1981 Omnibus Reconciliation Act – Policy Development*, Institute of Urban and Regional Research, The University of Iowa, for the Legislative Extended Assistance Group of the State of Iowa.

Lauria, M. (1983b). Programs for housing assistance, in *Iowa's Housing Prospects for the 1980s*, State of Iowa, Office of Planning and Programming, Des Moines.

Lauria, M. (1984). Implications of Marxian rent theory for community-controlled redevelopment strategies, *Journal of Planning Education and Research* 4 (1): 16–24.

Lauria, M. (1986). The internal transformation of community-controlled implementation organizations, *Administration and Society* 17 (4): 387–410.

Lauria, M. and Fisher, P. (1983). *Plant Closings in Iowa: Causes, Consequences, and Legislative Options*. Institute of Urban and Regional Research, The University of Iowa, for the Legislative Extended Assistance Group of the State of Iowa.

Lauria, M. Dufour, W., and Haughey, P. (1995). *The Holy Cross Neighborhood: Planning for Community Development*. Division of Urban Research and Policy Studies, College of Urban and Public Affairs, University of New Orleans, New Orleans, Louisiana.

Lauria, M., Dufour, D., Haughey, P. and Lamarque, J. (1996). *Citizen Planning for Community Development in the Lower Ninth Ward*. Division of Urban Research and Policy Studies, College of Urban and Public Affairs, University of New Orleans, New Orleans, Louisiana.

Minnesota Collective, The (1980) PLEASE SEND MONEY! *USG Newsletter* 5 (4) & 6 (1): 1.

Nord, D. P. (1976). Minneapolis and the pragmatic socialism of Thomas Van Lear, *Minnesota History*, 45: 2–10.

Olsson, G. (1975). *Birds in Egg*. Ann Arbor, MI: Department of Geography, University of Michigan.

Olsson, G. (1980). *Birds in Egg: Eggs in Bird*. London: Pion.

Pynchon, T. (1965). *The Crying of Lot 49*. Philadelphia: Lippincott.

Pynchon, T. (1972). *Gravity's Rainbow*. New York Viking Press.

Radcliff, S. (2017). Decolonizing geographical knowledges, *Transactions of the Institute of British Geographers* 42 (3): 329–333.

Sheppard, E. (1979). ANNOUNCEMENT: Formation of an AAG Group in Socialist Geography. *USG Newsletter* 5 (1): 6.

Sheppard, E. and Doherty, J. (1987). Editorial. *Antipode* 19(2): 93–94.

Steiner, G. (1971). *In Bluebeard's Castle: Some Notes toward the Redefinition of Culture*. New Haven, CT: Yale University Press.

Tuan, Y.-F. (1974). *Topophilia. A Study of Environmental Values and Perception*. Englewood Cliffs, NJ: Prentice Hall.

Tuan, Y. F. (1979). *Landscapes of Fear*. New York: Pantheon Books.

Watts, M. (2012). Of bats, birds and mice. In C. Abrahamsson and M. Gren, eds, *GO: On the Geographies of Gunnar Olsson*, pp. 143–154. Farnham: Ashgate.

Williams, R. (1973). *The Country and the City*. London: Chatto and Windus.

# 9

# Radical Geography Goes Francophone[1]

## Juan-Luis Klein

The Marxist version of radical geography that emerged in the late 1960s did not enjoy widespread support from the French-language universities in Canada.[2] Of course, some geographers were open and attentive to the cultural, ethnic, gender-oriented, and even epistemological questions posed by this form of radical geography, with some even incorporating Marxist elements into their analyses or teachings. But when they did, they did so as individuals rather than as part of a collective project. However, there was one exception, which is the topic of this chapter: the *Groupe de recherche sur l'espace, la dépendance et les inégalités* (GREDIN) (Research group on space, dependence, and inequalities) formed in 1975 at Université Laval in Québec City. Consisting of professors and graduate students from Laval's Department of Geography, the group implemented a Marxist-inspired geographical research program. It lasted until the 1980s and comprised both individual and joint research informed by a Marxist theoretical framework. It has been the only group of its kind in French-speaking Canada. This experiment to operationalize Marxist concepts within geography extended beyond the Université Laval with members of the group vigorously disseminating their work in forums and symposia outside the University.

The narrative of the GREDIN's research program I will provide is partly subjective. Of course, I rely on verifiable documents. Yet, the importance I give to these documents, the contexts in which I situate them, and the value I impute to their various theoretical and practical contributions, come from personal reflection. This chapter is informed

*Spatial Histories of Radical Geography: North America and Beyond*, First Edition.
Edited by Trevor J. Barnes and Eric Sheppard.
© 2019 John Wiley & Sons Ltd. Published 2019 by John Wiley & Sons Ltd.

throughout by my own experience, and very much tied to the objectives of this book.

In 1975, I began my Master's degree in geography, a few months after having left my home country, Chile. It was the same year that GREDIN formed. GREDIN provided me a bridge between my political activism experiences in Chile, and the academic work I did there between 1969 and 1972 in history and geography at the former Universidad Técnica del Estado (now Universidad de Santiago de Chile). During those years I fully participated in the Chilean revolution,[3] and which was distinctly Marxist. Of course, the *coup d'état* by General Augusto Pinochet[4] on September 11, 1973 put a violent end to all of that. It led me in June 1974 to come to Québec City as a political refugee and, several months later, to enroll at Université Laval as a graduate student in geography.

The move was traumatic, even a doubly traumatic. On the one hand, I was uprooted and displaced from my home country, with all that implied for a 23-year-old. On the other hand – and this is where the link with my experiences came in – I experienced an epistemological rupture. My vision of research and society did not align with the dominant vision of my new academic environment that was predominantly politically conservative. Within that context, my interactions with GREDIN allowed me to regain a linkage with the Marxism I formerly practiced in Chile. Better yet, that experience encouraged me to question the basic concepts and methods of the discipline of geography, something I had not done in my country. Indeed, in Chile, as in many other countries, and especially those where the *École française de géographie* predominated,[5] the discipline of geography, in contrast to history and other social sciences, remained very traditional, intellectually conservative, notwithstanding the social and even political commitment of many of its practitioners. I therefore saw GREDIN as an opportunity to make an epistemological contribution to Chilean geography.[6] These circumstances explain the intensity of my engagement with the group – an intensity that was further fostered by a special bond I shared both with Rodolphe De Koninck and Paul Y. Villeneuve, the two professors who assumed the group's academic and intellectual leadership, and with the other graduate students who participated in its creation.

Through different but convergent paths, De Koninck, Villeneuve, and the graduate students (both those who participated in the founding of the group and those who joined thereafter) all aimed to construct a geographical approach that both criticized the existing social reality as well as to transform it into into a better society better society. I situate my history of Francophone Canadian radical geography between 1975 and 1982, that is, between the inauguration of GREDIN and the publication of its most comprehensive work, the atlas *Développement inégal dans la région de Québec* [The development of inequality in the region of Québec] edited by De Koninck *et al.* (1982).

## Québec Geography Between Two Transitions

To understand fully the questions and orientations posed by GREDIN's research program in the mid-1970s, it is necessary to recall the broader Canadian, and especially Quebecois context in which the practice of geography took place.[7]

### The need for a critical approach

The discipline of geography in French Canada, institutionalized first at Université de Montréal and Université Laval, had its roots in the *École française de géographie*. It was largely influenced by the ideas of professors from France such as Raoul Blanchard and Pierre Desfontaines,[8] as well as the *Science de lieux* approach, and its main form, the regional monograph enunciated by Paul Vidal de la Blache. French Canadian geography was thus a descriptive discipline. By the 1960s, a period when social sciences questioned and modernized their theoretical frameworks, the descriptive method of French Canadian geography made it old fashioned, even anti-modernist.

The traditional orientation of Canadian Francophone geography derived from its original mission when first institutionalized in France and elsewhere, to train geography teachers. That was no longer true, however. The transformations experienced by Western societies as a result of changes in modes of governance altered the demands for specific geographical skills. Traditional geography became less relevant. Instead, required was an "active geography" capable of participating in the planning of territories, and providing solutions to economic and social problems. In response to the increasing role of public bodies in spatial planning, geography departments oriented themselves to training professionals capable of carrying out public policies around urban and regional planning and development. This, in turn, led to the establishment of an applied geography with a vocational orientation, but at the expense of the exploration and questioning of theory.

It was the theoretical vacuousness of the geography of that time that was criticized by several radical social movements in the 1970s. Ironically, the most important trend in applied geography was called "theoretical geography,"[9] which relied primarily on quantification and the application of mathematical tools in the analysis of geographical space.

Geographers such as William Bunge and David Harvey laid the foundation for a hypothetical-deductive theoretical spatial analysis that was to influence a number of geographers in Canadian universities, including those in French Canada.[10] In contrast, other geographers dissatisfied with theoretical geography proposed a critical analysis inspired by Marx and historical materialism. It was within this context that GREDIN was created. Its principal objective was to critique the transformative effects of contemporary capitalism in Québec.

## The modernization of Quebec as a result of the Quiet Revolution

What kind of *place* was Québec? Between 1936 and 1960, it was governed by one political party, *Union Nationale*, which identified with both the Catholic church and traditional rural elites. Headed by Maurice Duplessis, that party defended a very conservative nationalism, and was anti-union, anti-socialist and anti-progressive. This context changed dramatically in 1960 when a largely urban social coalition made up of business, organized labor and intellectuals, pushed Québec to economic and social modernization, and became the basis for a new Quebecois Liberal Party government. Taking power, the Liberals embarked on a set of critical changes known as the "Quiet Revolution" (Brunelle 1978).

The result was significant restructuring of government services, producing health reforms, improved accessibility to education, the creation of public enterprises concerned with the development of natural resources in peripheral regions, and major hydroelectric projects. This process also involved the administrative and functional modernization of the province. Specifically, the drive toward modernization instilled by these reforms led: to the creation in 1967 of administrative regions, the objective of which was to disperse government agencies to regional capitals; to establishing in 1968 a development and planning office (*Office de développement et de planification du Québec*, OPDQ); to forming new regional institutions that impacted territorial governance such as general and vocational colleges (*Collèges d'enseignement général et professionnel*, CEGEPs); and, above all, to founding in 1969 the Université du Québec network with the objective of democratizing access to higher education. Through these and other major planning operations, the Government of Québec sought to modernize cities and regions. As part of that effort, urban communities in Montréal, Hull, and Québec City were created and the government launched regional planning operations aimed at the economic revitalization of underdeveloped regions, especially remote rural areas.[11]

This process of modernization also affected various aspects of the management of Québec's land use. Urban renewal operations involved the relocation of citizens, particularly in urban inner districts and in rural areas. Those upheavals, in turn, produced protest movements (Klein *et al.* 2012) embodying a kind of "citizen unionization"[12] opposed to management operations, and their authoritarian implementation. The movement also led to forming citizen committees that defended the rights of the most disadvantaged segments of the population.

## The Development of a Marxist Approach to the Analysis of Cities

Planning and land-use operations carried out in Québec cities as part of the Quiet Revolution have been subject to various critical studies. For the most part they denounced the Quiet Revolution, suggesting that the aspiration of modernization was really aimed at realizing state monopoly capitalism; that is, the stage of capitalism in which financial capital, services, and technocracy combine with monopolies and big capital to maximize surplus value and profits. Among these critical works, ÉZOP-Québec's "Une ville à vendre" (A city for sale) was the best known.[13] It stood out for both its content and the influence it had on GREDIN. Initially published in 1972 as a booklet and re-issued in 1981 in book form, it gave a Marxist reading of the city in general, and the renewal of Québec City in particular. According to Villeneuve (1982: 279), "Une ville à vendre" was a milestone in the social changes sweeping through Québec society, with its authors conducting "the first detailed analysis of a concrete case of urban redevelopment carried out in Québec based on the conceptual framework proposed in the works of neo-Marxist authors such as Lefebvre, Althuser, Poulantzas, and Castells."[14]

The study consisted of four sections. The first presented the Marxist theoretical framework. The city was set within capitalism. As a result, urban planning favored the circulation of capital, not citizen wellbeing. The second described two Québec City neighborhoods that experienced transformative urban renewal. The third addressed political dimensions, and in particular, the role of the state in urban transformations. Here the report severely criticized all levels of the state for allowing the destruction of working-class neighborhoods, while at the same time allowing the expansion of individual property in the suburbs. The fourth analyzed the plan's ideological and social foundation.

The ÉZOP report had a considerable impact on the work of GREDIN on two levels. First, it situated the analysis of territorial modernization. *Une ville à vendre* inspired Marxist research on space in Québec, and influenced the work carried out by researchers associated with GREDIN.[15] Second, the authors of the report were actively involved in local (municipal and regional) politics. In 1976, their study inspired the creation of a left-wing municipal political party with the active participation of some of the researchers affiliated with GREDIN, especially Paul Y. Villeneuve.

There was a difference between the research carried out in Québec City, which addressed problems of a local nature, and the research carried out in Montréal, the perspective of which was certainly broader but less place-anchored. Was such locally based orientation of the research carried out in Québec City a sign of a bottom-up, political counter-practice and later disseminated in the province? asked Villeneuve (1982: 281). This question was the precursor of the action-oriented turn of 1980s spatial research that led researchers to focus on the local scene.

Indeed, in the wake of the crisis of Fordism in the late 1970s, and the deindustrialization it provoked, social actors already began to innovate socially. They developed a vision of economic development that was unique to the Québec social movement and that, following in the footsteps of initiatives in rural and urban areas, conceptualized the local as a basis for job creation, the social economy, and political participation. These were the early stages of what was to become 1980s local community economic development (Hamel 1991; Fontan 1992; Klein et al. 2014).

More generally, the radical geographical project that was GREDIN was at every turn shaped and directed by the specificities of a primarily Francophone Québec, with a distinctive history and geography. Place mattered. Of course, that radical geography drew on specifically Marxist concepts conceived outside of the province, but they were reworked and refashioned in GREDIN's analysis of the geographically historical peculiarities that made Québec so distinctive.

## The Construction and Implementation of a Research Program in Marxist Geography

This is an overview of the setting in which GREDIN emerged and operated. Of course, this group followed a similar trajectory to that of other radical geography groups elsewhere. All of them rejected what appeared to be an apolitical geography that was in the service of society's dominating power structures. What made GREDIN's critical approach distinctive, however, was the social context of Québec, and its ambitious modernization process. Specifically, Quebecois radical geography critically analyzed a top-heavy state apparatus on the one hand, and a large-scale territorial planning operation, on the other. As De Koninck (1978a: 304) wrote when introducing GREDIN, intellectual autonomy is

> all the more necessary as the society in which it unfolds has its own characteristics and therefore calls for a specific critique and targeted solutions – which in no way denies, it should be added, the importance of common ground with other societies. We only need to recall the sudden magnitude of this technocratic project, which in Québec effectively replaced the former theocratic project of the feudal elites, to convey the enormity of the task. Essentially, each spatial planning project oriented towards building profit, and thus involving the "organization of people" to generate capital, should be countered with a project dedicated to the revolution of space, in other words, the liberation of mankind.[16]

GREDIN's university roots were in Québec City, and the Québec administrative region.[17] This factor undoubtedly influenced the group's orientation and its specificity. Yet, even though the city of Québec is the capital of the province, it is Montréal, given its economic power and population size, which is the metropolis of the province.[18] Seen from the perspective of Québec City, this socio-spatial relationship of the province's cities appeared unbalanced and disadvantageous to the development opportunities of the regions other than Montréal, including the Québec administrative region. In that context, GREDIN researchers oriented their research toward redressing these regional inequalities, drawing from works on international unequal development (Cardoso and Faletto 1968; Amin 1973).

GREDIN consequently yielded a number of publications and Master's and doctoral theses (Table 9.1), culminating in the publication of the atlas *Développement inégal dans la région de Québec* (De Koninck *et al.* 1982).

## The conceptual construction of the GREDIN research program: Two founding publications

The first two GREDIN publications took the form of working papers and published as part of the *Notes et documents de recherche* (NDR) series of the Department of Geography at Université Laval. They accounted for issues NDR 9 and 10 (1978). NDR 9 was entitled *"Contributions à une géographie critique"* (Contributions to a critical geography), and NDR 10, *"Au sujet des exigences spatiales du mode de production capitaliste"* (On the spatial requirements of the capitalist mode of production).

**Table 9.1** Theses either completed or initiated as part of the GREDIN research program between 1975 and 1982*.

| Author and year of submission | Title (translated to English) | Level | Director |
|---|---|---|---|
| Dion, Marc, 1975 | Land ownership and land-use planning: The case of Île d'Orléans | M.A. | Rodolphe De Koninck |
| Piot, Monique, 1976 | Agriculture in the face of capital: Notes on the integration of agriculture into the capitalist market using the example of Quebec and Île d'Orléans | M.A. | Rodolphe De Koninck |
| Klein, Juan-Luis, 1978 | The regional question in late stage capitalism: The case of Quebec | M.A. | Paul Y. Villeneuve |
| Lavertue, Robert, 1980 | The industrialization of peripheral regions: The case of La Beauce | M.A. | Rodolphe De Koninck |
| Roy, Aimé, 1980 | The role of capital and landowners in the development of the city: The case of the Saint-Jean-Baptiste neighborhood in Quebec City | M.A. | Paul Y. Villeneuve |
| Klein, Juan-Luis, 1981 | Region, industry deployment and the cost of labor: Contribution to the analysis of manufacturing growth in the Québec administrative region | Ph.D. | Rodolphe De Konink and Paul Y. Villeneuve |
| Giguère, Jacques, 1981 | Mining monopolies and the Côte-Nord. Contribution to a critical regional geography | M.A. | Jacques Bernier |
| Pelletier, Lyse, 1982 | Local political power and social classes: The example of Quebec City | M.A. | Paul Y. Villeneuve |
| Nadeau, Jean, 1984 | From the submission of workers to the alienation of agricultural land: The case of Quebec agriculture | M.A. | Rodolphe De Koninck |

Table 9.1    (Continued)

| Author and year of submission | Title (translated to English) | Level | Director |
|---|---|---|---|
| Grasland, Loïc, 1986 | Small towns between the metropolis and the region | Ph.D. | Paul Y. Villeneuve |
| Lavertue, Robert, 1988 | The role of agriculture in the industrial development process | Ph.D. | Rodolphe De Koninck |

*I limited my selection of works to those completed under the research program on the Quebec regions, in particular, the Québec administrative region. Other theses and dissertations were also initiated or realized during this period, but with different foci.

Source: De Koninck et al. (1982), complemented by the "Projets, mémoires et thèses" section of the website of the Department of Geography at Université Laval: https://www.ggr.ulaval.ca/projets-memoires-et-theses.

NDR 9 was divided into four sections. The introduction, "*Changer la géographie: notes pour une discussion*" (Changing geography: notes for discussion), was written by De Koninck. He laid out the groundwork for a critique of both the École française de géographie, and the so-called "new geography" of English-speaking North America. His critique was formulated on principles of historical materialism, namely that ideas can only be understood by situating them within the material context in which they were generated (De Koninck 1978b: 6).[19] Specifically, DeKoninck criticized the *École française de géographie* for its inability to define rigorously the subject of the discipline, and for its complicity with the prevailing power structure, making it a "science of the status quo" (p. 11). However, he also finds fault with the new geography of the 1960s and 1970s, in particular, quantitative geography. While recognizing the progress it made in relation to French regional geography, he emphasized the static and simplistic nature of the models used by its main proponents (p. 15). Moreover, De Koninck underlined the contradictory nature of the models, and their obscuring of capitalist forces that determined the social and economic character of space or territory. At the same time, he criticized the fundamental propositions of cultural or phenomenological geography that dissociated geographical phenomena from their material context.

De Koninck critiqued assumptions of equilibrium and harmony, especially in urban and economic geography, and he also argued that that the relationship between humans and nature was determined by class struggle, the division of labor, and the entire antagonistic social relations inherent in the mode of production. In this way, he proposed a new geographical vision. Theory and methodology would derive from Marxism, but the discipline of geography would not be "encapsulated inside Marxism" (p. 41). This stance was to define the work of GREDIN researchers: Marxism would be a tool for understanding reality, but not doctrinaire. Theory would always be checked by fieldwork.[20]

The second section of NDR 9, "*Concepts fondamentaux à une première esquisse critique*" (Basic concepts for a first critical sketch), outlined a series of fundamental ideas first raised in the inaugural seminar held by GREDIN during the summer of 1976. They were also presented at a special session of the Canadian Association of Geographers (CAG) conference at Université Laval in 1976. Among the ideas discussed were: dialectical materialism, historical materialism, and mode of production (Piot 1978a); their connections with space (Klein 1978a); the specificity of these connections in relation to capitalist stages of production, especially imperialism and state monopoly capitalism (Naud 1978); and the unique forms that they took in urban contexts (Bélanger 1978).

Such a materialist approach did not deny the importance of ideas. Rather, it emphasized the contradictions that generated them. Further, these contradictions were seen to evolve according to their material supports, primarily economic. The contradictions, in turn, were linked to social relations of production, the spatial disparities that were consequently created, and the role of the state that managed those disparities. Social inequalities occur in space, taking the form of spatial inequalities. Caused by the action of capital and monopolies, with strong state involvement, the inequalities were part of the overall social and economic structure of state monopoly capitalism. Central spaces were integrated into the processes of capitalist accumulation, and the outside spaces dominated by those centers needed to be explained by this global framework.

The third section, "*Le développement, la région, la ville*" (Development, the region, the city), explored spatial inequalities in relation to: social classes (Klein and Naud 1978); uneven development (Piot 1978b); regional development strategies (Lavertue and Villeneuve 1978); and land-use planning (Beaumont and Guay 1978). The papers included in this section were written for the second GREDIN seminar held in the winter of 1977, and presented at the annual conference of the *Association canadienne française pour l'avancement des sciences* (ACFAS)[21] held in Trois-Rivières in May 1977. The works here used the basic concepts and notions of Marxism to examine empirically the Québec administrative region. Each sub-region was analyzed in terms of the various stages of capitalism that it had passed through (mercantile, industrial-competitive, monopoly). Each of those stages in turns was associated with a specific class structure, which created a particular local society, and conceptually interpreted as a Gramscian historical bloc. Also invoked was the idea of the articulation of spaces at different stages of capitalism.

The fourth section includes the case of a peripheral space analyzed by Pierre Mathieu (1978), Costa Rica. The study of the socio-historical evolution of Costa Rica allows the author to reveal the structure of the conflicting social relations in this country, in particular, dependency of the rural space and the national urban cores not only on one another but also on foreign monopolies and local ruling classes. The combined analysis of the siphoning of the surplus value produced by the agricultural sector, and the centralization of power in national capitals, illustrates the dynamics that govern the center-periphery relations in a dependent space.

In sum, NDR 9 provided the conceptual framework for the GREDIN research program. It was a program in which the conceptual and methodological grid of historical materialism was applied to the analysis of space by focusing on interregional and intraregional relations. It revealed the centrality of the production of surplus value and land rent, and the relations of the state to territory through its land-use policies and programs.

The second NDR (no. 10) was published in September 1978, "*Au sujet des exigences spatiales du mode de production capitaliste*" (About space requirements of the capitalist mode of production). This publication is divided into two sections. The first section includes four papers that were written for presentation at the Canadian Association of Geographers Congress in London, Ontario in May 1978. The second section includes five papers presented at the annual conference of ACFAS held in Ottawa also in May 1978.

Entitled "Notes d'approche" (Notes on the approach), the first part of NDR (no. 10) examines the regional issue from four angles: the division of labor (De Koninck 1978c), the role of the state (Piot and Hulbert 1978a), monopoly capital (Lavertue 1978a), and political power (Villeneuve 1978a). A central feature of this part is a discussion of the constraints on the revolutionary transformation of capitalism.

De Koninck's (1978c) chapter revisits a central aspect of regional analysis from a Marxist perspective, the town-country divide. He considers the separation of the city and the countryside as one of the main forms of the social division of labor. It involves separation of workers from the means of production, a split between peasants and workers, exploitation of rural spaces for the benefit of the city, monoculture in the countryside, and submission of the peasantry to the working class. Together, these processes result in an opposition between the city and the countryside (De Koninck 1978c: 8). For De Koninck, the dissolution of this opposition is central to the project of overcoming capitalism and building a socialist society. He thinks, though, this will be difficult given that the working class that will lead the revolution is primarily urban.

Piot and Hulbert's (1978a) chapter addresses the role of the state in the social division of labor. It highlights a central element of the Marxist analysis, the contradiction between the progressive socialization of labor and the means of production, on the one hand, and the privatization of the ownership of the means of production and the appropriation of surplus value, on the other. For Marxists this contradiction is the motor of the evolution of the relations of production and social relations (Piot and Hulbert 1978a). Piot and Hulbert (1978a) suggest that the state manages and deals with this contradiction by ensuring both a certain socialization of the ownership of the means of production, and the continuing accumulation of surplus value. Consequently, capital remains dominant, controlling both labor and the spaces where labor is implemented or deployed. The modalities of domination vary according to the stages of capitalism. The direct domination by industry in the early stages of capitalism is replaced later by an indirect centralized domination by corporate headquarters, financial capital, and the state. In this latter period, the processes of production and control are standardized while at the same time specialized spaces are produced to meet the interests of private and state-sector bourgeoisie, leading to uneven regional development.

Lavertue's (1978a) chapter examines the contradiction between the interests of monopoly capital and the dominant centralized classes (both private and public), and those of the local regional bourgeoisie. To illustrate the point, he uses the case-study of La Beauce.

Finally, Villeneuve's chapter (1978a) draws on an analysis of Québec nationalism. The author explains that two nationalist projects are underway. First is the mobilization of the Quebecois bourgeoisie in promoting a nationalist project that subscribes to state monopoly capitalism (Villeneuve 1978a: 44–45). Villeneuve argues that this project is essentially carried out by an alliance between the Québec bourgeoisie and the technocracy of the Quiet Revolution. The second project is revolutionary and should involve a coalition between workers and employees with the support of intellectuals.

The second part of NDR (no. 10) provides concrete applications of a radical geographical approach for the Québec administrative region (Klein 1978b; Piot and Hulbert 1978b; Lavertue 1978b; Roy and Beaumont 1978; Giguère 1978). It draws directly on the work of Jules Savarria (1976) on Québec uneven development. His analysis uses Samir Amin's (1973) ideas to suggest that Québec is a social formation (or sub-social formation) of the center, but also marked by the protracted presence of earlier stages of capitalism. The Québec administrative region is a patchwork quilt of differentiated spaces each at a different stage of capitalism but coexisting at the same time. Determining that patchwork quilt are different organic compositions of capital in different spaces, producing different amounts of surplus value. For example, in his analysis of the distribution of the branches of the Québec manufacturing sector, Klein (1978b) concluded that it was unevenly integrated into the relations of production of the dominant monopoly capitalism in North America, and as a result the region was relegated to the periphery.

Apart from the organic composition of capital, Piot and Hubert (1978b) demonstrated that the state also influenced the location of industrial sectors within Québec. They argued that Federal government intervention favored the development of a monopolistic bourgeoisie, while the provincial government favored the development of a middle bourgeoisie. More specifically, the implementation of these state policies allowed for the development of both productive manufacturing and service spaces, but in the process producing uneven development. This applied primarily to cities where the provincial government and the main services (university, health, commerce) are concentrated. They are Québec City and its suburbs, and the La Beauce area, a sub-region featuring a high concentration of SMEs and local bourgeoisie who assume a leadership role.

## The dissemination of GREDIN's work and collaboration with the Union of Socialist Geographers

Between 1976 and 1980, GREDIN's research program was presented at several thematic sessions held at annual conferences of scientific organizations, which explains its influence beyond Université Laval and the city of Québec. Those conferences included the Canadian Association of Geographers (CAG), and the geography section of the *Association canadienne française pour l'avancement des sciences* (ACFAS). In both cases, participants came from across Canada, even internationally (Table 9.2).

The 1976 CAG conference at Université Laval was especially important in that it brought together members of GREDIN and the Union of Socialist Geographers at a session on geography and dialectical materialism (see also Chapter 5).[22] That meeting led in 1978 to the publication of a special issue of *Cahiers de géographie du Québec* on historical materialism in geography. John Bradbury, one of the most active promoters of the USG in Canada, contributed a paper to that issue, "*La géographie et l'État*" (Geography and the state) (Bradbury 1978), as well as a short

**Table 9.2**   Presentations researchers affiliated with GREDIN between 1976 and 1980.

| Context, place and date | Theme | Participants; theme of the presentation |
|---|---|---|
| Canadian Association of Geographers (CAG), Quebec, May 1976. Collaboration with the USG | Dialectical materialism and geography | • Klein, J. L. and M. Piot; Some theoretical tools for a Marxist analysis of space<br>• Dion, M. Differential rent and social development<br>• Dion, M. and R. De Koninck; The state and development: Orléans, an island for sale<br>• De Koninck, R. and R. Lévesque; Culture, ideology and space<br>• Piot, M.; Agriculture in the stage of state monopoly capitalism |
| ACFAS, Trois-Rivières, May 1977 | Regional and spatial disparities | • Klein, J. L. and M.A. Naud; Regional development, a class problem<br>• Piot, M.; Uneven development and regional disparities: Reflections on the relationship between the mode of production and the space<br>• Mathieu, P.; Regional development and spatial inequalities<br>• Lavertue, R. and P. Y. Villeneuve; Regional development and polarization in Quebec<br>• Beaumont, L. and R. Guay; Urban planning in a capitalist environment |

**Table 9.2** (Continued)

| Context, place and date | Theme | Participants; theme of the presentation |
|---|---|---|
| ACFAS, Ottawa, May 1978 | Uneven development in the Québec administrative region | • Klein, J. L. and P. Villeneuve; Organic composition of capital and labor mobility at the regional level<br>• Lavertue, R.; The industrial vitality of La Beauce<br>• Giguere, J.; What are the foundations of the Baie Comeau Hauterive division?<br>• Roy, A. and L. Beaumont; Land monopoly and changes in the physical and social nature of neighborhoods<br>• Piot, M. and F. Hulbert; Développement régional et politique de subventions à l'entreprise privée (Regional development and the policy of subsidizing private enterprise) |
| Canadian Association of Geographers (CAG), London (Ont), May 1978 | The submission of regional spaces to the capitalist mode of production | • Villeneuve, P. and J. L. Klein; Uneven development at the regional level<br>• De Koninck, R.; Development and submission of regional spaces<br>• Lavertue, R.; Submission of the region to monopoly capital<br>• Piot, M. and F. Hulbert; The state, the monopolies and the region |

*(Continued)*

**Table 9.2** (Continued)

| Context, place and date | Theme | Participants; theme of the presentation |
|---|---|---|
| Canadian Association of Geographers (CAG), Victoria (BC), May 1979. Collaboration with the USG | Urban and regional problems in geography | • Klein, J. L.; Notes for a materialist theory of the region<br>• Piot, M. and J. Giguère; The role of the state in uneven regional development<br>• Lavertue, R.; The penetration of the dominant production relations in the regions: Notes on regional inequalities<br>• Carré, G.; Uneven regional development and social movements: The case of the Gaspé Peninsula<br>• Grasland, L.; Social utility and its link to the mode of production in geography<br>• Beaumont, L.; Urban planning and land ownership in Sainte-Foy |
| Canadian Association of Geographers (CAG), Montreal, May 1980 | For a critical geography | • Klein, J. L.; The notion of dependency and the region<br>• Piot, M.; The Québec space: Issues and outcomes of the class struggle<br>• Giguère, J.: The control of raw materials by multinationals: The case of the Côte-Nord ore<br>• Grasland, L.: The nature of recent economic development in the Québec administrative region |

*Source:* Based on data from NDR 9, pp. 159–160; Summaries of the presentations of GREDIN members at the 1979 CAG congress; and Résumés des communications/Abstracts of Papers CAG, UQAM 1980.

note with Bob Galois about the USG (Galois and Bradbury 1978). These collaborations continued. For example, in 1978, the USG Newsletter published Piot's text, "Uneven Development and Regional Disparities."[23] In 1979, at the University of Victoria a USG-sponsored session of the CAG was convened on the theme, *Problèmes urbains et régionaux en géographie* (Urban and regional problems in geography). And also that same year the USG Newsletter published (in English and French) a description of GREDIN written by De Koninck.[24]

Of course, these subjects were not the only ones addressed by the USG Newsletter on the situation in Quebec. Especially the Quebec referendum that took place in 1980 – the *Parti Québécois* had taken power in 1976 with the promise to hold a referendum on the sovereignty of Quebec – attracted the attention of a number of researchers involved with the USG Newsletter.[25]

## A focus field: The rural setting

There was also other, more targeted research carried out within GREDIN that helped clarify the geographic application of the Marxist approach to the Québec administrative region. Two complementary Master's theses focused on agriculture, and completed under the direction of Rodolphe De Koninck. Marc Dion's led to an article in *Cahiers de géographie de Québec*[26] (Dion and De Koninck 1976, "*L'Etat et l'aménagement: Orléans, une île à vendre*" (The state and land-use planning: Orléans, an island for sale). It should be noted that Île d'Orléans, located a few kilometers from Quebec City, was (and still is) symbolically important. The study addressed the role of the state in the commodification of the Island's natural environment brought about by spatial planning regulations. The analysis of the process of appropriation of the island by an urban bourgeoisie presented empirical evidence of how space is subject to the interests of capital at the expense of local residents and family farming. The second M.A. thesis was by Monique Piot, "*L'agriculture face au capital*" (Agriculture versus capital). It was summarized in an article published in the journal *Anthropologie et sociétés* (1977: 71–88), "*Agriculture et capitalisme au Québec: l'agro-industrie et l'État*" (Agriculture and capitalism in Quebec: Agro-industry and the state). It analyzed agricultural production in Quebec. Production was mainly carried out by family farms governed by non-capitalist production relations, but nevertheless integrated into a capitalist agri-food chain. The study showed that this process was supported by state intervention within a context of state monopoly capitalism.

## Asserting the marxist approach to geography in academia

The role of GREDIN in shaping radical geography was bolstered by the publication of a special issue, "*Le matérialisme historique en géographie*," edited by Rodolphe De Koninck in *Cahiers de géographie du Québec*. To my knowledge, this has been the only issue ever of a recognized geography journal published in Canada explicitly devoted to Marxist analysis. The issue included articles by R. De Koninck, R. Peet, P. Villeneuve, J.-L. Klein, B. Bernier, M. Bruneau, and P. Forcier, as well as notes by C. Raffestin and J. Bradbury, and short descriptions of the USG (Galois and Bradbury) and GREDIN (De Koninck). The issue also included reviews of relevant publications focused on a Marxist approach to territorial planning.

In a modified version of the text "*Changer la géographie: notes pour une discussion*," with the somewhat more powerful title "*Contre l'idéalisme en géographie*" (Against idealism in geography), De Koninck (1978d) set the tone for the objective of the issue: a critical geography of the relations of production and class struggle (De Koninck 1978e: 118).[27]

The basic concepts of such a geography were further elaborated by Richard Peet (1978) in his article, "Materialism, Social Formation and Socio-Spatial Relations: An Essay in Marxist Geography." He proposed and outlined what he called the "spatial dialectic" (pp. 154–155). Bernier (1978), for his part, contributed to the issue with a focus on the city. Villeneuve (1978b) laid out methodological tools that were applied to Canadian social class. Klein (1978c) examined regional inequalities in the Québec administrative region. Finally, Bruneau (1978) and Forcier (1978) analyzed two Asian social formations: Thailand and Cambodia. The last two contributions showed the socio-spatial contradictions that arose as a result of the spatial expansion of the capitalist mode of production, which encroached on spaces dominated by pre-capitalist modes of production.

## The final output: The publication of an atlas on uneven regional development

The final output of GREDIN was a major publication, the atlas, *Développement inégal dans la région de Québec* (De Koninck *et al.* 1982). It was intended as an analytical and cartographic contribution; analytical because it applied the uneven development approach, and cartographic because it was a computer-generated atlas.[28] Moreover, it successfully combined two types of expertise: a cartographic expertise developed by a quantitative geographical research group, the *Laboratoire d'analyse spatiale et de cartographie automatique régionale* (LASCAR);

and radical geographical knowledge on uneven development developed within GREDIN. The atlas showed that the approaches of theoretical geography and historical materialism were compatible.

The atlas was divided into an introduction, which is about how the book was produced, and four sections presenting the results. In total, it covers 40 topics grouped under the overarching themes of population, agriculture, industry, and employment. The four sections are supplemented by an exhaustive bibliography on uneven development. As a whole, the atlas shows the constraining characteristics of uneven development in the Québec administrative region. There is a core urban region, with a concentration of population, and peripheral rural areas much more sparsely populated; location sites of economic production are often not interconnected; several areas have weak productive structures; there is an exodus of the population from the most impoverished areas; and there are differences in the workforce across all areas. Overall, the spatial structure of Quebec as conveyed by the atlas is of a core-periphery. The administrative core, Quebec City, dominated the rest of the region by well-defined tentacles of management that was concentrated in a skilled and above all masculine workforce. The core controlled and drained the vitality of the periphery, the locus of production and execution, and made up of an often female and unskilled workforce (De Koninck et al. 1982: 9).[29] The criticism of this uneven distribution of the male and female workforce in the structuring of the regional space is particularly important. It led to a more advanced gendered analysis of work division, power inequality and income inequity.

## The conceptual components of the research program

Historical materialism as mobilized through GREDIN's work is based on the fundamental assumption that space is a social product that acts both as a condition and a constraint for the development of social relations of production within the capitalist mode of production. This key concept of capitalism was refined by GREDIN with the help of the concept of stages of capitalism. GREDIN examined the capitalist mode of production with reference to four stages: (1) primitive capitalism, in which the emerging bourgeoisie seizes the means of production, and merchant capital is subject to industrial capital; (2) competitive capitalism, in which capital expands; (3) monopolist capitalism, in which the bourgeoisie separates into two fractions, monopolistic and non-monopolistic; and (4) state monopoly capitalism, in which the state injects capital, assumes the costs that the enterprise does not assume, and organizes social relations to favor the growth of capital and profits.

It is the fundamental premise that space is a social product that leads to the two overarching themes of analysis of GREDIN's research program. First, that regional inequalities are produced by uneven exchange; and second, spatial planning stems from state monopoly capitalism. The main assumption of this program was that the combination of stages of capitalism, that is, State monopoly capitalism, Monopolistic capitalism, Competitive capitalism, and Precapitalism, within specific spaces determined their role in the division of labor and differentiated them both within and between social formations. These differences explained the uneven exchange between places and regions and the capacity of their social classes, in conjunction with the state, either to put into action social relations of domination-submission or to fight against them.

Later the concept of social class was revised by including the notion of social class fraction, which corresponded to stages of capitalism. The notions of the upper, middle, and petty bourgeoisie were thus applied, as well as the hypothesis of a fundamental contradiction between the interests of the upper bourgeoisie and those of the middle and petty bourgeoisie. Further, the concept of a "historical bloc" was mobilized to propose, if not strongly advocate, an alliance between non-monopolistic bourgeoisie (that is to say, petty and middle) and working classes. The analysis postulated a convergence of interests between petty bourgeoisie, middle bourgeoisie, and working classes because each belonged to a territory, which differentiated them from the upper bourgeoisie, the ruling class. The latter represented the interests of big capital and exercised its domination at the national and supranational levels, thereby imposing the conditions for uneven exchange between socio-spatial formations.

## Conclusion

The 1970s were a period of transition in the evolution of geography in the West. The recourse to Marxism represented a step forward in the construction of the scientificity of the discipline. Marxism was a method of studying reality through both the formulation of general laws of the evolution of society, and the identification of its driving forces. The GREDIN team was inspired by Marxism to understand, explain and transform society. The method was: (1) materialistic, because it saw the ideological sphere as a reflection of its material base; (2) historical, because it analyzed an evolving society; and (3) scientific, because it conceived of society as an ordered whole, governed by laws that could be objectively studied and transformed (Klein 1981: 16). Marxism thus provided an explanation of the forces that induced the evolution of society and its spatial configuration. These forces were considered the result of the contradiction between the social relations of production and the productive forces, which include the capacities and knowledge that a society possesses and puts into action within the production process. Overall, the Marxist method comprised a paradigm that opened the door to the construction of a critical geography based on theoretically powerful conceptualizations that proposed a vision of history anticipating the end point of socialism.

In the case of GREDIN, its framework favored the development of a research program focused on the region, informed by findings from the Québec administrative region based on fieldwork. GREDIN's approach did not subject the region to theoretical laws, instead it used them to clarify its analysis, to take Québec as a place seriously. Its method spread academically and to territorial actors because it revealed the social structures in which spatial inequalities were embedded and provided actors with the tools to substantiate their claims and criticisms. Subsequently, GREDIN's agenda was echoed in other regions and in other research groups, notably in peripheral regions (Eastern Québec or Saguenay–Lac-Saint-Jean).[30] Their uptake was not unrelated to the importance that the notion of territory had taken as a key concept in Québec's social geography (Rose and Gilbert 2005: 273). From that perspective, a territory is shaped by social and ideological economic forces that divide or coalesce social actors on the basis of their economic, social and political interests, as well as their identity (Rose and Gilbert 2005).

This approach was re-adjusted in the 1980s in the wake of the "crisis of grand narratives" (Lyotard 1979). Among them was a narrative about the path to socialism. The consequence was an actionist turn to analyses that privileged social movements rather than structures. Those analyses were less teleological and unilineal but nevertheless retained the resolve to fight inequalities. In that context, geography became largely oriented to a vision of development centered on actors (Klein 2010). Such a vision postulates that the local territory is a generative framework of social links and collective actions. The local reference of social actors, that is, their territorial identity, may lead them to accomplish collective actions that have economic objectives and are inspired by the sense of belonging to a local territory, thereby conciliating the economy and society. This gives rise to a community-based approach that goes under different names (social economy, solidarity-based economy, community economy, popular economy), which actually refers to forms of wealth production in which the community participates in a direct way by applying associative modalities with regard to the production of goods or services as well as to the sharing of benefits. Further, this new geography fosters an understanding of space and territory as tools to solve citizen's problems, and to equip social actors with the collective capacities necessary to change debilitating trends. From this perspective, I believe that the legacy of the radical geography of the 1970s was crucial for the development of a new approach. At its center is social, even socio-territorial innovation driven by civil society, and aimed at social trans-formation (Klein and Harrisson 2007).

## Notes

1   I would like to thank Damaris Rose who provided me many documents and helpful suggestions. I also wish to thank Trevor Barnes who made a gorgeous editing of a first version of this text. Finally, I dedicate the chapter *to the memory of Samir Amin who inspired the GREDIN's intellectual journey.* He was many things, including a great radical geographer.

2   My reference is primarily Canadian French-language universities with programs of study in geography. These are essentially universities in the province of Quebec – Université Laval, Université de Montréal, Université du Québec à Montréal, Université du Québec à Trois-Rivières, Université du Québec à Chicoutimi, Université du Québec à Rimouski – and the University of Moncton in the province of New Brunswick and the Univer-sity of Sudbury in the province of Ontario. I also included the University of Ottawa, which is partly French-speaking.

3   I am referring, of course, to the election of the government of Salvador Allende, leader of the left-wing coalition of the Popular Unity in 1970, and the revolutionary process that this government put in place to move toward realizing a socialist society. My university in Santiago was intensely involved in that process.

4   Besides managing to become commander-in-chief of the Chilean army, Pinochet also taught geopolitics, which have led some to identify him as a geographer.

5   This was generally the case in Latin American universities.

6   Like the majority of Chilean refugees at that time, I was convinced that the dictatorship of Pinochet would not last and that my exile would be no longer than three years.

7   See the special issue of the *Cahiers de géographie du Québec* about the 1970s geographical transitions (Klein 2010).

8   See Hamelin and Hamelin (1986)

9   Named after the seminal eponymous book by William Bunge, published in 1962.

10  For example, the research unit *Laboratoire d'analyse spatiale et de cartographie automatique régionale* (LASCAR) at Université Laval was founded to pursue this perspective.

11  For a more about these reforms and the social and territorial changes they brought about, see Klein *et al.* 2012.

12 "Syndicalisation citoyenne" in French; see Favreau (1989), Lévesque (1984), Bélanger and Lévesque (1992), Klein *et al.* (2014).

13 ÉZOP: Étude des zones prioritaires d'intervention dans la ville de Québec (Study of priority areas for intervention in Quebec City).

14 Author's translation.

15 The fact that the article by Dion and De Koninck (1976) was produced within the framework of GREDIN was called "Orléans: une île à vendre" is no coincidence.

16 Author's translation

17 In 2000, the region was renamed *Région de la Capitale-Nationale* (Capital-National region to assert the national self-understanding of the province of Quebec. Because my chapter refers to the period prior to this name change, I will retain the former name of the region.

18 In fact, until well into the 1950s, Montreal was the metropolis of all of Canada. However, starting at the end of that decade, economic and political power gradually became concentrated in Toronto (Fontan *et al.* 2005).

19 This principle is essential at the epistemological level. Marx and Engels assert: "The mode of production of material life conditions the general process of social, political and intellectual life." (Marx and Engels 1976: 530). This means that all the relationships of society and the state, all religious and legal systems, and all theoretical views, can only be deduced and understood from the conditions of material life.

20 This is very important because it distinguished the approach adopted by GREDIN from other groups which, although sharing similar orientations, developed a macro-social and abstract approach that placed little emphasis on the materiality of geography. Geography for them served only illustrative purposes.

21 In 2001, the name of this association was changed to *Association francophone pour le savoir*, but the acronym ACFAS has been retained.

22 *USG Newsletter*, Vol. 1, no. 4–5, 1976, p. 11.

23 *USG Newsletter*, Vol. 4, no. 1, 1978, pp. 41–45.

24 *USG Newsletter*, Vol. 4 no. 3, 1979, pp. 17–19.

25 Note the following texts on Quebec: *USG Newsletter* (Vol. 2, no. 2, February 1977): E. Wadell "Quebec 15 November 1976"; B. Richardson, "A different view of the Quebec situation"; S. Ruddick, "The Montreal citizens movement-Potential of a Plurist Party". *UGG Newsletter* (Vol. 3, no. 4, April–May, 1978). J. Bradbury, "Genesis and Exodus. Capital and Population Movements and the Quebec 1976 Election". Moreover, in 1981 the *USG Newsletter* (Vol. 6, no. 2, pp. 31–34) published in French a text signed by J. Aubin and L. Seguin providing a perspective on the 1980 referendum held in Quebec.

26 The journal changed its name in 1978 from *Cahiers de géographie de Québec* to *Cahiers de géographie du Québec* to reflect its mission of addressing the geography of the whole of Quebec and not only Quebec City. See *Direction de la revue* (1978) "Les *Cahiers de géographie du Québec au service de la géographie québécoise*." Vol. 22, no. 55, pp. 5–8.

27  Our translation.
28  The production of the atlas was financially supported by the "Formation des chercheurs et d'action concertée" program of the Quebec Ministry of Education (today named: Fonds de recherche du Québec-Société et culture) as well as other grants obtained from the same ministry and from Université Laval (De Koninck *et al.* 1982, 9).
29  Our translation.
30  In particular, the *Groupe de recherche sur le développement de l'Est du Québec* (GRIDEQ) at Université du Québec à Rimouski and the *Groupe de recherche et d'intervention régionales* (GRIR) at Université du Québec à Chicoutimi.

## References

Amin, S. (1973). *Le développement inégal.* Paris: Éditions de Minuit.

Beaumont, L. and Guay, R. (1978). La planification urbaine en milieu capitaliste monopoliste. In GREDIN, ed., *Contributions à une géographie critique*, pp. 125–136. Québec, département de géographie de l'université Laval. NDR 9.

Bélanger, F. (1978). Pour une approche critique de la planification urbaine. In GREDIN, ed., *Contributions à une géographie critique*, pp. 87–93. Québec, département de géographie de l'université Laval. NDR 9.

Bélanger, P. R. and Lévesque, B. (1992). Le mouvement populaire et communautaire: de la revendication au partenariat (1963–1992). In G. Daigle and G. Rocher, eds, *Le Québec en jeu. Comprendre les grands défis*, pp. 713–747. Montréal: Presses de l'Université de Montréal.

Bernier, B. (1978). Les phénomènes urbains dans le capitalisme actuel. *Cahiers de géographie du Québec* 22 (56): 189–216.

Bradbury, J. (1978). La géographie et l'État. *Cahiers de géographie du Québec.* 22 (56): 293–300.

Bruneau, M. (1978). Évolution de la formation sociale et transformation de l'espace dans le nord de la Thaïlande (1850–1977). *Cahiers de géographie du Québec* 22 (56): 217–263.

Brunelle, D. (1978). *La désillusion tranquille.* Montréal: HMH.

Cardoso, F.-E. and Faletto, E. (1968). *Dependencia y desarrollo en América latina.* México: Siglo 21.

De Koninck, R. (1978a). Le Groupe de recherche sur l'espace, la dépendance et les inégalités. *Cahiers de géographie du Québec* 22 (56): 303–305.

De Koninck, R. (1978b). Changer la géographie: notes pour une discussion. In GREDIN, ed., *Contributions à une géographie critique*, pp. 3–50. Québec, département de géographie de l'université Laval. NDR 9.

De Koninck, R. (1978c). À propos de la division du travail, des hommes et des espaces: notes sur la question ville campagne. In GREDIN, ed., *Au sujet des exigences spatiales du mode de production capitaliste*, pp. 3–14. Québec, département de géographie de l'université Laval. NDR 10.

De Koninck, R. (1978d). Contre l'idéalisme en géographie. *Cahiers de géographie du Québec* 22 (56): 123–145.

De Koninck, R. (1978e). Le matérialisme historique en géographie: note liminaire. *Cahiers de géographie du Québec* 22 (56): 117–122.

De Koninck, R., Lavertue, R., Raveneau, J. with the collaboration of Fortin, R., Gosselin, J., Grasland, L., Guenet, M., Klein, J.-L., Pelletier, L., Perrotte, R., and Villeneuve, P-Y. (1982). *Le développement inégal dans la région de Québec*. Québec: Presses de l'Université Laval.

Dion, M. and De Koninck, R. (1976). L'État et l'aménagement: Orléans, une île à vendre. *Cahiers de géographie du Québec* 20 (49): 39–67.

ÉZOP-QUÉBEC (1981). *Une ville à vendre*. Laval: Éditions coopératives Albert Saint-Martin.

Favreau, L. (1989). *Mouvement populaire et intervention communautaire*. Montréal: Centre de formation populaire.

Fontan, J.-M. (1992). *Les corporations de développement économique communautaire montréalaises. Du développement économique communautaire au développement local de l'économie*. Ph.D. thesis in sociology. Université de Montréal.

Fontan, J.-M., Klein, J.-L. and Tremblay, D.-G. (2005). *Innovation socioterritoriale et reconversion économique. Le cas de Montréal*. Paris: L'Harmattan.

Forcier, P. (1978). Croissance de la population et stagnation de l'agriculture au Cambodge: essai sur les conditions permissives d'un processus révolutionnaire. *Cahiers de géographie du Québec* 22 (56): 265–277.

Galois, B. and Bradbury, J. (1978). L'union des géographes socialistes. *Cahiers de géographie du Québec* 22 (56): 301–302.

Giguère, J. (1978). Les fondements de la division Baie-Comeau/Hauterive. In GREDIN, ed., *Au sujet des exigences spatiales du mode de production capitaliste*, pp. 87–98. Québec, département de géographie de l'université Laval. NDR 10.

Hamel, P. (1991). *Action collective et démocratie locale. Les mouvements urbains Montréalais*. Montréal: Presses de l'Université de Montréal.

Hamelin, L. E and Hamelin, C. (1986). Les carrières canadiennes de Raoul Blanchard et Pierre Deffontaines. *Cahiers de géographie du Québec* 30 (80): 137–150.

Klein J.-L., Tremblay, D.-G., and Fontan, J.-M. (2014). Social actors and hybrid governance in community economic development in Montreal. In N. Bradford and A. Bramwell, eds, *Governing Urban Economies: Innovation and Inclusion in Canadian City Regions*, pp. 37–57. Toronto: University of Toronto Press.

Klein, J.-L. and Naud, M.-A. (1978). Le développement régional, un problème de classes. In GREDIN, ed., *Contributions à une géographie critique*, pp. 97–103. Québec, département de géographie de l'université Laval. NDR 9.

Klein, J.-L. and Harrisson, D. eds (2007). *L'innovation sociale*. Québec: Presses de l'Université du Québec.

Klein, J.-L. (1978a). Le matérialisme historique et l'espace. In GREDIN, ed., *Contributions à une géographie critique*, pp. 69–77. Québec, département de géographie de l'université Laval. NDR 9.

Klein, J.-L. (1978b). Division spatiale du travail et développement régional inégal dans la région de Québec. In GREDIN, ed., *Au sujet des exigences spatiales du mode de production capitaliste*, pp. 49–61. Québec, département de géographie de l'université Laval. NDR 10.

Klein, J.-L. (1978c). Du matérialisme historique aux inégalités régionales. Le cas de la région de Québec. *Cahiers de géographie du Québec* 22 (56): 173–187.

Klein, J.-L. (1981). *Région, déploiement du capital et cout du travail. Contribution à l'analyse de la croissance manufacturière dans la région de Québec.* Ph.D. thesis en Geography. Université Laval.

Klein, J.-L. (2010). Les changements de paradigme en géographie et l'aménagement du territoire au Québec: les années 1970. *Cahiers de Géographie du Québec* 54 (151): 133–152.

Klein, J.-L., Fontan, J.-M. Harrisson, D. and Lévesque, B. (2012). The Quebec System of social innovation: a focused analysis on the local development field. *FINISTERRA Revista Portuguesa de Geografia XLVII* (94): 9–28.

Lavertue, R. and Villeneuve, P.-Y. (1978). Développement régional et polarisation au Québec. Le rapport HME revu et corrigé. In GREDIN ed., *Contributions à une géographie critique*, pp. 113–123. Québec, département de géographie de l'université Laval. NDR 9.

Lavertue, R. (1978a) La soumission de la région au capital monopoliste. In GREDIN, ed., *Au sujet des exigences spatiales du mode de production capitaliste*, pp. 23–31. Québec, département de géographie de l'université Laval. NDR 10.

Lavertue, R. (1978b) La croissance industrielle de La Beauce. In GREDIN, ed., *Au sujet des exigences spatiales du mode de production capitaliste*, pp. 77–86. Québec, département de géographie de l'université Laval. NDR 10.

Lévesque, B. (1984). Le mouvement populaire du Québec: de la formule syndicale à la formule coopérative? *Coopératives et Développement* 16 (2): 43–66.

Lyotard, J.-F. (1979). *La condition postmoderne.* Paris: Éditions de minuit.

Marx, K. and Engels, F. (1976). Feuerbach, l'opposition de la conception matérialiste et idéaliste. In *Œuvres choisies (T.1)*, pp. 10–81. Moscou: Editions du Progrès.

Mathieu, P. (1978). Développement régional et inégalités spatiales: le cas du Costa Rica. In GREDIN, ed., *Contributions à une géographie critique*, pp. 139–156. Québec, département de géographie de l'université Laval. NDR 9.

Naud, M.-A. (1978). L'impérialisme et ses contraintes spatiales. In GREDIN, ed., *Contributions à une géographie critique*, pp. 79–86. Québec, département de géographie de l'université Laval. NDR 9.

Peet, R. (1978). Materialism, Social Formation, and socio-spatial relations: an essay in Marxist geography. *Cahiers de géographie du Québec* 22 (56): 147–157.

Piot. M. (1977). Agriculture et capitalisme au Québec: l'agroindustrie et l'État. *Anthropologie et Sociétés* 12: 71–88.

Piot, M. (1978a). Les concepts fondamentaux du matérialisme dialectique et historique. In GREDIN, ed., *Contributions à une géographie critique*, pp. 53–68. Québec, département de géographie de l'université Laval. NDR 9.

Piot, M. (1978b). Développement inégal et disparités régionales: réflexions sur le rapport du mode de production à l'espace. In GREDIN, ed., *Contributions à une géographie critique*, pp. 105–111. Québec, département de géographie de l'université Laval. NDR 9.

Piot, M. and Hulbert, F. (1978a). L'État, les monopoles et la région. In GREDIN, ed., *Au sujet des exigences spatiales du mode de production capitaliste*, pp. 15–22. Québec, département de géographie de l'université Laval. NDR 10.

Piot, M. and Hulbert, F. (1978b). Développement régional et politique de subventions à l'entreprise privée dans la région 03 de Québec. In GREDIN, ed., *Au sujet des exigences spatiales du mode de production capitaliste*, pp. 63–76. Québec, département de géographie de l'université Laval. NDR 10.

Rose, D. and Gilbert, A. (2005). Country report: Glimpses of social and cultural geography in Canada and Québec at the turn of the millennium, *Social & Cultural Geography* 6(2): 271–298, DOI: 10.1080/14649360500074725.

Roy, A. and Beaumont, L. (1978). Concentration de la propriété foncière et réaménagement urbain à Québec et à Ste-Foy. In GREDIN, ed., *Au sujet des exigences spatiales du mode de production capitaliste*, pp. 99–117. Québec, département de géographie de l'université Laval. NDR 10.

Savarria, J. (1976). Le Québec est-il une société périphérique? *Sociologie et sociétés* VII (2): 115–128.

Villeneuve, P.-Y. (1978a). Développement régional et pouvoir politique au Québec: PME ou CME? In GREDIN, ed., *Au sujet des exigences spatiales du mode de production capitaliste*, pp. 33–46. Québec, département de géographie de l'université Laval. NDR 10.

Villeneuve, P.-Y. (1978b). Classes sociales, régions et accumulation du capital. *Cahiers de géographie du Québec* 22 (56): 159–172.

Villeneuve, P.-Y. (1982). Ezop-Québec (1981) Une ville à vendre. *Cahiers de géographie du Québec* 26 (68): 279–281.

# Part II

## Radical Geography beyond North America

# 10

# Japan

## *The Yada Faction versus North American Radical Geography*

Fujio Mizuoka[1]

## The Long Heritage of Marxist Geography in Japan

While it may seem surprising, critical or radical geography appeared in Japan as early as the 1930s. Notwithstanding the totalitarian Japanese state's suppression of academic freedom, Marxism attained its zenith in pre-World War II Japan in this period. Several Soviet and German publications were translated into Japanese, including those of Wittfogel (1929). Yet Japanese authorities suppressed this trend as the war intensified, as evidenced by Keishi Ohara's arrest,[2] accused of participating in the Research Group on Materialism, and subsequent resignation from Yokohama Commercial College in 1942 due to his critical position toward the aggressive policy of the Japanese militarist regime (Takeuchi 2000: 72).

With the arrival of freedom of speech during post-war democratization, critical geographers organized themselves. This is likely the first attempt worldwide to establish a radical "counter institution" to mainstream Geography: Geographers

> turned their back on the existing authority and order, and ... struggled to reform the Association of Japanese Geographers [AJG, the national school] by campaigning for candidates [advocating democratic reforms] as councillors; while they simultaneously poised themselves for the foundation of the JAEG [Japan Association of Economic Geographers] in 1954. (Kazamaki 1998: 72).

*Spatial Histories of Radical Geography: North America and Beyond*, First Edition.
Edited by Trevor J. Barnes and Eric Sheppard.
© 2019 John Wiley & Sons Ltd. Published 2019 by John Wiley & Sons Ltd.

For these rebels, the term "economic geography" was a surrogate for "critical" or "Marxist" geography. The 1954 inaugural issue of *The Annals of the Japan Association of Economic Geographers* (*AJAEG*) showed a strong critical and theoretical inclination, publishing such articles as "Materialism and Geography" and "The Allocation of Agricultural Productive Forces in the Soviet Union". The second issue contained a paper by Tetsuro Kawashima[3], later to become seminal, establishing a new agenda for economic geography. Formulating a Marxist theory concerning the "spatial distribution of economic phenomena and localities" and their "development and demise" (Kawashima 1955: 9), he concludes: "Both the overcoming of localities and the transcendence of class are the major targets that humans must and can achieve" (p. 17).

## The Progress of Critical Geography in Japan: 1950s–1960s

During the JAEG's early years, its members worked hard to establish their conceptual foundations with Soviet ideology as their inspiration. Kamozawa (1954: 271) proclaimed: "Upon the gradual transition from socialism to communism, the Soviet academic circle, in drawing upon the epoch-making paper of Stalin, achieved publication of papers that further promoted economic geography," asking Japanese geographers to "study the works of Stalin thoroughly." From the late 1960s to the early 1970s, JAEG members strove hard to create concepts unique to their own critical geography. A series of articles containing creative conceptualizations of critical geography also appeared in the 1966 volume of the *Geographical Review of Japan* (*GRJ*), the organ of the Association of Japanese Geographers.

During this brief interlude, critical geography in Japan enjoyed its heyday (perhaps along with its counterpart in France) relative to geographers taking a critical line worldwide, even though Japanese critical geography had little global reach. Critical geographers attempted to establish concepts of society-space or space-place interfaces, a staple theme decades later among radical geographers in North America and other English-speaking countries (Mizuoka *et al.* 2005: 458–459). The work of Noboru Ueno is of particular note. In his influential *The Ultimate Origin of Chorography*, Ueno (1972) attempted to integrate Marxist and humanistic approaches into critical geographic theory, a significant theoretical achievement within critical Japanese geographical scholarship. Drawing upon interpretations of Heideggerian phenomenology, Ueno explained the objective as intersubjective in interpreting produced nature (Mizuoka *et al.* 2005: 459–461).

## Advent of the Yada Faction: 1973 and After

1973 marked a major turning point for radical geographies in both Japan and North America, but in totally contrary directions. In Japan, Toshifumi Yada (1973) published "On Economic Geography." Born in 1941, as an undergraduate Yada engaged in a settlement house movement. Under the strong influence of the Japanese Communist Party (JCP), this movement sought to empower neighborhood people living in poverty, while training the participating students to become radical intellectuals. As a graduate student, Yada became a Marxist geographer specializing in resource economics, especially the demise of domestic coal mining, denouncing the government's policy of scrapping the domestic coal industry in favor of petroleum controlled by multinational capital. He claimed that the government sought to "support the large enterprises and eliminate small companies" (Yada 1967: 19), damning "monopoly capital exploiting and abusing domestic resources on the pretext of 'regional development' and 'urbanization'". Using a pseudonym, he demanded self-sufficiency in energy through exhaustive deployment of domestic coal mining and the direct trading for oil from socialist and other oil-producing countries. This demand followed the JCP's revolutionary program to establish a "Democratic Coalition Government" (Akiba 1976: 299).

In 1973 Yada published a paper, later to become the conceptual intro-duction to *The Regional Structure of Post-War Japanese Capitalism* (Moritaki and Nohara 1975), castigating earlier Japanese critical geogra-phers, including Kamozawa, Kawashima, and Ueno, for subscribing to the "economic chorography school." He put forward an alternative that he named the "regional structure concept"*(chiiki kōzō ron)*. According to his definition (Yada 1990: 15–16), *chiiki kōzō* is "the system of the regional division of labor of a national economy," which is "in principle determined by the industrial structure, or the system of social division of labor. The "[l]ocation of various sectors and functions that constitute the industrial structure" and "regional [economic] circuits that unfold based on these locations" (Yada 2000: 300) work together to configure a "*chiiki kōzō.*" Yada then set out to identify relatively autonomous "regional economic circuits" within Japan (2000: 301). For this purpose, he organized the Chiiki Kōzō Kenkyūkai (CKK: Research Group for Regional Structure), working with the conventional industrial geographer Kitamura as figure-head, and with a core group of a score of other economic geographers.

In pursuit of identifying *chiiki kōzō*, critical concepts were tacitly replaced with neoclassical ones, raising skepticism about how the identification of "regional economic circuits" could be the agenda for a radical economic geography. Identifying regional economic circuits is more homologous with central-place theory or shift-share analysis: a better fit to neoclassical economic geography. Associating *Chiiki kōzō* with critical scholarship was thus imputed barely to Yada's previous scholarship and his association with the JCP.

Yada went about quite secretly building a faction. While a graduate student at Hitotsubashi University I was invited to a "workshop," held in early 1977 in a suburban Tokyo, on the pretext of reviewing *The Regional Structure of Post-war Japanese Capitalism*. In the workshop, various tasks for "strategic academic politics" were assigned to, and mutually agreed upon by the participants. At the workshop's conclusion, I was told not to disclose it to anyone else, a method of organizing a faction behind the scene that resembled a Leninist-Stalinist-type communist party.

The JAEG commemorated its 40th anniversary by devoting its 1993 annual meeting to the topic "Society and Space," drawing heavily upon North American radical geography. Yet this was the last flash of this critical legacy within the JAEG. Yada school pushed through a "constitutional amendment" to secure hegemony over the JAEG. Kenji Yamamoto proposed the amendment in 1999, intended to abolish the free and direct election of the JAEG executive.[4] Its members were to be chosen instead "by consultation" with JAEG's elected councilors, who played only a nominal role in the daily operation of the JAEG, in other words under the table. During a three-hour debate at the JAEG general assembly, on May 23, 1999, Mitsuo Yamakawa, another close ally of Yada who assumed chairpersonship of the general assembly, attempted to reject the oppositional group's right to propose a more democratic alternative to the original amendment bill. As the administration began to distribute ballots to members on the floor, Kitamura (the former head of CKK) demanded that the vote be taken by a show of hands, forcing everyone to reveal their positions openly. Then JAEG president Keiichi Takeuchi regarded this amendment to the JAEG constitution as just enhancing the feasibility and efficiency of its administration, thereby demonstrating tacit support for the Yada faction's takeover of the JAEG.[5]

Under the amended constitution, the autumn 1999 election outcome was "certainly very disappointing, although by now expected" (Smith 2000). Almost across the board, most of the economic geographers who had been active since the 1960s and 1970s outside the CKK, having served as JAEG councilors for many years, lost their seats. Younger generation critical geographers closely collaborating with scholars of radical geography in North America also were defeated. Aono, who had attempted in the 1970s to organize a study group for Richard Peet's edited *Radical Geography* book (Peet 1977), ran in 1999 as a JAEG presidential candidate but lost by a wide margin to Yada.

A group of the Yada faction's handpicked members formed the JAEG executive board, with Yada as president. Immediately thereafter, Toshio

Nohara, who had once collaborated with Yada by editing *The Regional Structure of Post-war Japanese Capitalism*, quit the JAEG altogether.[6] In short, while North American critical geographers strove hard "to construct a new, philosophical base for human geography" (Peet 1977: 20), their Japanese counterparts were stifled through this sort of political play for domination in academic circles.

Yada's accusations against the "economic chorography school" were more a political maneuver to establish his faction than a serious academic endeavor. Indeed, there was no tangible substance of such a "school." Yet Yada's personal ambition undermined the heritage of a radical geography that Japanese geographers should feel proud about at least until 1972.

## Attempts to Associate with North American Radical Geography

I was an undergraduate studying Marxian economics when Ueno's work was published. Yada became more prominent during my graduate study. Reading *Capital* (Marx 2012–2013 edn) and *Theories of Surplus Value* (Marx 2000 edn) in graduate seminars, I joined Yada's study group, but became increasingly skeptical of its critical nature. At a 1979 seminar organized by the National Association of Geography Graduate Students, I therefore offered a friendly criticism of Yada's work, arguing that his *chiiki kōzō* concept was a rehash of ideas presented by Japanese regional economists in the 1950–1960s, and explaining that Yada's "concept cannot establish originality in the social sciences at large" (Zenchi-inren 1980: 151). Much to my surprise, Yada vehemently replied, claiming that "Mizuoka [is] bent on pointing out obsoleteness to me, who is striving for the systematic concept of economic geography, ... without putting forward any of his own concepts" (Yada 1980: 156).

While I was a visiting lecturer at the University of Hong Kong (1979–1981), responding to Yada's critique I sought to establish a stronger conceptual foundation to replace *chiiki kōzō*. I aspired to pursue a Ph.D. by presenting a systematic conception of critical economic geography, and the radical geography then emerging in North America appeared useful for this purpose. Thus, financed by a full Fulbright scholarship for study in the United States, I moved to Clark University for my Ph.D. as it was a base of the North American radical geography movement through its publication of *Antipode*.

My doctoral dissertation, titled *Annihilation of Space* (Mizuoka 1986), presented my own critical economic geography in response to Yada's provocation. My argument focused on the concept of space subsumption, drawing upon Marx's formal and real subsumption of labor (Saito and Mizuoka 2008: fig. 1). Peet was my supervisor, and David Harvey volunteered to be a member of my dissertation committee.

After conferral of my doctoral degree, I was appointed in 1986 as an Associate Professor of economic geography in the Faculty of Economics of Hitotsubashi University, where Marxian economics still had a considerable influence. Harvey kindly wrote me a strong letter of recommendation to the Dean. Here, I worked to introduce North American radical geography to Japan. I organized a group of Marxist economists and geographers to translate Harvey's *The Limits to Capital* (1982) and *Urbanization of Capital* (1985), as well as Allen Scott's *Metropolis* (1988).

Nohara had been the first Marxist geographer to introduce *The Limits to Capital* to a Japanese audience, before the translation was available. In his comprehensive review, Nohara (1986) emphasized Harvey's concepts of the built environment, spatial configuration, and the first to third cuts of crisis theory. Examining how these could be applied to the analysis of localized industries, he positioned Harvey's concepts as one of the *chiiki kōzō* models (Nohara 1986: tables 1–4), which he defined as the fundamental contextualized concept of place. In spite of this attempt to associate Harvey with *chiiki kōzō*, Yada showed little interest in Nohara's argument.

Harvey's work aroused more interest in Japan among sociologists and Marxian economists than geographers. Although *Social Justice and the City* (1973) was translated into Japanese by geographers in 1980, *Condition of Postmodernity* (1992) was translated by a group of sociologists conducting research on the concept of the urban built environment (Yoshihara, translator-in-chief 1999), although it contained a considerable number of mistranslations and wrong interpretations. Recent works of Harvey on *Capital* have been translated by Marxian economists, notably Seiya Morita (who also has translated some works of Trotsky).

North American radical geography has had influence outside academia as well. A movie director who had studied architecture as an undergraduate in Tokyo went to UC Berkeley in 2004, yet was disappointed by what he saw as its low-spirited armchair liberals: "university staffs were rich and in a safety zone unlike the liberals risking one's life in 1960s, and kept talking over wine what the liberal might be" (Ishikawa 2015). Quitting UC Berkeley and transferring to New York City University, she attended the high-spirited lectures of Harvey and others there, which presumably inspired her movie production later on.

## Academic Seclusion from Global Trends in Radical Geography

In the meantime, Yada and his faction of geographers carefully avoided adopting the radical geography of North America and other Anglophone countries. Allowing the penetration of more sophisticated foreign radical geographies would have surely undermined the hegemony of *chiiki kōzō* among Japanese geographers. Yada also tacitly shifted his political position from the JCP revolutionary line to align with Japan's neoliberalizing state, as evidenced by his participation in government councils on national spatial policies.

Although North American radical geographies were diffusing globally, Galapagos-like academic seclusion ensued among the Japanese economic geographers closely associated with the JAEG. Hiroshi Matsubara (2000), a Yada disciple who had once showed enthusiasm for Harvey's geography, confessed that he was no longer interested in Harvey's built environment concept.[7] Some JAEG members worked to translate Lloyd and Dicken's (1990) third edition of *Location in Space* into Japanese (Ito, translator-in-chief, 1997). This edition was unique in drawing heavily upon critical concepts of space developed in Anglophone countries, especially in Part II. Yet this translation failed to properly render the text into Japanese. For instance, the Marxist term "mode of production" was dropped entirely in its discussion of the conditioning factor of capitalist social relationships (Lloyd and Dicken 1990: 362), and heterogeneity(-nous) in relation to the spatial (Lloyd and Dicken 1990: 83, 100) was mistranslated as "homogeneity."

Neoliberal concepts came to permeate the JAEG, replacing critical concepts. Yamakawa organized the JAEG's 1998 annual conference on "Deregulation and Regional Economy." Yada gave an uncritical briefing on the main features of recent national development projects, emphasizing "National Land Axes". He commented (Yada 1998: 102–103):

> We should no longer use the concept of balanced growth to legitimize mere redistribution of public investment and income. ... There is no need to provide every single local state with uniform sets of services. With transportation and other networks well-equipped, those who value proximity to a city and enjoyment of urban services should opt to live in the city while those who prefer to indulge themselves in nature with only occasional trips to the city might opt for living in a "multi-natural living zone. ... The residents are then left to their own choices...

In short, Yada adopted a hypothesis akin to Tiebout's (1956) "voting with one's feet" with clear contempt for egalitarian spatial planning ideals.

Yada also nourished a wishful self-image, creating a chart titled "The Principal Theories of Economic Geography in the World" (Yada 2000: 287–288). He placed a collection of scholars along its "spatial axis," including the "aspatial" Marx, Daniel Bell and Freeman, Wallerstein, and Lipietz's "spatial system of world economy," continuing to Pred and Castells' "spatial system of informational economy," Massey and Porter's "spatial system of corporate economy," and Scott and Markusen's "spatial system of regional economy." He culminated the list with the geographer who had put forward "the spatial system of national economy and its restructuring" – Yada himself.

## Beyond Japan: The East Asian Regional Conferences in Alternative Geography

Japan's pre-war critical geography did not diffuse to its former colonies. Shortly after Korea's independence and Taiwan's handover to China, these states were placed under political dictatorships where dissident thought, including Marxism, was brutally oppressed. The British in Hong Kong also detested Communist thought.

In this Cold War political climate, it was not until the 1980s, when South Korea and Taiwan became democratic, that social movements were ignited and scholars gained academic freedom. The work of North American radical geographers was readily available to geographers there, who recognized the significance of critical and Marxist thought. Unlike Japan, there was no past legacy of critical geography to hinder the adoption of radical geography. In South Korea, for example, a geographer addressed the "tasks for alternative geography in Korea" (Choi 2007: 237–240).

In December 1999, a group of Korean and Japanese critical geographers met in Kyongju, South Korea to hold the first meeting of the East Asian Regional Conferences in Alternative Geography (EARCAG), where Neil Smith delivered the keynote lecture. Hong Kong critical geographers joined EARCAG soon thereafter, and Wing-Shing Tang at Hong Kong Baptist University hosted its second meeting in Hong Kong in 2001, with Richard Peet delivering the keynote (Tang and Mizuoka 2010). EARCAG's geographical coverage expanded into Southeast Asia in 2012 when Amriah Buang at the National University of Malaysia hosted the 6th meeting. Over two decades, EARCAG has grown into a regional centerpiece of the radical geography movement, informally associated with the achievements of North American radical geography yet also striving to develop the critical geography concepts unique to East Asia. The current task of EARCAG should be to maintain a high critical spirit comparable to North American radical geography in 1970s into the future.

Some EARCAG members had contributed to the organization of the 6th Global Conference on Economic Geography (GCEG), to be held in Tokyo in 2021, on the theme "global, struggle/conflict, and symbiosis" (The GCEG2021 Organising Committee, 2017). In drawing upon radical scholarship's achievements in North America and across the world, the GCEG2021 Organising Committee had aimed to make this conference distinct from previous more bourgeois-oriented GCEGs, focusing on critical orientations to support grass-roots and subjugated peoples.

## Conclusion

Notwithstanding early forays into Marxist economic geography, dating back to the 1930s, it has been a major struggle for radical economic geographers trained in the North American Marxist tradition to gain a foothold within a Japanese economic geography dominated by a Yada faction that is dismissive of such work. In response, such radical geographers have networked with other East Asian colleagues via EARCAG. In Japan, concepts of radical geography have been penetrating into critical scholarship more broadly, not just in Geography, as manifest in the enthusiasm to publish the second edition of Harvey's *Limits to Capital*, translated into Japanese based on the latest Verso edition. There is also a move to deepen critical geographical theories, including my plan to publish a book on a comprehensive theory of spatial struggle based on the concept of subsumption of space.

At the same time, and outside economic geography, Japanese cultural and political geographers have begun to create their own spaces for practicing radical and critical geography within Japan, such as the annual journals *Space, Society, and Geographical Thought* (for more details, see Mizuoka *et al.* 2005) and *City, Culture and Society*.

## Notes

1   This chapter is a heavily edited version of the author's previous paper "The Demise of a Critical Institution of Economic Geography in Japan," in *Critical and Radical Geographies of the Social, the Spatial and the Political, 2006* and Mizuoka (1996). Earlier drafts of this chapter were presented at the AAG Annual Meeting in Honolulu in March 1999 and at the Global Conference on Economic Geography in Singapore in December 2000.
2   Keishi Ohara (1903–1972), an economist specialising in North American economies, taught economic geography at Yokohama Commercial

College during the pre-war period published a book on critical geography (Ohara 1950). He served as president of the Japan Association of Economic Geographers (1963–1974).

3  Tetsuro Kawashima (1918–2002), an economist originally specialising in Marxist ground rent theory, taught economic geography at Osaka City University until 1982 and served as president of JAEG (1979–1988).

4  Yada's successor at Hosei and Kyushu Universities, the JAEG secretary-general, and a member of the Advisory Board of the 4th and 5th Global Conference on Economic Geography

5  The author's direct observation of the general assembly held in Chukyo University, Nagoya.

6  Toshio Nohara (1930–), a Marxist economic geographer closely associated with the JCP, specializes in localized industry and cooperative movements. Having initially supported Yada's *chiiki kōzō*, he took up the coeditorship of *The Regional Structure of Post-war Japanese Capitalism* (Moritaki and Nohara 1975). He was also once active in the JAEG, e.g. serving as the president of its Nagoya division.

7  Matsubara, replying to my question at his presentation at the 3rd Global Conference on Economic Geography (GCEG) held in Seoul from June 28–July 2, 2011.

8  All literature by Japanese authors cited here is in Japanese, unless otherwise noted.

# References[8]

Akiba, S. (1976). On the new policies towards coal industry under the "energy crisis": a history of the post-war coal policies and critique of the 6th Report, 4, *Keizai*, 142: 288–299.

Choi, B. D. (2007). Beyond developmentalism and neoliberalism: development process and alternative visions for Korean geography. *Journal of the Korean Geographical Society*, 42(2): 218–242.

The GCEG2021 Organising Committee (2017). *Official Website*. www.gceg2021.org/ (accessed on September 15, 2017).

Harvey, D. (1973) *Social Justice and the City*. London: Edward Arnold. (Translated into Japanese by K. Takeuchi and M. Matsumoto, Tokyo: Nippon Buritanika, 1980).

Harvey, D. (1982) *The Limits to Capital*. Chicago: University of Chicago Press. (Translators-in-chief, K. Matsuishi and F. Mizuoka, Tokyo: Taimeido, 1989–1990).

Harvey, D. (1985). *The Urbanization of Capital*. Baltimore: The Johns Hopkins University Press. (Translator-in-chief, F. Mizuoka, Tokyo: Aoki Shoten, 1992).

Harvey, D. (1992) *The Condition of Postmodernity*. New York: Wiley-Blackwell (Translator-in-chief, N. Yoshihara, Tokyo: Aoki Shoten, 1999).

Ishikawa, T. (2015). An interview: the direction of borderless crowd, *Traverse: Kyoto University Architectural Journal*, 15. https://www.traverse-architecture.com/interview15–2-1 (accessed on January 18, 2018).

Ito, Y. (translator-in-chief) (1997). *Ritchi to Kūkan: Keizaichirigaku no kisoriron*: Vols. 1 and 2. Tokyo: Kokon Shoin. [Original: Lloyd, P. and Dicken, P. (1990). *Location in Space: Theoretical Perspective in Economic Geography*. 3rd edn, New York: HarperCollins].

Kamozawa, I. (1954). Znacheniye Trula I. V. Stalina 'Ekonomicheskie Problemi Sochializma v SSSR' Dlya Ekonomicheskoi Geografii (O. A. Konstantinov), *Geographical Review of Japan* 27(6): 269–271.

Kawashima, T. (1955). On the economic region, *Annals of the Japan Association of Economic Geographers* 2: 1–17.

Kazamaki, Y. (1998). Prehistory of the Japan Association of Economic Geographers and Our Meetings, *Annals of the Japan Association of Economic Geographers* 44(1): 65–73.

Marx, K. (2012–2013 edn). *Das Kapital, Volumes 1–3*. Berlin: Dietz Verlag.

Marx, K. (2000 edn) *Theorien über den Mehrwert, Marx-Engels Werke, Volume 26 Teil 1–3*. Berlin: Dietz Verlag.

Matsubara, H. (2000). David Harvey: spatial and urban theories of capital. In T. Yada and H. Matsubara, eds, *Gendai Keizai Chirigaku: Sono Genryū to Chiiki Kōzō Ron [Contemporary Economic Geography: Its Origin and the Theory of Regional Structure]*, pp. 2–20. Kyoto: Minerva Shobo.

Mizuoka, F. (1986). *Annihilation of Space: A Theory of Marxist Economic Geography*. Ph.D. Dissertation, Clark University.

Mizuoka, F. (1996). The disciplinary dialectics that has played eternal pendulum swings: Spatial theories and disconstructionism in the history of alternative social and economic geography in Japan, *Geographical Review of Japan, Series B*, 69(1): 95–112 (in English).

Mizuoka, F., Mizuuchi, T., Hisatake, T., Tsutsumi, K., and Fujita, T. (2005). The critical heritage of Japanese geography: its tortured trajectory for eight decades, *Society and Space* 23(3): 453–473 (in English).

Moritaki, K. and Nohara, T. eds (1975). *Sengo Nippon shihonshugi no chiiki kōzō [The regional structure of post-war Japanese capitalism]*. Kyoto: Chobunsha.

Nohara, T. (1986). *Gendai Chiiki Sangyo [The Contemporary Localised Industries]* Tokyo: Shin-Hyoron.

Ohara, K. (1950). *Shakai Chirigaku no Kiso Riron [Fundamental Theories of Social Geography]*. Tokyo: Kokon Shoin.

Peet, R. ed. (1977). *Radical Geography*. London: Methuen.

Saito, A. and Mizuoka. F. (2008). Fujio Mizuoka and the theory of space subsumption. In R. Kitchin and N. Thrift, eds, *International Encyclopedia of Human Geography, Volume 6*, pp. 4–10. Amsterdam: Elsevier (in English).

Scott, A. J. (1988). *Metropolis: From Division of Labor to Urban Form*, Berkeley: University of California Press (Translator-in-chief, Mizuoka. Tokyo: Kokon Shoin, 1996).

Smith, N. (2000). A message as to the election result of the JAEG. Quoted in Keizai Chiri Gakkai – A Counter Website. http://hit-u.ac/gakkai/ (accessed on September 15, 2017).

Takeuchi, K. (2000). Japanese Geopolitics in the 1930s and 1940s. In K. Dodds and D. Atkinson, eds, *Geopolitical Traditions: A Century of Geopolitical Thought*, pp. 72–92. London: Routledge. (in English).

Tang, W-S and Mizuoka, F. eds (2010). *East Asia: A Critical Geography Perspective*. Tokyo: Kokon Shoin.

Tiebout, C. M. (1956). A pure theory of local expenditures, *Journal of Political Economy* 64(5): 416–424.

Ueno, N. (1972). *Chishigaku no genten [The ultimate origin of chorography]*. Tokyo: Taimeido.

Wittfogel, K. A. (1929). Geopolitik, geographischer Materialismus und Marxismus, in *Unter dem Banner des Marxismus*, 3 (1, 4 and 5). Translated into Japanese by Kawanishi, 1935, as *Chirigaku Hihan [A critical appraisal of geography]*. Tokyo: Yukosha. Also translated into English by G. L. Ulmen, *Antipode* 17(1): 21–71.

Yada, T. (1967) The abandonment of poor coal seams caused by the rationalization of the coal-mining industry in the Joban coal field, *Annals of the Japan Association of Economic Geographers* 13(1): 1–19.

Yada, T. (1973). On economic-geography, *Keizai-Shirin (The Hosei University Economic Review)* 41(3, 4): 373–410.

Yada, T. (1980). On the criticism of Fujio Mizuoka, *Chiri* 25(9): 154–156.

Yada, T. (1990). *Chiiki Kōzō no Riron [The Theory of Regional Structure]*. Kyoto: Minerva Shobo.

Yada, T., Shimohirao, I., Tsuboi, T., Araya, K., and Suzuki, H. (1998). Land management for the 21st century. *Annals of the Japan Association of Economic Geographers* 44(4): 355–364.

Yada, T. (2000). Contemporary economic geography and *Chiiki Kōzō-ron*. In T. Yada and H. Matsubara, eds, *Gendai Keizai Chirigaku: Sono Genryū to Chiiki Kōzō Ron [Contemporary Economic Geography: Its Origin and the Theory of Regional Structure]*, pp. 279–312. Kyoto: Minerva Shobo.

Zenchi-inren (1980). The Zenchi-inren Seminar Report. *Chiri* 25(7): 138–151.

# 11

# The Rise and Decline of Radical Geography in South Africa

Brij Maharaj

Over the past two decades, there have been many reviews of radical geography (e.g., Castree 2000). However, there is little reference to any radical work outside the Anglo-American network. Reflecting on such "parochialism," Robinson (2003: 275) warned ominously that "a geography whose intellectual vision is limited to the concerns and perspectives of the richest countries in the world has little hope of effectively participating in the debates that will matter in the twenty-first century." This chapter analyzes the rise and decline of radical geography in South Africa, an appropriate juncture as the country celebrates the centenary of the discipline at the tertiary level, whose roots can be traced back to 1916 when colonialism was dominant (Visser *et al.* 2016). Most branches of the geography discipline in South Africa were "imported" from the Anglo-American world, albeit after a considerable time lag, and radical geography was no different. Similar to its Anglo-American impact, the influence of radical geography in South Africa was responsible for generating a great deal of scholarly debate and introspection about the relevance of the discipline, as well as more theoretically informed research influenced by social, economic, and political realities. Prominent issues included imperialism, inequalities, welfare, justice, environmental degradation, gender/race oppression, labor exploitation, neoliberalism, and human rights violations.

The argument in this chapter proceeds in five stages. The ferment in apartheid geography is discussed in the first section. Apartheid, as an example of state-directed socio-spatial structuring, held a "special

*Spatial Histories of Radical Geography: North America and Beyond*, First Edition.
Edited by Trevor J. Barnes and Eric Sheppard.
© 2019 John Wiley & Sons Ltd. Published 2019 by John Wiley & Sons Ltd.

fascination for the geographer ... Indeed, it would not be too much of an exaggeration to describe apartheid as the most ambitious contemporary exercise in applied geography" (Smith 1982: 1). In the 1980s a new generation of radical scholars who were trained in North America began to challenge the Eurocentric hegemony, calling for decolonization and the development of indigenous geography, and this is the theme of the second section. The fragmentation of apartheid society was reflected in the academic world, and the geography fraternity was no exception. Black[1] geographers made a critical contribution to the restructuring of geography, and were very clear about its implications for empowerment, emancipation and transformation. This is discussed in the third section. The fourth section discusses how radical geographers who set the critical agenda in the 1980s were increasingly seduced by the neoliberal agendas of the post-apartheid state, many migrating to the lucrative, consulting industry. The final section points to how a radical, critical, dissenting geography agenda is being revived, however, in the post-apartheid era. While this paper draws primarily from geographical sources, it is also sensitive to critical literature in the social sciences.

## Complicity in Legitimizing Apartheid

Research and teaching in South African geography followed scholarly trends in the western world, although with a significant time lag. Similar to most colonial societies, the origins of geography as an academic discipline was intended to serve the imperial agenda. Indeed, Wellings (1986: 121) critiqued "expatriate models and methodologies, often years past their prime and quite irrelevant to African settings."

Environmental determinism was the dominant paradigm, legitimating the superiority of the European colonizers and the associated racist discourse on a supposedly scientific basis (Wesso 1994). This was replaced by the spatial science approach in the late 1960s and early 1970s, which by then radical Anglo-American geographers were challenging. Although impressive in terms of quantitative sophistication, such studies were unable to unravel or explain processes influencing the geography of apartheid (Rogerson and Browett 1986). The spatial science approach promoted a "utilitarian geography" that "buttressed a society based on inequalities of race, class and gender" (Wesso 1994: 333). Indeed, with their uncritical, descriptive studies geographers were accused of being complicit in legitimizing apartheid: "As segregation hardened into apartheid after 1948, geographers scarcely missed a beat. Geographical skills were bent to the service of the new state and to the entrenchment of race in space" (Crush 1991: 9).

The research focus was largely on the spatial interaction of white people and their environment, with black communities and their problems appearing to be largely invisible to the geographical imagination. By 1980, only two articles about black townships had been published in the *South African Geographical Journal* (Beavon 1982). McCarthy (1982: 53) warned ominously that unless geographers adopted more critical, radical approaches, the discipline would "be judged by history as having been totally incapable of addressing the fundamental issues of its times." It had become increasingly apparent that there was an urgent need to decolonize South African geography.

## Decolonization and Progressive Approaches

In a scholarly context, decolonization has two implications. First, there had to be a shift away from western paradigms toward the development of more indigenous approaches to understand socio-spatial processes. Secondly, there was a need to reduce the gap between ivory towers and communities, in order to empower the poor and the disadvantaged (Crush 1992).

The June Soweto 1976 uprisings served as a catalyst for geographers to become more aware of the need to investigate problems experienced by the black underprivileged and exploited masses. As a result, geographers began to attend increasingly to the housing and employment problems of blacks, also adopting a more critical stance toward explanation (Davies 1976; Beavon 1982; Mabin and Parnell 1983). As historian Bickford-Smith (2008: 288) has emphasized, "the Soweto uprising of 1976 confirmed to most observers that the anti-apartheid struggle (in contrast to anti-colonial struggles in many other parts of Africa) would be largely urban in character."

Almost simultaneously, geography also came under the influence of a "new school" of neo-Marxist historiography developed by social historians, sociologists and political scientists. From this perspective, South African society could not be explained purely in terms of race, but required a more nuanced analysis, "informed by historical materialism, on the themes of class, capitalism, racial domination and the institutions of labour repression" (Rogerson and Beavon 1988: 85). Such an analysis revealed that "South Africa's landscapes of social control and domination are imbued with a history of interaction and struggle between black and white, labour and capital, class and race" (Crush 1986: 3).

Significantly, social historians were also influenced by geography, as evident, for example, in the work of Bonner and Lodge (1989: 1) who maintained that apartheid segregation in South Africa made the contestation of space one of the central political struggles.... A central theme in the history of black urban communities in South Africa was ... their attempts to create and defend illegal space. A central thrust of state urban policy was equally to close down such communities and to quarantine them in localities selected by the state where they could be more effectively regimented and controlled.

This period also was characterized by a methodological and epistemological shift in urban research toward the political economy paradigm (Beavon and Rogerson 1981), influenced by a new generation of progressive South African geographers who had trained in North America. Prominent names include Jeff McCarthy and Mike Sutcliffe (both trained at Ohio State University under Kevin Cox); Jonathan Crush (who grew up in southern Africa, with a university education in the UK and a Ph.D. from Queen's University); Chris Rogerson (who got a Ph.D. from Queen's University); and Alan Mabin (Ph.D. Simon Fraser). From academic positions at liberal white institutions (especially the University of Witwatersrand and the University of Natal), many of these scholars had links with, and provided support for, trade unions and community and civic organizations. This momentum accelerated in the 1980s with the publication of a number of articles, both locally and internationally, critically evaluating the development of geography and geographers in south and southern Africa (e.g., Beavon and Rogerson 1981; Crush and Rogerson 1983; Wellings and McCarthy 1983; Wellings 1986).

While some geographers had Marxist leanings (e.g., McCarthy 1982; Sutcliffe and Wellings 1985; Mabin 1986), the majority followed a more liberal tradition. Associated with this trend, the difficulties of "living under apartheid" became a major theme in geographic research (Smith 1982). Yet a major concern was that questions of empowerment, emancipation and transformation were not addressed (Wellings 1986). Furthermore, black academics were largely invisible and remained on the periphery of the discipline (Magi 1990). The state of the discipline report commissioned the Society of South African Geographers (2003) concluded that this "exclusion either by careless omission or intent ... stands to the shame of the discipline as a whole."

## Empowerment, Emancipation, and Transformation

Institutionally, academic geography in the apartheid era was organized and administered by the liberal English South African Geography Society (SAGS) and the conservative Afrikaans Society for Geography. The few black academic geographers were based mainly in ethnic universities, pejoratively referred to as tribal or bush colleges. Mamdani (1999: 131) commented that while "white universities were islands of privilege ... black universities ... were designed more as detention centres for black intellectuals." Mabin (1989) argued that the invisibility of black geographers in the publishing arena was an indictment of the development of the discipline.

Although they played a leading role in exposing the "inhuman geography of apartheid" in the 1980s, very few progressive geographers were involved in the debate on democratic transformation and empowerment. Black geographers played an important role, however, in addressing these issues. The transformation of the discipline was initiated at the Conference of the Society for Geography held at the University of Pretoria in 1989. Harold Wesso, from the University of Western Cape (for colored students, which since the mid-1980s developed a national and international reputation as a home for the left), called for the development of a people's geography, one that is opposed to all forms of oppression and exploitation and promotes "the values and the ideal of a truly just, non-racial and democratic society" (Wesso 1989: 10). At the Quadrennial Conference of the South African Geography Society in Potchefstroom in 1991, Dhiru Soni (from the University of Durban-Westville for Indians, a seedbed of anti-apartheid activities since its inception in 1972), similarly argued that "both geography and geographers need to substantially redefine their roles if they intend to be relevant and recognised as progressive forces in the transformation towards a democratic, non-racial and non-sexist society" (Soni 1991: 10).

The issues raised by black geographers at Pretoria in 1989 and Potchefstroom in 1991 set in motion a series of events that culminated in the disbandment of both the SAGS and the Society for Geography, and the launch of a new, non-racial Society of South African Geographers (SSAG). This occurred in 1995 at the former University of Durban-Westville (not surprisingly, many doyens of North American radical geography attended, including Jim Blaut, Neil Smith, and Ben Wisner).

In spite of numerous challenges, black geographers also were becoming nationally and internationally visible in the publishing arena (e.g., Khosa 1989; Magi 1989; Seethal 1991; Soni and Maharaj 1991), under the influence of mentors from abroad. Dhiru Soni and Meshack Khosa took courses with David Harvey (at Johns Hopkins and Oxford, respectively), Abdi Samatar advised Cecil Seethal's Ph.D. at The University of Iowa. I interacted with David M. Smith from Queen Mary College (University of London) and Jeff McCarthy was my Ph.D. advisor at the University of Natal.

## Post-Colonial Era: Radicals in Retreat

In the 1990s, South Africa embarked on the long journey toward reconstruction, development and planning in the post-apartheid era. Success would depend on a sensitive understanding of the "geographical legacy of apartheid and the scars it has left behind, and also to the complex local, regional and environmental diversity that characterizes the South African whole" (McCarthy 1991: 23).

As the post-apartheid transition proceeded, however, many of the progressive geographers who had set the critical agendas of the 1980s increasingly became complacent conformists in the post-apartheid era, notwithstanding glaring and increasingly apparent socio-spatial inequalities. Three reasons can be identified for what can be described as both a decline in radical geography and the retreat of intellectuals in the post-apartheid era: The corporatization and commodification of the academy; the silence and connivance of academics; and the policy/consultancy turn.

### *Corporatization and commodification*

In the post-apartheid era, universities went through a process of restructuring that took many forms. There was a shift from discipline-based offerings to school-based programs that were more responsive to market needs. Almost simultaneously, there was an increasingly centralized top-down, authoritarian managerial style, in contrast to the traditional collegial approach (Duncan 2007). Southall and Cobbing (2001: 1) discuss the rise of "corporate authoritarianism" in South Africa, whereby academics are "increasingly subordinated to administrators, who in turn are becoming increasingly intolerant of robust internal dissent." In short, there was a neoliberal corporatization and commodification of higher education, in response to the needs of global capital (Mama 2004) and paralleling a pro-market turn within the African National Congress.

The ostensible reason for the shift to managerialism and corporatization was to make universities more responsible and accountable (Webster and Mosoetsa 2002), with disciplines and scholarship increasingly judged in terms of market value and technical quality, rather than societal relevance. As David Harvey (1998: 115) has argued, in an era of commodification and corporatization "the quite proper demand that the university be accountable gets translated into the reductionist idea that everything is simply a matter of accounting."

## Academic silence and connivance

There has been disquiet about the silence of intellectuals on critical public policy issues in democratic South Africa. According to Ramphele (1999: 205) the "silence of academics on questions of redistribution of wealth and opportunity ... is tantamount to connivance." Possible reasons for this silence include "euphoria over the transition, the uncertainties accompanying all major political transitions, the emphasis on reconciliation to stabilise our democracy, a perceived ANC leadership 'intolerance' of debate on major issues, (and) a somewhat 'dislocated' mass movement" (Nzimande 2005: 1).

President Nelson Mandela (1997) expressed concern that academics who played a critical role in challenging apartheid were now losing their way. Yet the state's response to critique of its policies was to label protagonists as racists, reactionaries or as "ultraleftists." Critique was stigmatized and marginalized as being "unpatriotic" (Desai and Bohmke 1997), which can "only undermine democracy" (Ramphele 2000: 105). Further, there is a view that black intellectuals were expected to be more patriotic and sympathetic toward the government because they were "supposed to understand the struggles and support the projects of the emerging state" (Jansen 2006: 190).

Indeed Jonathan Jansen (2006: 197) has been scathing about black intellectuals, who allow themselves to be coopted by the "machinery of government where compliance and conformity are more highly valued, even to the point of dishonesty and self-denial." This is reminiscent of Cornel West, writing about US black intellectuals as appearing "too hungry for status to be angry, too eager for acceptance to be bold, too self-invested in advancement to be defiant" (West 1993: 38).

## Policy-consultancy nexus

Geographers were well equipped to analyze many of the problems emerging during the phase of post-apartheid restructuring and reconstruction (Rogerson and McCarthy 1992; Parnell 2007), on such issues as the redrawing of municipal and regional boundaries, land reform and restitution, urban and rural development, environmental hazards, and access to basic needs. Indeed, many progressive geographers became policy makers and consultants to the new government, albeit "designing and implementing policy quite uncritically" (Oldfield *et al.* 2004: 293). As sociologist Seekings (2000: 835) put it, the "combination of new decision-makers in the state and well connected academics outside of it has resulted in a booming but largely uncritical policy studies industry."

According to Parnell (2007: 111) "consulting, policy or applied research is a ubiquitous feature of geography departments, [and] we have been tardy, or perhaps reluctant, to open the conversation about the implications of the way many of us now work." As Bekker (1996: 20) has argued, however, consultants and "policy makers are not free agents. Their activities are confined both by their political masters and by budgetary allocations that oblige then to prioritize often against their better judgement."

Maano Ramutsindela (2002) argued that increased opportunities for consultants threatened to "erode the availability of the critical intellectual mass in the discipline" (p. 7), expressing the fear that uncritical engagement in the policy arena "will most probably lock the discipline into the same trends that obtained in the past" (p. 8). Lindsay Bremner (2000: 87) offered another indictment against policy makers and consultants, opining that very few of their policies have "challenged the apartheid geography or have come to terms with the changing social patterns of post-apartheid South African cities," and have in fact reinforced race-class inequalities.

## Advancing Critical Scholarship and Activism

Notwithstanding a certain retreat of radical geography for the reasons discussed previously, some seminal intellectual projects led by a younger generation of scholars are advancing critical scholarship and activism. As socioeconomic and spatial inequalities widened in the post-apartheid era, there were calls for more radical critiques of reconstruction strategies, especially in terms of impacts on the poor and disadvantaged (Bond 2000a; Maharaj et al. 2010). There was a view that if geography is to be meaningful and progressive in the "context of the challenges and opportunities engendered by transformation in South Africa, then the subject and its practitioners should adopt a critical and confrontational posture while continuously exploring the possibilities for positive change" (Soni 1992: 86). Khosa (2002: 21) expanded on this contention, maintaining that critical geographers should be grappling with the following questions:

1   To what extent did the transition to democracy improve the state of the people in South Africa?
2   How is the new constitutional democracy contributing to social justice?
3   What are the social, economic, and political processes leading to historical continuity and change in the post-apartheid period?
4   What is the changing nature and direction of civil engagement in the emerging democratic South Africa?

Lupton (1992: 106) advocated for a "new interpretation from the Marxist tradition ... to advance the intellectual and political project of poor and working people struggling for a new and better world" (Lupton 1992: 106). Patrick Bond, a student of David Harvey based in South Africa since the late 1980s, provided the intellectual lead in this direction, contending that although many radical scholars of the 1980s were seduced by neoliberalism in the 1990s, leftist intellectuals should not despair: Urban and rural "South Africa offers fertile case study material in the application of political-economic theory to contemporary policy and practice" (Bond 1995: 149).

Bond (2000a, 2000b) was referring to the neoliberal departure from the basic needs and social justice orientated Reconstruction and Development Programme (RDP), which occurred with the 1996 adoption of the Growth, Employment, and Redistribution strategy (GEAR) emphasizing fiscal discipline, debt reduction, and privatization of public services, which had far-reaching geographical implications. Under apartheid, access to services had a distinct spatiality: Townships were inadequately serviced, if at all, while the racially privileged enjoyed access to services comparable to those in the first world (Turok 1994). Privatizing basic services militates against the aim of building an inclusive society, and the provision of a minimum level of service to disadvantaged areas reemphasizes apartheid boundaries in the geography of service distribution (Bakker and Hemson 2000). Further, as the responsibility for poverty was shifted from an apartheid government to market forces, the poor became liable for their own circumstances thereby masking the structural legacy of exploitation and oppression (Gibson 2001).

The immediate response of the poor has been mass action. In recent years, there have been about 8000 service delivery protests annually as the poor mobilize and militate against the neoliberal agenda (Mottiar and Bond 2012). Desai (2002) and Ballard, Habib, and Valodia (2006) offer remarkable insights into how poor people are responding to landlessness, homelessness, evictions, privatization, and water and electricity disconnections. Organization and mobilization around these issues have provided the catalyst for new social movements to emerge, filling a vacuum created by the absence of a political party promoting the welfare of poor people beyond rhetoric.

## Conclusion

While radical geographers in the 1980s exposed the historical, political, and ideological forces structuring apartheid social space, there was a deafening silence as class cleavages and socio-spatial inequalities were reproduced in the democratic era. This can be attributed to the "rise of the disciplinary university". There is thus an urgent need for a new breed of radical scholars to critically analyze the socio-spatial impacts of the neoliberal agenda, and to consider alternate paradigms for reconstruction and development. This is beginning to emerge. Emerging critical themes include: Environmental justice (Bond 2002; Patel 2009); displacement and livelihoods (Maharaj 2017); sexuality (Visser 2003; Dirsuweit 2006); gender (Dirsuweit and Mohamed 2016); human rights (Maharaj 2005); social justice (Visser 2001); impacts of mega events (Maharaj 2011, 2015); xenophobia, violence, refugees (Peberdy and Jara 2011; Bollaert and Maharaj 2018); privatization (Narsiah 2011); race and ethnicity (Ballard 2010); land reform and restitution (Ramutsindela 2007); and geographies of corruption (Maharaj 2016).

Given South Africa's repressive apartheid legacy and Africa's postcolonial record, the need to sustain a critical, independent intellectual prospectus in geography in particular, and academia in general, cannot be overemphasized,

## Note

1   Any study of the South African social formation cannot avoid reference to race and ethnic divisions, even in the non-racial democratic era. "Black" refers to Africans, "coloureds," and Indians. Use of such terminology does not in any way legitimate racist ideology.

## References

Bakker, K. and Hemson, D. (2000). Privatising water: BoTT and hydropolitics in the new South Africa. *South African Geographical Journal* 82: 3–12.
Ballard, R. (2010). "Slaughter in the suburbs": Livestock slaughter and race in post-apartheid cities. *Ethnic and Racial Studies* 33 (6): 1069–1087.
Ballard, R., Habib, A., and Valodia, I. eds (2006). *Voices of Protest – Social Movements in Post-Apartheid South Africa*. Pietermaritzburg: UKZN Press.
Beavon, K. S. O. (1982). Black townships in South Africa: Terra incognita for urban geographers. *South African Geographical Journal* 64: 3–20.
Beavon, K. S. O. and Rogerson, C. M. (1981). Trekking on: Recent trends in the human geography of Southern Africa. *Progress in Human Geography* 5: 159–189.
Bekker, S. (1996). The policy making predicament. *Indicator SA* 13: 17–20.
Bickford-Smith, V. (2008). Urban history in the new South Africa: Continuity and innovation since the end of apartheid. *Urban History* 35: 288–315.
Bollaert, E. and Maharaj, B. (2018). Experiences of a hidden population: Life stories of refugees in Pietermaritzburg. *African Geographical Review* 37: 146–158.

Bond, P. (1995). Urban social movements, the housing question and development discourse in South Africa. In D. B. Moore and G. J. Schmitz, eds, *Debating Development Discourse – Institutional and Popular Perspectives*, pp. 149–177. London: Macmillan.

Bond, P. (2000a). *Elite Transition: From Apartheid to Neoliberalism in South Africa*. London: Pluto Press.

Bond, P. (2000b). *Cities of Gold, Townships of Coal: Essays on South Africa's New Urban Crisis*. Trenton: Africa World Press.

Bond, P. (2002). *Unsustainable South Africa: Environment, Development and Social Protest*. Pietermaritzburg: University of Natal Press.

Bonner, P. and Lodge, T. (1989). Introduction. In P. Bonner, I. Hofmeyr, and D. James, eds, *Holding their Ground – Class, locality and Culture in 19th and 20th Century South Africa*, pp. 1–17. Johannesburg: Ravan Press.

Bremner, L. (2000). Post-apartheid urban geography: A case study of Greater Johannesburg's rapid land development programme. *Development Southern Africa* 17: 87–104.

Castree, N. (2000). Professionalisation, activism and the university: Whither "critical geography"? *Environment and Planning A* 32: 955–970.

Crush, J. (1986). Towards a people's historical geography for South Africa. *Journal of Historical Geography* 12: 2–3.

Crush, J. (1991). The discourse of progressive human geography. *Progress in Human Geography* 15: 395–414.

Crush, J. (1992). Beyond the frontier: The new South African historical geography. In C. Rogerson and J. McCarthy, eds, *Geography in a Changing South Africa – Progress and Prospects*, pp. 10–73. Cape Town: Oxford University Press.

Crush, J. and Rogerson, C. (1983). New wave African historiography and African historical geography. *Progress in Human Geography* 7: 203–231.

Davies, R. J. (1976). *Of Cities and Societies: A Geographer's Viewpoint*, Inaugural Lecture, University of Cape Town.

Desai, A. (2002). *We are the Poors – Community Struggles in Post-Apartheid South Africa*. Monthly Review Press, New York.

Desai, A. and Bohmke, H. (1997). The death of the intellectual, the birth of a salesman. *Debate* 3: 10–34

Dirsuweit, T. (2006). The problem of identities: The lesbian, gay, bisexual, transgender and intersex social movement in South Africa. In R. Ballard, A. Habib and I. Valodia, eds, *Voices of Protest – Social Movements in Post-Apartheid South Africa*, pp. 325–347. Pietermaritzburg: UKZN Press.

Dirsuweit, T. and Mohamed, S. (2016). Vertical and horizontal communities of practice: Gender and geography in South Africa. *South African Geographical Journal* 98: 531–541.

Duncan, J, (2007). Rise of the disciplinary university. https://mg.co.za/article/2007–05–22-rise-of-the-disciplinary-university, accessed May 12, 2016.

Gibson, N. (2001). Transition from apartheid. *Journal of Asian and African Studies* 36: 65–85.

Harvey, D. (1998). University Inc. *The Atlantic Monthly*, October: 112–116.

Jansen, J. (2006). The (self-imposed) crisis of the black intellectual. In M. Alexander, ed., *Articulations: A Harold Wolpe Memorial Lecture Collection*, pp. 189–199. Trenton: African World Press.

Khosa, M. (1989). "Dipalangwang": Black commuting in the apartheid city. *African Urban Quarterly* 4: 251–259.

Khosa, M. (2002). (Dis)Empowerment through social transformation. *South African Geographical Journal* 84: 21–29.

Lupton, M. (1992). Economic crisis, deracialisation, and spatial restructuring: Challenges for radical geography. In C. Rogerson and J. McCarthy, eds, *Geography in a Changing South Africa: Progress and Prospects*, pp. 95–110. Cape Town: OUP

Mabin, A. and Parnell, S. (1983). Recommodification and working-class home ownership: New directions for South African cities? *South African Geographical Journal* 65: 148–166.

Mabin, A. (1986). Labour, capital, class struggle and the origins of residential segregation in Kimberley, 1880–1920. *Journal of Historical Geography* 12(1): 4–26.

Mabin, A. (1989). Does geography matter? A review of the past decade's books by geographers of contemporary South Africa. *South African Geographical Journal* 71: 121–126.

Magi, L. M. (1989). Cognition of recreation resources through photographic images. *South African Geographical Journal* 71: 67–73.

Magi, L. M. (1990). *Geography in Society: The Site of Struggle*. Inaugural Lecture, University of Zululand.

Maharaj, B. (2005). Geography, human rights and development: reflections from South Africa. *Geoforum* 36: 133–135.

Maharaj, B. (2011). 2010 FIFA World Cup: (South) "Africa's time has come"? *South African Geographical Journal* 93: 49–62.

Maharaj, B. (2015). The turn of the South? Social and economic impacts of mega events in India, Brazil and South Africa. *Local Economy* 39: 983–999.

Maharaj, B. (2016). Corruption – A Geographical Perspective: South African Reflections. Presentation at the *Centenary Conference of the Society of South African Geographers, Stellenbosch University*, September 25–28.

Maharaj, B. (2017). Contesting displacement and the struggle for survival: The case of subsistence fisher folk in Durban, South Africa. *Local Economy* 32: 744–762.

Maharaj, B., Desai, A., and Bond, P. (2010). *Zuma's Own Goal – Losing South Africa's "War on Poverty"*. Trenton, NJ: African World Press.

Mama, A. (2004). *Critical capacities: Facing the challenges of intellectual development in Africa*. Inaugural address, Institute of Social Studies, The Hague, April 28.

Mandela, N. (1997). Address by President Nelson Mandela on receiving honorary degree from University of Pretoria, Pretoria, December 4. www.mandela.gov.za/mandela_speeches/1997/971204_up.htm

Mamdani, M. (1999). There can be no African renaissance without an Africa-focused intelligentsia. In M. W. Makgoba, ed., *African Renaissance: The New Struggle*, pp. 125–136. Sandton: Mafube.

McCarthy, J. J. (1982). Radical geography, mainstream geography and southern Africa. *Social Dynamics* 8: 53–70.

McCarthy, J. J. (1991). *Theory, Practice and Development.* Inaugural lecture, University of Natal, Pietermaritzburg.

Mottiar, S. and Bond, P. (2012). The politics of discontent and social protest in Durban. *Politikon* 39: 309–330.

Narsiah, S. (2011). The struggle for water, life and dignity in South African cities: The case of Johannesburg. *Urban Geography* 32(2): 149–155.

Nzimande, B. (2005). The contribution of public intellectuals in defining public interest in South Africa. Paper presented at the *Harold Wolpe Memorial Seminar*, February 16.

Oldfield, S., Parnell, S., and Mabin, A. (2004). Engagement and reconstruction in critical research: Negotiating urban practice, policy and theory in South Africa. *Social and Cultural Geography* 5: 285–299.

Parnell, S. (2007). The academic – policy interface in postapartheid urban research: Personal reflections. *South African Geographical Journal* 89: 111–120.

Patel, Z. (2009). Environmental justice in South Africa: Tools and trade-offs. *Social Dynamics* 35 (1): 94–110.

Peberdy, S. and Jara, M. K. (2011). Humanitarian and social mobilization in Cape Town: Civil society and the May 2008 xenophobic violence. *Politikon* 38: 37–57.

Ramphele, M. (1999). The responsibility side of the academic freedom coin. *Pretexts: Literary and Cultural Studies* 8: 201–206.

Ramphele, M. (2000). Critic and citizen: A response. *Pretexts: Literary and Cultural Studies* 9: 105–107.

Ramutsindela, M. (2002). The philosophy of progress and South African geography. *South African Geographical Journal* 84: 4–11.

Ramutsindela, M. (2007). The geographical imprint of land restitution with reference to Limpopo Province, South Africa. *Tijdschrift voor Economische en Sociale Geografie* 98: 455–467.

Robinson, J. (2003). Postcolonialising geography: Tactics and pitfalls. *Singapore Journal of Tropical Geography* 24: 273–289.

Rogerson, C. M. and Beavon, K. S. O. (1988). Towards a geography of the common people in South Africa. In J. Eyles, ed., *Research in Human Geography – Introductions and Investigations*, pp. 83–99. Oxford: Basil Blackwell.

Rogerson, C. M. and Browett, J. G. (1986). Social geography under Apartheid. In J. Eyles ed., *Social Geography in International Perspective*, pp. 221–250. London: Croon Helm.

Rogerson, C. and McCarthy, J. (eds.) (1992). *Geography in a Changing South Africa –Progress and Prospects.* Cape Town: Oxford University Press.

Seekings, J. (2000). Introduction: Urban studies in South Africa after apartheid. *International Journal of Urban and Regional Studies* 24: 832–840.

Seethal, C. (1991). Restructuring the local state in South Africa: Regional Services councils and crisis resolution. *Political Geography Quarterly* 10: 8–25.

Smith, D. M. ed. (1982). *Living under Apartheid*. London: George Allen and Unwin.

Society of South African Geographers (2003). GEOGRAPHY: The State of the Discipline in South Africa: A Survey, 2000/2001 Report. www.ssag.co.za/PowerCMS/files/State%20of%20the%20Discipline.pdf (accessed July 2, 2018).

Soni, D. (1991). Knocking at the door: The future of geography and geographers in South Africa. Paper presented at the *Conference of the South African Geographical Society, University of Potchefstroom*.

Soni, D. (1992). Human geography in the 1990s: Challenges engendered by transformation. In C. Rogerson and J. McCarthy, eds, *Geography in a Changing South Africa: Progress and Prospects*, pp. 86–94. Cape Town: OUP.

Soni, D. and Maharaj, B. (1991). Emerging rural urban forms in South Africa. *Antipode* 23: 47–67.

Southall, R. and Cobbing, J. (2001). From racial liberalism to corporate authoritarianism: The Shell Affair and the assault on academic freedom in South Africa. *Social Dynamics* 27: 1–42.

Sutcliffe, M. and Wellings, P. (1985). Worker militancy in South Africa: A sociospatial analysis of trade union activism in the manufacturing sector. *Environment and Planning D: Society and Space* 3: 357–79.

Turok, I. (1994). Urban planning in the transition from apartheid Part 1: The legacy of social control. *Town Planning Review* 65: 243–259.

Visser, G. (2001). Social justice, integrated development planning and post-apartheid urban reconstruction. *Urban Studies* 38: 1673–1699.

Visser, G. (2003). Gay men, tourism and urban space: reflections on Africa's "gay capital". *Tourism Geographies* 5: 168–189.

Visser, G., Donaldson, R., and Seethal, C. eds. (2016). *The Origin and Growth of Geography as a Discipline at South African Universities*. Stellenbosch: Sun Press.

Webster, E. and Mosoetsa, S. (2002). At the chalk face: managerialism and the changing academic workplace 1995–2001. *Transformation* 48: 59–82.

Wellings, P. (1986). Editor's introduction – geography and development studies in southern africa: A progressive prospectus. *Geoforum* 17: 119–131.

Wellings, P. A. and McCarthy, J. J. (1983). Whither southern African Human Geography. *Area* 15: 337–345.

Wesso, H. (1989.) People's education: Towards a people's geography in South Africa. Paper presented at the Society for Geography Conference, 3–6 July.

Wesso, H. (1994). The colonisation of geographic thought: the South African Experience. In A. Godlewska and N. Smith, eds, *Geography and Empire*, pp. 316–332. Oxford: Blackwell.

West, C. (1993). *Race Matters*. Boston: Beacon Press.

# 12

# The Geographies of Critical Geography
## The Development of Critical Geography in Mexico

Verónica Crossa

In the early 1990s, when I was exploring options to study my under-graduate degree in geography in Mexico City, I visited what seemed the obvious place in search for information and orientation: The National Autonomous University of Mexico (UNAM), the biggest public university in Mexico (and Latin America) and the only institution offering a degree in geography at that time in the capital city. Although I was not sure exactly what I wanted to focus on, my interest was primarily in human geography. I had a few ideas, albeit vague, of what I hoped a geography degree would offer: a more profound understanding of the spatiality of inequality, a reading on poverty, and a concern over social problems, particularly through a geographical lens. In a way, and without really knowing it, I was searching for a geography degree that offered a critical perspective on the relationship between space and society. Unfortunately, I found none of this in the UNAM. Instead, I encountered an extremely traditional curriculum, inundated with physical geography classes, and a human geography program that overemphasized descriptive and regional analyses. This lack of a critical perspective was one of the main reasons why I pursued most of my geography career outside of Mexico.

This original misfortune always puzzled me and sparked much of my curiosity over what Harvey, many years later, called the geographies of

*Spatial Histories of Radical Geography: North America and Beyond*, First Edition.
Edited by Trevor J. Barnes and Eric Sheppard.
© 2019 John Wiley & Sons Ltd. Published 2019 by John Wiley & Sons Ltd.

critical geography (Harvey 2006). It was difficult for me to understand how and why a career in human geography in a university like the UNAM (especially its Faculty of Philosophy and Letters), an institution known for its political position relative to a number of social issues experienced in the country, was so distant from a critical vocabulary that was already widespread in Mexico in such disciplines as sociology, anthropology, and even political science. It is not that a critical vocabulary was non-existent within the social sciences in Mexico, but it was certainly absent from the field of Geography.

This short chapter seeks to address some of the reasons behind this absence. I will pose some ideas as to why a critical and even radical vocabulary, so present in certain academic and institutional spaces, was at the same time so absent in geography in Mexico City even during the 1990s. Having said this, and as I will elaborate throughout this chapter, the landscape of critical human geography is rapidly changing in Mexico, entering new realms in the last two decades with the emergence of new geography departments, new interdisciplinary approaches, and new faculty members, many of whom studied abroad and who have brought with them more novel and critical ideas for how to approach the relationship between space and society. The chapter is structured as follows: the next section briefly provides some general contextual national issues to understand the development of geography as a discipline in Mexico. The second section explores some of the reasons that help explain the limited development of critical geography in Mexico. The concluding section explores some essential aspects of the current state of affairs within Mexican academia and potential for the growth of a critical geographical landscape.

## A Critical Perspective: A Few Preambles

The trajectory of ideas, theories, or epistemological approaches is almost always a study of the interrelationship between the history of people, institutions, and practices, one that is sometimes difficult to reconstruct. Part of the difficulty has to do with the fact that people and institutional practices are not always distinguishable, especially when looking at certain processes historically. It is hard to know, for instance, if a limitation in the development of particular ideas in a geography university curriculum is linked to a sort of departmental inertia, if it has to do with individualized conflicts between particular faculty members, or both. Similarly, it is difficult to decipher the extent to which personal professional trajectories define the line of thought in a particular department, or if the institutional composition shapes individual trajectories. In the case of critical geography in Mexico, I would argue that it is both. Rather than provide a "chicken-and-egg" account of the evolution of the discipline, I hope to give a more nuanced account of the underlying reasons why critical geography failed to be present during most of the twentieth century within Mexican human geography.

Throughout this essay, I use the term "critical" rather than "radical" for multiple reasons. Traditionally, the use of "radical" as a prefix to geography is practically absent in academic debates within the field in Mexico. When discussed among academic circles, the notion of radical geography is described as something "out there," a perspective that while present in international circles and debates, remained marginal in Mexico (Aguilar 1994). Mexican scholars who make reference to radical geography have done so through a sort of geographical-historiography approach, fed by a curiosity to understand how ideas in the field have evolved, but not necessarily making any substantive claims for the value of those ideas in thinking through context-specific problems in Mexico. The lack of use of the term radical to describe a branch in human geography is hardly coincidental. It is partly an issue of language and translation but perhaps more importantly it is a political matter, especially when concerned with geography in Mexico. As I will discuss, geography played a part in processes of state-formation in the post-revolutionary period, but contributed very little to the actual revolutionary project prior to the formation of the national political party that ruled Mexico for more than 70 years (from 1929 to 2000). During this time, the term radical was used to describe particular actions, ideas or practices, primarily from outsiders who greatly influenced the political ambience of a country that was in the process of reinventing itself. Radicals were, for example, Mexicans and foreigners who in 1919 founded the Mexican communist party. During the start of the twentieth century, the term radical offered a way of characterizing those that were opposed to the most profound forms of autocracy consolidated during the time of Porfirio Díaz (1876–1911). The term acquired multiple meanings during the Mexican revolution, from those who fought for a secular state, to those who fought for the eradication of the last vestiges of Spanish colonialism in all its manifestations. More recently, however, particularly in the last decade, the term critical geography has figured more largely in debates within the field among geographers who have, in most cases, studied abroad (primarily in the U.S. and U.K.), bringing back with them many of the concerns cultivated during their years as graduate students.

As I will discuss in the next section, the landscape of critical geography in Mexico is nascent, a very recent development that marks a significant shift in the trajectory of the discipline. Mexico's political and economic landscape has been changing rapidly, particularly since the second half of the twentieth century, intense transformations that provide an extremely rich laboratory for the involvement of scholars interested in the formation of a more just, more equitable society. During this period, many scholars within the social sciences, primarily within sociology, anthropology, and urban planning, became concerned with phenomena as diverse as modernization, economic development, land tenure, indigenismo, agriculture and agrarian reform, education, health, and urban growth. For the geographers present during this period, however, their engagement acted mostly in accordance with the requirements and desires of the party in power, the PRI. As Ramírez (2003: 566) argues, "very few geographers, at least known, contributed to the revolutionary project. On the contrary, it seems that they adopted a conservative position instead of seeking to contribute to the changes needed to solve poverty within rural areas."

It is essential to highlight, however, that while a critical approach was a distant reality in Mexican geography, critical geography was flourishing in other parts of Latin America; particularly in Brazil with the influence of the Brazilian geographer Milton Santos, who was trained in France. Santos was among the first Latin American geographers to stress the practice of geography as based on a commitment toward social change. His work explored the Latin American reality as a product of a colonial past that shaped the socio-spatial relations of the region. Accordingly, hunger, poverty, inequality, and racism, to mention just a few, were processes that geographers had the responsibility not only of describing through ideographic approximations but of influencing and transforming. Santos' work in Mexico eventually was well received, but only among certain geographers who were starting to think through the limitations of what was fundamentally an empirical and descriptive discipline.

By the 1960s, scholars in other social sciences were playing an important role in questioning many of the consequences experienced as a result of policies set forth during the development of the post-revolutionary Mexican state. By the 1970s, for example, the revolutionary program of agrarian reform had become stagnant and was seriously challenged by modernist projects prioritizing industrial growth at any cost, including at the expense of what originally was at the heart of the revolutionary discourse: the reality of the nation's peasantry and the future of rural Mexico. As anthropologist Lomnitz highlights,

> [...] the redeeming program of the Mexican State, which sought, using the Cardenista formula, to "convert the indigenous populations into Mexicans", was seen as a project of indigenous subjugation by the new generation of anthropologists [...] This is why anthropologists of the '68 generation sought to move away from State policies and were harshly critical of

earlier anthropologists who had achieved the modernist glorification of anthropology. Instead, they searched for a new light within spaces of inter-mediation between the peasantry – who were carefully watched by the market and the State – and the State itself (Lomnitz 2016: 290).[1]

During this period, special attention also was given to Latin American urban contexts. Cities were experiencing unprecedented levels of growth, reflecting, inter alia, high levels of industrialization triggering large rural-urban migratory flows especially to capital cities. In Mexico City, the percentage of people living in irregular housing increased from 14% in 1956 to 60% in 1990 (Gilbert 2004: 58). Interestingly, while much of the work that originated from contexts outside of the region drew on a culture of poverty approach to understand inequality, segregation, and the rise of informal settlements, Mexican scholars were highly critical of such ideas and instead proposed a different, critical reading. During these decades, the work of many Mexican anthropologists, sociologists and urban planners, while valuing the study of marginality and poverty influenced by urban scholars from the Chicago School, rejected many of the conclusions arrived at in this scholarship. For example, the now clas-sic text by Larissa Lomnitz (1975) on the survival mechanism among marginalized sectors of the urban population elegantly challenged many of the essentialist features of work by scholars like Oscar Lewis and his approach to poverty and culture in Mexico. Many of Lewis' ideas stem largely from Park's understanding of culture as a self-perpetuating pro-cess developed particularly in marginalized neighborhoods in Chicago. For Park, the behavior of human beings living under impoverished con-ditions takes a life of its own, creating a specific way of being and a social order that become cultural traits for a specific population. For Lewis, the notion of "culture of poverty" described culture as a list of identifiable behaviors and practices conducted by marginalized popula-tions (in Mexico). For Lomnitz (1975), culture should be understood as the context through which certain behaviors and social practices can be explained, not a list of ingredients that inevitably blend into naturalized characteristics of a population.

Another very important contribution to the development of critical ideas within the Mexican social sciences, but also more broadly in Latin America, was an engagement with what became known as new depen-dency theory during the 1970s. Whereas many Anglo-American Marxists derive inspiration from French regulation theory, Marxist literature in Latin America was often discussed in terms of new dependency theory (Cardoso and Faletto 1979). Differences in the way capitalism operates in distinct contexts help explain why Marxist scholars from different places derive inspiration from different theoretical perspectives (Ettlinger 1999). According to Cardoso and Faletto (1979: 172), theoretical schemes concerning the formation of capitalist society in present-day developed countries are of little use in understanding the situation in Latin American countries. Not only the historical moments but also the structural conditions of development and society are different.

New dependency theory developed as an intellectual project among many Latin American social scientists, shaping the political ideology of a generation of Latin American leftists throughout the 1960s and 1970s.

The new dependency theory, as a critical approach within the Mexican social sciences, permeated numerous studies across a wide range of disciplines. The work of scholars like Manuel Castells (in Chile), Anibal Quijano (Peru and Chile), Jorge Hardoy (Argentina), and Paul Singer (Brazil), concerned with capitalist social relations in Latin America and their spatialities in urban contexts, greatly influenced the development of a critical vocabulary that sought to capture emerging forms of rural disenfranchisement, the growth of irregular settlements in peripheral urban areas, and the creation of new forms of urban employment – the informal economy. Quijano's *Dependency, Social Change, and Urbanization in Latin America* (Quijano 1967), Schteingart's *Urbanization and Dependency in Latin America* (Schteingart 1973), and Castells' (1973) *Imperialism and Urbanization in Latin America* are still considered classic texts concerned with the connection between imperialism, colonialism, and urbanization in the region. These last two were published in the same year as David Harvey's (1973) *Social Justice and the City*, a book with limited circulation in Latin America until its eventual translation into Spanish in 1977.

The critical vocabulary developed in these other disciplines skipped over geography, raising the question of why this vocabulary was so alien within geographical circles. The next section seeks to explore this question. By focusing on the limited development of critical geography in Mexico, I will show how particular institutional and socio-political arrangements within the academic sphere helped shape a specific type of engagement with a critical narrative within other social sciences, but not within geography.

## Critical Geography in Mexico

It is difficult to pinpoint exactly when critical geography began to have some sort of presence in Mexican geography. Indeed, some would claim that it still has no presence. I argue here that its presence is limited but emergent, especially in the last decade or so. The short and recent history of critical geography in Mexico can be understood by looking more closely at the institutionalization of geography as a discipline, particularly within the Mexican higher education system. While geography has a long history in Mexico (Moncada 1994), it was institutionalized in university circles only in the early twentieth century (Chías et al. 1994). Two noticeable aspects of this process have been critical in shaping much of the Mexican trajectory of the discipline. The first has to do with normative visions constructed around the purpose and value of the discipline as one centered on the practicalities of studying the spatial arrangement of objects in space. This vision was influential at a time when Mexico was in the midst of its national modernization project. The second aspect has to do with the particular constellation of institutional arrangements established in the UNAM, which created a demarcated separation between teaching and research. I discuss each in turn.

Geography in Mexico during the twentieth century emerged as a descriptive science that initially harnessed, as its primary imprint, the development of cartographic-technical skills (Ramírez et al. 2009). Geographers were prepared and hired primarily as mapmakers, to spatially locate multiple features of Mexico's post-revolutionary territorial reality. In many cases, geographers were in charge of instructing particular branches of the state about the distribution of the country's resources, transformations in land tenure, where land was fertile and thus where potentially communal territories were located (in the context of agrarian reform), and where land was barren. Geography was first institutionalized in the 1930s in the UNAM as a hard science, located within the School of Engineers. It was not until the 1940s that geography was transferred to the humanities – within the Faculty of Philosophy and Letters. This relocation entailed a slight modification of the objectives of the discipline, from one based mostly on cartography to one that also provided the descriptive tools necessary for having a say over the territorial organization of the country. Based on the original curriculum, the main objective of the UNAM Department of Geography was to contribute to the social and economic development of the nation, through the transmission of spatial knowledge (Chías et al. 1994: 77–78). So, geography emerged as a map-making discipline that, with itseventual location in the humanities, slowly incorporated a descriptive approach for describing regionalism for the purpose of economic development.

Three scholars, Ángel Bassols (1925–2012), Jorge Vivó (1906–1979), and Jorge Tamayo (1912–1978), became the pillars of geography in Mexico during the 1940s, profoundly influencing the subsequent trajectory of the discipline. Tamayo's work is perhaps less cited than the first two scholars, but his contribution to the field cannot be ignored.

His work was illustrative of a post-revolutionary geography defined as an applied science, concerned primarily with territorial mapping and spatial organization. One of his most emblematic studies was the General Geography of Mexico (Tamayo 1962), an empirically rich descriptive study of Mexico's territorial composition (natural, economic, social, and industrial) published between in 1962.

Some contemporary geographers see in the work of Bassols and Vivó the first traces of a critical approach within the field (Delgadillo and Torres 2010; Hiernaux 2010), but this has been a subject of much dispute (Ramírez 2003). Their work certainly sought to develop a discipline that moved beyond cartographic methods to foreground the descriptive and analytical qualities of the geographic eye. In that sense, their work, especially Bassols', can be compared to that of William Bunge, particularly with respect to their influence on a number of themes such as the value placed over expeditions and fieldwork as part of the identity of the discipline, their influence over the rootedness of geography as a spatial science, and their general concern over issues of social justice. Vivó's scholarship was also significant, but his major contribution was to the formation of an important generation of professional Mexican geographers. Furthermore, Vivó and Bassols' thinking was formed under intense contexts of political change. Bassols studied economic geography at Lomonosov Moscow State University, graduating in 1949. Vivó, a Cuban native, actively participated in the political struggle against Batista, leading to his political exile in 1929. These circumstances, while different in nature, played an important role in shaping their politics. Yet while their political views were very much what we would describe as radical, some would argue that these ideas remained located mostly within their personal life and were not actively transferred to their academic.[2] It is worth noting that while Bassols' influence in geography is undeniable, he was a faculty member in the School of Economics at the UNAM. Most of his work reinforced the importance of incorporating applied techniques to study socioeconomic and regional development, particularly as a way of understanding some of the great challenges facing Mexico.

While some contemporary geographers (Moncada and López 2015) argue that critical geography in Mexico was born from the work of individuals like Bassols in the 1950s, others locate a more robust emergence of a critical vocabulary in the late 1970s, when a group of university students and professors joined many of the emergent political movements and participated in one of the largest and longest university strikes experienced by the UNAM (Ramírez et al. 2009: 107). This was the point when many geography scholars united to express concern over the need to open the discipline to a critical understanding of social problems, through a commitment toward justice and social change. This concern materialized in a series of efforts, including creating an independent association known as the Union of Progressive Geographers of Mexico in 1978,

> organized to promote the development of geography as a scientific discipline oriented toward the value of conducting research and to end the isolation of the discipline by linking it with other sciences, specifically the social sciences and to Marxism, a theoretical approach that had remained absent from geographic thought in Mexico (Ramírez *et al.* 2009: 107).

A few years later, in 1983, the group founded *Posición*, a journal that provided space for reflections around a critical analysis of sociospatial relations. The journal became a safe space in which to voice a number of concerns around the discipline in Mexico. While most of the articles published were substantive in nature, others openly voiced a concern for the way in which geography was taught in university circles. Many called for the urgent need to revise a university curriculum that had remained untouched since its creation in the 1940s. Many of these pedagogical concerns arose from a frustration, felt by faculty and students alike, toward a discipline that seemed stagnant when compared to discussions and debates within the field in other parts of the world. The journal published a total of eight issues on an irregular basis, until it was discontinued in 1999. The exact reasons behind the demise of the journal and the termination of the Union are unknown. Many of the individuals involved in its development recall happenstance incidents, linked to personal differences, as the root cause of the dissolution of the group and the subsequent paralysis of what seemed like a promising path toward a pedagogical change. This critical approach in geography remained very much within the academic sphere, but arguably did not translate into challenging hierarchical power relations within the institutional spaces of the university structure. In addition to a polarization of positions among faculty members in geography at the UNAM, others argue that the nascent critical geography developed through the creation of the Union and the journal was undermined by the opening of geography departments elsewhere in the country, eroding the centrality of many efforts that had been built within specific spaces of the UNAM (Ramírez 2003).

Notwithstanding this opening of new geography departments, the development of a critical perspective remained quite sluggish within the discipline as a whole. This is partly linked to the legacy of how the discipline began in Mexico and a need, among geographers themselves, to position geography as a legitimate science. A primary pedagogical objective for many of the new departments was developing a critical mass of spatial-scientists who would then work in government offices like the INEGI (National Institute of Geography, Statistics and Informatics), created in 1983. Geography was differentiated from sociology, economics, anthropology or political sciences fundamentally on the basis of its spatial sensitivity, understood at the time as spatial analysis. Furthermore, many of the individuals who had started to create a critical mass of critical geographers during the 1970s moved to other disciplines (planning, urban studies, sociology, and landscape architecture), or left the country to do their postgraduate degrees (mainly to the U.K., U.S., and France).

## Conclusion: What Might the Future of Mexican Critical Geography Look Like?

The landscape of geography in Mexico is changing rapidly. Today, it would be fair to say that a body of work exists within the discipline that explores a multiplicity of issues with a critical approach. Part of this transformation undoubtedly has to do with the role of particular individuals who continued to push forward a critical agenda in geography, in terms of both teaching and research.[3] Beyond individualized serendipitous cases, two important processes help explain the slow, albeit persistent growth of a critical perspective within Mexican geography. First, was the creation of CONACYT (the National Council for Science and Technology) in 1971, triggering a sudden infusion of resources and opportunities for students to travel abroad for a semester, or to pursue their graduate studies. This abruptly opened the discipline up beyond the national boundaries of Mexico, facilitating the movement of people and ideas: The study of geography suddenly seemed limitless. CONACYT provided grants for students to pursue graduate degrees abroad, under the condition that they return to Mexico for a few years to work in the public sector, as a way of recompensing the nation for what it gave them. Thus, an entire generation of geographers, trained elsewhere (mostly in the U.S., France, and England), returned to Mexico with new ideas, new epistemologies, and new understandings of the discipline and its possible future. Many pushed for these new agendas within geography, others joined different departments, and yet others pushed for the opening of new geography departments in the country. This triggered the second factor: the multiplication of geography departments across universities, both in Mexico City and nationwide. For example, a geography department was created in Guadalajara in 1980, a human geography department emerged in the Universidad Autónoma

Metropolitana-Iztapalapa in 2002, and a socio-territorial studies department was founded in UAM-Cuajimalpa in 2005. Even though some geography departments remain relatively traditional in their pedagogical approaches, individuals certainly are infusing the curricula with critical perspectives. This diffusion of geography departments across the country is certainly an important consideration.

Despite this promising scenario for critical geography, new generations of geographers nonetheless face particular structural challenges. First, in a context of increasing austerity measures that affect most aspects of public life, it is important to acknowledge what many of us know, have experienced, or have fought against, an academic model based on productivity under market logics. While the Mexican academic scenario remains *relatively* distant from a rampant logic that exists elsewhere (for instance, in many U.S. and U.K. academic institutions), it is certainly moving in that direction. In 1984, CONACYT developed what became known as the National System of Researchers (SNI in Spanish), a national evaluation body that places and promotes academics across the country. It utilizes a quantitative system of performance measures, based on a number of somewhat vague and sometimes discretional benchmarks that include publications (individual, not co-authored(?)), teaching, and postgraduate supervision. It is considered prestigious to join the SNI, whose members receive a monthly monetary remuneration that, depending on their performance, can provide a very generous monthly allowance helping many get by relatively comfortably. Needless to say, this system is facilitating an individualized publish-or-perish model dragging with it similar consequences to those experienced elsewhere: An entrepreneurial spirit where those who play by the rules of the game dictate both those rules and who is allowed in, and a hyper-production of articles that must be cutting-edge. Yet the definition of cutting-edge research stresses publication in Anglophone journals ranked highly on international performance indices. Other forms of knowledge production, which might figure largely in, for example, alternative community development agendas or indigenous forms of knowledge production, are simply ignored. The professional and economic pressures to engage in particular forms of fast-track knowledge production accentuates a type of research that on occasions can drift away from socio-spatial research agendas that are slower, uncertain, and open to multiple possibilities, something that critical geographers have noticeably voiced concern over (Fuller and Kitchen 2004).

These are issues that of course exist elsewhere. In the Mexican context, we can add the question of language. Articles published in English are judged more positively (implicitly, but in some cases overtly), considered higher standard and thus "count more" than those appearing in national journals. That is where a colonial element overlaps with an already uneven terrain of hyper-competitiveness, point-counting, and decimal fighting. In an academic political economy of this nature, the future of critical perspectives within geography (or other social sciences for that matter) can appear bleak. However, many young critical geographers, especially those who have returned to Mexico having had the opportunity to study elsewhere (which is an issue in and of itself), have developed important international networks that place them in favorable positions relative to the diffusion of critical work at the national scale. This is the case both in terms of publications (collaborating with scholars elsewhere) and in the realm of teaching. Geography graduate and undergraduate students in Mexico today have more opportunities to engage with interesting national and international critical debates that open up avenues for new possibilities and perhaps new critical geographies.

## Notes

1   Translated by author. Original quote: Por otra parte, el programa redentor del Estado Mexicano, que buscaba, usando la fórmula cardenista, "convertir a los indios en mexicanos", comenzó a ser visto por las nuevas generaciones de antropólogos como un proyecto de sometimiento del indio – es decir, de aquel fetiche adorado, ya, por la antropología como el talismán y objeto de su éxito –. Por eso, la antropología de la generación del 68 quiso alejarse de las políticas públicas del Estado y criticó duramente a la generación de antropólogos que había conseguido la apoteosis modernista de la antropología, buscando en vez nueva energía en el espacio de intermediación entre el campesinado – asechado por las fuerzas del mercado y del Estado – y el propio Estado. (Lomnitz 2016: 290)
2   Interview with Professor José Omar Moncada Maya, from the Instituto de Geografía at the Universidad Nacional Autónoma de México, by Verónica Crossa, August 4, 2017.
3   Scholars like: Julie-Anne Boudreau, Georgina Calderón, Gian Carlo Delgado Ramos, Javier Delgadillo, Jerónimo Díaz, Efraín Hernández, Blanca Ramírez or Luis Salinas, to mention only a few names that have emerged in various conversations.

## References

Aguilar, A. (1994). La radicalización de la geografía. Nuevas direcciones en el debate. In A. Aguilar, A. and Moncada, O. eds, *La geografía humana en México: institucionalización y desarrollo recientes*, pp. 38–54. México: Universidad National Autónoma de México y Fondo de Cultura.
Cardoso, F. and Faletto, E. (1979). *Dependency and Development in Latin America*. Translated by Marjory Mattingly Urquidi. Berkeley: University of California Press.

Castells, M. (1973). *Imperialismo y Urbanización en América Latina*. Barcelona: Gustavo Gili.

Chías, L., Cruz, J., and Malcón, A. (1994) Desarrollo de la geografía universitaria: el colegio de Geografía. UNAM y sus planes de estudio. In A. Aguilar and O. Moncada, eds, *La geografía humana en México: institucionalización y desarrollo recientes*, pp. 76–91. México: Universidad National Autónoma de México y Fondo de Cultura.

Delgadillo, J. and Torres, F. (2010). La geografía regional en México: aproximaciones a la obra y sus autores. In, D. Hiernaux, ed., *Construyendo la geografía humana: El estado de la cuestión desde México*. México: Anthropos & Universidad Autónoma Metropolitana.

Ettlinger, N. (1999). Local trajectories in the global economy. *Progress in Human Geography* 23: 335–357.

Fuller, D. and Kitchen, R. (2004). *Radical Theory/Critical Praxis: Making a Difference Beyond the Academy*. Canada: Praxis Press.

Gilbert, A. (2004), Love in the time of enhanced capital flows: Reflections on the links between liberalization and informality. In Ananya Roy and AlSayyad, eds, *Urban Informality: Transna- tional Perspectives from the Middle East, Latin America, and South Asia*, pp. 33–65, Nueva York: Lexington Books.

Harvey, D. (1973). *Social Justice and the City*. Oxford: Blackwell Publishers.

Harvey, D. (2006). Editorial: The geographies of critical geography. *Transactions of the Institute of British Geographers* 31: 409–412.

Hiernaux, D. (2010). *Construyendo la geografía humana: El estado de la cuestión desde México*. México: Anthropos & Universidad Autónoma Metropolitana.

Lomnitz, C. (2016). *La nación desdibujada: México en trece ensayos*. Mexico: Malpaso.

Lomnitz, L. (1975). *Cómo sobreviven los marginados*. México: Siglo XXI editores.

Moncada, O. (1994). La geografía en México: institucionalización académica y professional. In A. Aguilar and O. Moncada, eds, *La geografía humana en México: institucionalización y desarrollo recientes*, pp. 57–75. México: Universidad National Autónoma de México y Fondo de Cultura.

Moncada, O. and López A. (2015) *70 años del Instituto de Geografía: historia, actualidad y perspectiva*. Mexico: Universidad National Autónoma de México.

Quijano, A. (1967). *Dependencia, Cambio Social y Urbanización en Latinoamérica*. Santiago: División de Asuntos Sociales CEPAL

Ramírez, B. (2003). Geographical practice in Mexico: the cultural geography project, *Social & Cultural Geography* 4(4): 565–578.

Ramírez, B., Montañez, G., and Zusman, P. (2009). Geografías críticas latinoamericanas. In M. Chávez and M. Checa, eds, *El espacio en las ciencias sociales: geografía, interdisciplinariedad y compromiso*, pp. 103–128. Mexico: El Colegio de Michoacán.

Schteingart, M. ed. (1973). *Urbanización y dependencia en América Latina*. Buenos Aires: Ediciones S.I.A.P.

Tamayo, J. (1962). *Geografía general de México (4 tomos y atlas)*. México: Instituto Mexicano de Investigaciones Económicas.

# 13

# "Let's here [sic] it for the Brits, You help us here"[1]

## North American Radical Geography and British Radical Geography Education

Joanne Norcup

The purpose of this chapter is to show that radical geographical ideas from North America complemented radical education aspirations in Britain, albeit not necessarily those found in only universities. In particular, those ideas played a significant role in inspiring a group of concerned academics, educators, activists, and artists who formed the Association for Curriculum Development in Geography (ACDG). During the 1980s, the Association produced a journal, *Contemporary Issues in Geography and Education* (*CIGE*), which argued for fundamental changes both to school geography education and to geographical knowledge-making at universities.

Studying the journal helps illuminate both the broader context that produced geography's radical turn, and the supportive transatlantic networks of exchange that existed between radical humanist geographers working in North America and Britain, contributing to sustaining the movement. The *CIGE* archive, and later the oral history associated with it (discussed in Norcup 2015a), tell a story of the rise of British radical geography during the 1980s. Both sources reference authors publishing

*Spatial Histories of Radical Geography: North America and Beyond*, First Edition.
Edited by Trevor J. Barnes and Eric Sheppard.
© 2019 John Wiley & Sons Ltd. Published 2019 by John Wiley & Sons Ltd.

in both *Antipode* and the *Union of Socialist Geographers*. They also document a continued relationship between the editors and William Bunge. Bunge's handwritten note to Ian Cook provides this chapter's title, affirming Bunge's admiration for both *CIGE* and British radical geography education. The archive and oral histories further suggest that North American radical works from the 1960s and 1970s aligned with anti-racist, post-colonial, and feminist politics that emerged in Britain during the 1980s, creating space for developing critical thinkers to articulate radical geographical knowledge.

## Contemporary Issues in Geography and Education

### Part 1: Sparks from North America

In 1983, Oxford University Press published an edited collection of essays reflecting on the state of geography education in Britain. Edited by the British geography educator, John Huckle, *Geographical Education: Reflection and Action* gathered a range of academic writers and geography educators to comment on the state of geographical education. The motivation for the collection were the "personal costs" experienced by teachers and students of the new geography espoused during the 1970s that favored scientific rationalism and mathematical modeling of geographical phenomena at the expense of a humanistic approach that took seriously people and their rich and complicated interior and corporeal existence. The new geography asked its practitioners:

> to discount private everyday experiences and to think of themselves as [both] insignificant inhabitants of a world in which truth and knowledge are static features of external reality, and ... victims of natural and social forces over which they have little control. In such circumstances human values and potentials are denied, and geography lessons lose much of their appeal (Huckle 1983: 1).

Divided into three sections, contributors to *Geographical Education* explored the recent history of teaching geography and new perspectives on it. Chapter 7 was written by Ian Cook. He discussed the histories of radical geography, and argued for meaningful and practical ways that radical ideas could be deployed to invigorate the discipline as "a live subject concerned with relevant social and environmental issues" (Cook 1983: 75). Cook promoted "two pillars" of critique of the new geography of positivism. First, were the power of geographical expeditions initially carried out by Bill Bunge in Detroit, Toronto, and Windsor (Ontario). For Cook (1983: 75), they "... marked a practical and activist focus which was regarded as being more important than the development of a deep theoretical understanding of the problems with which they dealt." Second, Cook signposted the importance of creating publishing space through which radical theoretical discussions and practical strategies could circulate, exemplified by *Antipode*. Cook also noted that for activism and theoretical ideas to develop across the subject as whole networks needed to be constructed to ensure interaction and solidarity. Many, Cook noted "… find solidarity in the Union of Socialist Geographers" (Cook 1983: 75). Cook's chapter later applied radical ideas to the critique of a school syllabus. He concluded his chapter with a rallying cry to geographers:

> Adoption of a radical type of curriculum will not be easy … It requires the teacher to have a wider and deeper knowledge of social processes than is customary at present; it requires much of the pupil; and it will provoke considerable reaction from the establishment. Nevertheless, to me it seems imperative that geographers [make the] attempt … The alternative is for the subject to wither in its refusal to face the deep and real issues of modern society. (Cook 1983: 81).

Cook's chapter and Huckle's book marked a significant publishing moment for radical geography education in Britain, introducing geography educators to the activities of radical geographers in North America, sharing their ideas. At the time, Cook was a lecturer at Liverpool Polytechnic where the focus on teaching was as vital to his work as any research he might undertake. For Cook, practical radical ideas utilized in educating students should be at the very heart of political and professional practice. As a subscriber to the newsletters of the Union of Socialist Geographers (see Peake, this volume), Cook's own radicalism was nurtured and enlarged. Becoming a series co-editor of *CIGE*, he would later influence members within the UK and in North America.

## Part 2: Shared trajectories and convergence of effort

On the April 23, 1983, Dawn Gill wrote her first letter to Ian Cook.

> I'm interested in learning more about the Union of Socialist Geographers mentioned in your chapter of John Huckle's book. Please will you send me details. Perhaps you will be interested in some of the work which is going on in London at the moment. A group of teachers has set up an "Association for Curriculum Development in Geography" with the purpose of running a magazine entitled Contemporary Issues in Geography and Education… Perhaps you'd like to work with us in some way[2]

Dawn Gill was head of geography at Quinton Kynaston Secondary School in North West London.[3] Gill had made national news after writing a report criticizing a popular school geography curriculum she saw as inherently racist. Gill's radicalism evolved from a commitment to anti-racist education, to advocating equal access to education. These ends were informed by both a variety of personal and professional geographies and education networks, along with a global/local community politics based on where she lived in Hackney, a long-time immigrant and working-class inner-city district in North East London. Key to Gill's own radical geography education was extra-curricular reading David Harvey's (1973) *Social Justice and the City* and Richard Peet's (1977) *Radical Geography* while completing a Master's degree in geography education at the Institute of Education, London University, during the very early 1980s. Inspired by Harvey, Gill's Master's dissertation examined the socio-spatial inequalities of access to education. Her later report was based on themes from her Masters' thesis. It was particularly timely given it coincided with the UK urban riots during the summer of 1981. Focused on a popular school geography curriculum, Gill concluded in her report that the curriculum was complicit in perpetuating both racial stereotypes and reductive representations of people and places (Gill 1982; and see also Norcup 2015b).

Connecting with Ian Cook proved crucial for Gill's publishing ambitions for the ACDG. By May 1983, Cook joined the ACDG along with a growing number of members (such as Roger Lee and John Huckle), all of who found intellectual inspiration from *Antipode* and the Union of Socialist Geographers. In a letter to Gill dated July 13, 1983, Cook suggested they co-edit the journal *CIGE*.[4] Cook brought both practical experience of writing for academic journals, as well as his knowledge of radical geography from his involvement with the Union of Socialist Geographers (USG). Between May and November 1983, correspondence between Roger Lee, Ian Cook, and Dawn Gill discussed such issues as the tone, content, style, and ethics of the journal.[5] *CIGE* would contain accessibly written but academically rigorous articles with reference lists, suggestions for further reading, and resources for use in teaching geography education across schools, colleges, and universities. At the heart of *CIGE* were nine aims that mirrored concerns of the political left in Britain.

## Part 3: "An Emancipatory Geography", CIGE – aims and themes

Contemporary Issues in Geography and Education seeks to open up or broaden areas of debate and to examine current controversies and disseminate ideas and materials which help to develop a critical approach to the learning and teaching of geography. The journal seeks to promote an emancipatory geography; it seeks, in other words, to promote the idea that the future is ours to create – or to destroy – and to demonstrate that education bears some responsibility for building a world responsive to human needs, diversity, and capabilities. (*CIGE* 1983: 1)

*CIGE*'s nine aims implicitly anticipate the politics of intersectionality (Figure 13.1; Crenshaw 1989). They attempt to reconstruct geographical education to recognize compounding power geographies of inequality generated by social processes of race, gender, and class. *CIGE* wanted to foster an interchange of ideas among researchers, students, and educationalists to examine critically the ideological content of geography education curricula, and at the same time to validate humanist and radical ideas in geographical teaching and research. A thematic approach was chosen to facilitate diverse critical explorations around a topic. Those themes were *Geography for a Multicultural Society Volume 1 and 2*, which focused on geographies of race and racism (*CIGE* 1.1 1983; *CIGE* 1.2 1984); *The Global Economy: Trade, Aid and Multination*als (*CIGE* 1.3 1984); South Africa: Apartheid Capitalism (*CIGE* 2.1 1985); Confronting the Ecological Crisis: *Environmental Geographies* (*CIGE* 2.2 1986); *War and Peace* (*CIGE* 2.3 1988); *Gender and Geography* (CIGE 3.1 989); and finally, *Anarchism and Geography* (*CIGE* 3.2 1990).

CIGE's editorial group was initially formed by a range of geography educators (John Huckle, Frances Slater, Julian Agyeman), students, social and environmental activists, and academic researchers. Over time, however, the makeup of the editorial group became dominated by academic geographers who were either lecturers (Peter Jackson, David Pepper), or who were doctoral students (Sarah Whatmore, Linda Peake). Gill and Cook worked alongside theme editors (often members of the editorial group) to solicit contributors for respective issues. For the most part, contributions came from their wider network of politically aligned contacts from North America, the UK and beyond.

## Part 4: Solidarity and support: contributors and circulation of CIGE

Eight issues of *CIGE* were published between 1983 and 1991. They illustrate the geographical scope of radical ideas coming from North American geography as well as critical and radical ideas such as anti-racism, postcolonialism, and feminism that were evolving and gaining cultural purchase in direct response to changes in British society during the 1980s.

Contemporary Issues in Geography and Education   Vol. 1, No. 1, Autumn 1983

# An Introduction to 'Contemporary Issues in Geography and Education'

## Editorial

'*Contemporary Issues in Geography and Education*' seeks to open up or broaden areas of debate and to examine current controversies within the discipline, at all levels. The journal aims to collect and disseminate ideas and materials which help to develop a critical approach to the learning and teaching of geography. The journal seeks to promote an emancipatory geography; it seeks, in other words, to promote the idea that the future is ours to create — or to destroy — and to demonstrate that education bears some responsibility for bulding a world responsive to human needs, diversity and capabilities.

Individuals in many schools, colleges and universities are beginning to develop learning materials and styles of working which encourage more active participation in decision-making on the part of the learner. Making choices and commitments in a real world context should develop from, and be linked with, decision-making in the classroom. Especially relevant to geographical education are techniques and materials which may encourage involvement in matters of local, national and global concern. The journal will help educators to share ideas on these issues.

Although the journal recognises the importance of other branches of the subject, it is focused primarily upon human geography. Some articles will, however, raise questions of a more general nature, relevant to the teacher as educator as well as subject specialist.

Many recent developments in geography have great potential to provide an education which is critical and constructive. As yet, however, these developments are scarcely reflected in school curricula, and have not been adopted in syllabuses. In this context the objectives of '*Contemporary Issues in Geography and Education*' are:

- to develop a critique of current curricula

- to explore the assumptions underlying much of geographical education and to make these assumptions explicit

- to examine the ideological content of geographical education in relation to its political context

- to demonstrate the relevance and importance of humanist and radical ideas for teaching and research in geography

- to promote an interchange of ideas between researchers, students and educationalists in geography

- to encourage dialogue between geographers and the various groups and organizations concerned with major issues in education. We envisage that these would include groups involved in world studies, peace studies, human rights education, environmental education, development education, multi-cultural and anti-racist education, anti-sexist education, urban studies and community education, education for equality and education for political awareness and participation

- to facilitate the exchange of ideas on learning materials and classroom strategies

- to foster a geographical education which is more relevant to the present and future everyday lives of ordinary people and the communities in which they live

- to encourage the realisation of the links between critical understanding and the active transformation of the world in which we live.

1

**Figure 13.1**   *CIGE*'s nine aims, found in every journal issue and used as editorial and writing guide. *Source: CIGE* 1983 (p. 1).

Publishing enabled *CIGE* to exemplify the capacious possibilities for radical humanist approaches. It legitimized theoretical debates and created practical educational resource materials to carry these ideas into diverse learning spaces. The first two issues of *CIGE* dealt specifically with geographies of race and racism within geography. While the launch issue focused attention largely on a British context, the following issue discussed international exemplars. Cook's article, "Colonial past: Postcolonial present: Alternative perspectives in geography" was an early example of a radical geographer discussing the unequal landscapes of colonialism and its socio-spatial and educational legacies. It weaved radical geography discussions on colonial legacies with radical and humanist discussions from North America as well as drawing on post-colonial writers such as Franz Fanon (Cook 1984: 9) to produce an article explaining key theoretical terms and suggestions for teaching. International articles challenged the legacies of colonialism across the education system were included by way of example. John Fein's article "Structural silence," examined the invisibility of first nation people in Australian geography teaching and how this perpetuated colonial ideas and furthered their marginalization in Australian society. The reproduction of William Bunge's 1965 *CRISIS* magazine article, "Racism in geography," alongside an edited 1983 republishing of Gwendolyn Warren's 1971 Detroit Geographical Expedition and Institute essay, "No ratwalls on Bewick," gave recent historic and contemporaneous insights to the racist geographies in the United States. Writing "An appreciation" at the end of Bunge's article signaled *CIGE*'s support of the radical geographer.

Solidarity for critical radical geographers was not limited to the Anglo-American academy. In *The Global Economy* issue (1.3), *CIGE* published a letter alerting readers to the political imprisonment of Dr. Kamoji Wachiira, a former geography lecturer at the University of Kenya, whose criticisms of Kenyan authorities about their economic and agricultural environmental policies led to his detention. Including this letter and address details for readers to write in solidarity illustrated *CIGE*'s commitment to post-colonial politics by encouraging practical action. This was further exemplified in the issue on *South Africa: Apartheid Capitalism*. Reproduction of the African National Congress's (ANC) Freedom Charter at the back of the issue was a significant political statement. Publishing this alongside articles from anti-apartheid campaigners, South African academics and British academics researching the theme (including Phil O'Keefe who was editor of *Antipode* at the time), demonstrated *CIGE*'s global political solidarity. Gill's own anti-racist work in London afforded her political insights from education activists across the Black Caribbean and Pan-African diaspora. Resource centers such as the John La Rose Centre (now George Padmore Institute) and New Beacon

Books gave *CIGE* access to more nuanced insights into the creation of anti-racist geography education (see Fairless Nicholson 2017). *Confronting the Ecological Crisis* issue afforded space to consider the geopolitical uses and abuses of natural environmental resources and the ecological consequences of capitalism. David Pepper, Andrew Sayer, and Greg Richards all contributed articles that echoed closely their writing in The London Group of The Union of Socialist Geographer's 1983 publication, *Society and Nature: Socialist Perspectives on the Relationship between Human and Physical Geography* (London Group of the Union of Socialist Geographers 1983). The *War and Peace* issue reiterated themes of global power and colonialism, and Bunge was supported again with the publication of a review essay (Bunge 1988b), while his *Nuclear War Atlas* was reviewed by David Pepper. Pepper's own article in the issue focused on ways to teach about nuclear war. The *Gender and Geography* issue was largely edited by members of the Women and Geography Studies Group of the Institute of British Geographers. Drawing on emerging feminist ideas (and citing writing by North American geographers Wilbur Zelinsky, Janice Monk, and Susan Hanson) indicated a commitment to sharing transatlantic feminist ambitions. Linda Peake, Jo Foord, Liz Bondi, Jo Little, and Sarah Whatmore all contributed articles. The *Anarchism and Geography* issue drew strongly from British and American anarchist geographers, with articles by Ian Cook and David Pepper (who jointed edited this final issue), British anarchist Colin Ward, and North American geographer Myrna Breitbart.

More broadly, *CIGE*'s archive hints at its larger aspirations. In the minutes of the editorial group, there is reference to other theme issues in production, including an issue on Central America edited by Doreen Massey.[6] Resource challenges impeded the ambition, however. An aborted attempt by Gill to meet with Richard Peet about a possible *Antipode/CIGE* collaboration was *never* followed up. Nevertheless, the space for shared discussions and mutual support for the ideas of radical humanist geographies galvanized practitioners and enabled ideas to germinate and gain currency across human geographers in the UK.

## Part 5: "How delightful a letter, just full of "things to do" which is my kind of geography ..." Transatlantic solidarity 2: The Bunge Letters[7]

There are documents in the remaining *CIGE* archive that connect with the work of radical North American geographers. Significant among them are five letters from William Bunge.[8] It pays to consider the details of these letters in relation to the context in which they were written. They show that *CIGE* gave solidarity and a publishing outlet to a radical geographer who was then finding it difficult to find publishing space in mainstream Anglo-American journals[9] (Bergmann and Morrill 2017). Bunge's first three letters were sent within a month of each other just before, during, and immediately after the launch of *CIGE* between October and November 1983. The final two letters were dated January 16, 1985 and June 17, 1986. These remaining letters coincided with Cook and Gill soliciting materials for the theme issues in which Bunge's work was eventually reproduced. The letters hint that there was more contact than only those five remaining letters. Both Cook and Gill recalled receiving letters from Bunge but only the letters addressed to Cook remain, a product of Cook's stable employment tenure at Liverpool Polytechnic where he kept the correspondence. Gill's employment was more peripatetic, but she confirmed she was in contact with Bunge both by letter and telephone conversation. Both Bunge and Gill had strong reason to believe their telephones were tapped, so Gill recorded those conversations as evidence.[10]

   Cook and Gill contacted Bunge before the launch of *CIGE* to request permission to reproduce some of his work. They wanted it to connect to the journal's discussions of anti-racist geography education, and the histories of such endeavors in geography. Such a connection not only benefited *CIGE* in aligning the journal with a significant figure in the history of radical geography, but produced mutual benefit. Bunge's first letter is enthusiastic at receiving their requests, attested by his quote beginning his letter that forms part of the title for this sub-section.[11] Bunge agreed to let them reproduce his 1965 "Racism in geography" article ("by all means reprint it, reprint it"), but he also included further materials which he encouraged them to publish too, attaching his review of George A. David and O. Fred Donaldson's 1975 book, *Blacks in the United States: A Geographic Perspective*. Bunge described it to Cook as "a great book which evidently the authors paid for with their lives...."[12] Bunge continued his letter by discussing the first imprint of his *Nuclear War Atlas,* and signed off by asking Cook to gather peace maps from Europe "or at least English peace maps ... They will go into the second edition of the *Atlas.* Peace, Bill Bunge" Less than a month letter, Cook received a second, much shorter letter from Bunge,[13] Dated November 14, 1983, Bunge begins, "I am vague as to what material you might find useful so I have enclosed a list of everything I have written, ... hope this is of some further help." Four pages of photocopied reference lists are attached, listing 90 articles under the heading, "How to Explore and Save Humanity". Two days later, Bunge sent Cook an annotated photocopied letter that began, "'Dear friends of Peace and The Nuclear War Atlas, Christmas is coming, and so is Nuclear War."[14] The letter is a general call for assistance in selling and circulating copies of his *Nuclear War Atlas*. At the top of his letter in red ink, Bunge wrote, "Dear Prof

Cook, Let's here [sic] it for *the Brits*, you help us here. Respectfully, Bill."
Bunge's final two letters concern his *Nuclear War Atlas*, while also
enthusing radical left British politics ("What is the name of that man who
leads the coal miners? My, oh, my, I only wish I had his guts …"). The final
letter concerns *CIGE's* publication of Bunge's book review, "Geography
without Geographers," a book review of Norman Myer's *GAIA: An Atlas
for Planetary Management*.[15] Bunge agrees, instructing Cook to contact
Karl Raitz for permission ("I have agreed to have only three of sixteen
pages published by Raitz in *The Professional Geographer* under the
condition that there is no censorship, and how he can accomplish that by
throwing out thirteen-sixteenths of the work beats me!")[16] Bunge's letter
discusses themes of genocide, nuclear war, and saving children before concluding with two requests. The first was to ask Cook to help in starting

> … a petition among English geographers asking Wayne State University
> Board of Governors and President to stop their relentless persecution
> of me in my academic freedom fight … sometime timing is everything.
> So my trial is October 9th…. The British address knocks out all the
> hicks over here including me.

The second was to advocate for the *Nuclear War Atlas* with Alan
Halsworth at the publishing company Basil Blackwell: "Your enthusiasm
for me helps me with him." Bunge's letters repeatedly illustrate the potency
of *CIGE* in circulating his ideas and in offering radical collegiate support.

## Part 6: Bunge's significance in CIGE and the significance of CIGE to Bunge

The inclusion of Bunge's writings in *CIGE* during the 1980s are
noteworthy. In a decade where Bunge's writing and work were rarely
republished in established Anglo-American academic journals, finding
publication space in a journal that was subscribed to by many UK and
international university libraries for the purpose of use in seminars
and tutorials ensured a degree of readership, and, in turn, a potential
supportive and collegiate network for him (Norcup 2015a). Bunge's
letters indicate the extent to which he lent on the editors to affirm his
work and circulate his ideas in an international radical publication.
This is not to suggest that Bunge was either unquestioningly supported
by *CIGE* nor that *CIGE* was widely read, but it maintained his presence
in print. Bunge's candid letters to the editors of *CIGE* strongly hint at
him being *persona non grata* in mainstream North American radical
geography by this point, and his own admission of the political trials and

tribulations he personally experienced. Nevertheless, his work remained inspiring to UK-based early-career academics such as Peter Jackson, who returned to the UK from doctoral research in North America enthused to make a similar attempt to forge fieldwork "expeditions," recalling a short-lived attempt made by himself and Susan Smith at Oxford that was met with disdain (Norcup 2015a). For the most part, attempts at inculcating a Detroit-like expedition approach linking community activism with the academy did not take root in the UK during the decade. This could be attributed largely to the formal institutional structure and pedagogy of UK university geography, especially at sites like Oxford. Nevertheless, precisely because of CIGE's dissemination of Bunge's work, Jackson and Smith were able to develop their new academic trajectories, as found in their edited book *Exploring Social Geography* (Jackson and Smith 1984). Bob Colenutt's attempt to create a London Expedition found more success, but it was short-lived given that by the mid-1980s teachers and college lecturers were feeling increasingly under attack by the Conservative government and mainstream news media that characterized them as the "Looney Left" in education (Norcup 2015b).

Nevertheless, the inclusion of Bunge's work in *CIGE* kept alive the possibility of a different framing of geographical education, fieldtrips, and curricula. It opened up the idea that education innovation and geographical knowledge-making need not simply trickle-down from the academy, but rather ripple across and through diverse formal and informal sites and spaces of learning. Jackson (1989) hints that in his paper about anti-racist education. He suggests that new geographical curricula reported by *CIGE*, and found in "school magazines" and statements by school educators, should be taken up by academic lecturers and teachers.

Bunge's notoriety and "cult of personality," something to which he was latterly frustrated by, also served to keep his earlier work in the folk memory of the discipline in the U.K. By the early 1990s, early-career U.K.-based academics were returning to Bunge's work in an effort to explore more deeply and understand more fully the microscales of social relationships that went into making radical and critical geographies of knowledge production, expanding the academic sub-field (Merrifield 1995). Bunge's field expeditions have endured, forming an important template for geographers of what it means to be an "activist-scholar."

It is sad then that despite all Bunge's words of appreciation for the editors in his correspondence ("Dawn Gill, she has a beautiful soul"),[17] there remains a palpable misogyny in Bunge's (lack of) accurate citational reference to both Gill and to Warren's *CIGE* article. In the bibliography of the second edition of Bunge's *Nuclear War Atlas* he references Gwendolyn Warren's "No ratwalls on Bewick article" from 1971 as being in Field notes, number 6, and "in the UK, copies are available from Dawn Gill, 29

Barretts Grove, London" (Bunge 1988a: 199). Providing Gill's then personal home address rather than the actual journal reference is at best disrespectful, serving to reduce Gill and as mere secretarial support rather than acknowledge the larger radical project she was undertaking. That Bunge fails to give Warren a full citation either from the pages of *CIGE* or from the original Fieldnotes publication effectively diminishes and obscures women radical geographer and publications (see also Warren *et al.*, this volume). Over time, the compounding effect of such citational slippage served to diminish the radical geographical activism of Gill, Cook, Warren, and *CIGE*, instead perpetuating the idea of Bunge as the lone radical geography maverick. It boosted further the cult of personality around him, making Bunge appear heroic, rather than recognizing the networks and solidarities in schools and community-based institutions from which he gained inspiration for much of his work.

Just like *CIGE*, what might appear as a fleeting spark (Walford 2001), can ignite contemporaneous reflections on and performances of geographical knowledge. Bunge's presence in *CIGE* gave kudos to the journal and a publishing space through which networks and solidarities were galvanized, and radical ideas forwarded. *CIGE* gained much from Bunge's name, not least the confidence and inspiration to circulate ideas, to recommend readings, to provide histories of radical geography that countered the educational lean to the political Right, and to create publication space to forge international solidarities for the future.

## Conclusions

This brief overview of the publishing life of the British-based journal *CIGE* illustrates how it created a vital space of commonality for radical and critical geographers who worked for it, contributed to its pages, and read it. For North American radical geographers as well as geographers in other parts of the world, *CIGE* provided and linked with past radical endeavors, offering future possibilities for radical geography education. Recovering the correspondences of Bunge in what remains of the *CIGE* correspondence archive illustrates how radical geographical ideas circulated internationally, helping to constitute social and intellectual networks between Britain and North America.

## Notes

1  Quote from a handwritten annotation on a letter to Ian Cook from Bill Bunge, November 16, 1983 (*CIGE* archive 1983/000034). Bunge's misspelling of hear as "here" is in the original letter. The *CIGE* archive contains documents

amassed during doctoral research spanning a decade of interviewing and recovering primary materials from the time. In large part, the letters discussed in this chapter come from the correspondence file of Ian G. Cook who was co-series editor of *CIGE* alongside Dawn Gill, founder of the Association for Curriculum Development in Geography (*ACDG*) of which *CIGE* was its journal. Further details can be found in Norcup (2015a).

2   Letter to Ian Cook from Dawn Gill, April 23, 1983. *CIGE* archive 1983/000001.

3   The British equivalent to American high school.

4   Letter from Ian Cook to Dawn Gill, July 13, 1983; *CIGE* archive 1983/000013.

5   See letters and minutes of *CIGE* editorial group meetings. *CIGE* archive 1983/000006 – 1983/000032.

6   Minutes of *CIGE* editorial group meetings *CIGE* archive 1983/00007, 1986/000012).

7   Letter from Bunge to Cook (*CIGE* 1983/000020a).

8   *CIGE* archive 1983/000020a, 1983/000033, 1983/000034; 1985/0000001, and 1986/0000015.

9   *CIGE* 1986/0000015.

10  Unfortunately, these are missing presumed lost. However other *CIGE* members recall and corroborate this (Norcup 2015a).

11  *CIGE* 1983/000020a.

12  This is not reprinted in *CIGE*.

13  *CIGE* 1983/000033.

14  *CIGE* 1983/000034.

15  *CIGE* 1986/000015.

16  Raitz does not. The complete unedited review is published in Issue 2.3 of *CIGE* with an editorial note explaining, "Important books like this one take more space than trivial ones, and therefore cannot be fitted into some mechanically predetermined format" (*CIGE* 1988: 78).

17  *CIGE* archive 1986/000012.

18  The entire series of *CIGE* is freely accessible online at Dr. Jo Norcup's production company, Geography Workshop Productions Limited. Established in 2015, the company takes its motto "The Future is Ours to Create" from *CIGE*'s initial editorial. The production company exists to produce media resources, workshops and fieldtrips aligned with the intersectional ambitions and shared spirit of *CIGE*. The complete journal series was made available to honor the memory of Dawn Gill who died in 2017. See http://geographyworkshop.com/2018/03/29/archive-afterlives-digitising-cige/.

# References[18]

Bergmann, L. and Morrill, R. (2017). In Memorium: William Wheeler Bunge: Radical Geographer (1928–2013). *Annals of the Association of American Geographers* 108 (1): 291–300.

Bunge, W. (1983). Racism in geography. Geography and education for a multicultural society: Part 2, *Contemporary Issues in Geography and Education* 1 (2): 10–11.

Bunge, W. (1988a). *Nuclear War Atlas*. 2nd Edn. Oxford. Basil Blackwell.

Bunge, W. (1988b). A geography without geographers: A book review of GAIA: An atlas of planetary management by Norman Myers (ed.) In War and Peace issue of *Contemporary Issues in Geography and Education* 2 (3): 78–83.

Cook, I. G. (1983). Radical geography. In J. Huckle, ed., *Geographical Education Reflection and Action*, pp. 74–82. Oxford: Oxford University Press.

Crenshaw, K. (1989). Demarginalizing the intersection of race and sex: A black feminist critique of antidiscrimination doctrine, feminist theory and antiracist politics. https://heinonline.org/HOL/LandingPage?handle = hein. journals/uchclf1989&div = 10 accessed September 30, 2017.

Fairless-Nicholson, J. (2017). Learning anti-racism: Preliminary snap-shots from PhD research on the historical geographies of anti-racist education in London, 1960–1990. Paper given at *Royal Geographical Society Annual Conference, August 2017*. [Contact author for copy].

Gill, D. (1982). Secondary School Geography in London: its contribution to multicultural education. In D. Gill, ed., *Geography for the Young School Leaver: A Critique and Secondary School Geography in London: An Assessment on its Contribution to Multicultural Education*. Working Paper No. 2, Centre for Multicultural Education.

Harvey, D. (1973). *Social Justice and the City*. London: Edward Arnold.

Huckle, J. (1983). *Geographical Education Reflection and Action*. Oxford: Oxford University Press.

Jackson, P. A. and Smith, S. J. (1984). *Exploring Social Geography*. London: George Allen and Unwin.

Jackson, P. (1989). Challenging racism through geography teaching. *Journal of Geography in Higher Education* 13 (1): 5–14.

Merrifield, A. (1995). Situated knowledge through exploration: reflections on Bunge's "Geographical Expeditions". *Antipode* 27 (1): 49–70.

Norcup, J. (2015a). *Awkward Geographies? An historical and cultural geography of the journal Contemporary Issues in Geography and Education (CIGE) (1983 – 1991)*. Doctoral Thesis, University of Glasgow. http://theses. gla.ac.uk/6849/Last accessed September 30, 2017.

Norcup, J. (2015b). Geography education, grey literature and the geographical canon. *Journal of Historical Geography* 49: 64–74.

Peet, R. (1977). *Radical Geography: Alternative Viewpoints*. London: Methuen.

The London Group of the Union of Socialist Geographers. 1983. *Society and Nature: Socialist Perspectives on the Relationship Between Human and Physical Geography*. Thames Polytechnic Print Unit. London.

Walford, R. (2001). *Geography in British Schools 1850 – 2000*. London: Woborn Press.

Warren, G. (1983). No ratwalls on Bewick. *Contemporary Issues in Geography and Education* 1 (2): 14–18.

# 14

# "Can these words, commonly applied to the Anglo-Saxon social sciences, fit the French?" Circulation, Translation, and Reception of Radical Geography in the French Academic Context

Yann Calbérac

> *For the observer of French geography, the identification of its "radical" or "critical" currents immediately poses a basic problem: can these words, commonly applied to the Anglo-Saxon social sciences, fit the French? (Lévy 1985: 9)*

More than 30 years after Jacques Lévy asked it, his question remains germane. How possible is it to transfer words from one intellectual and linguistic context to another? Further, can concepts and rules brought from one context organize and structure an intellectual community in a different one? Central to my chapter is the question: Is it possible to take *radical geography* that is deeply linked to the intellectual and political context of the United States of the 1960s and 1970s and translate it both

*Spatial Histories of Radical Geography: North America and Beyond*, First Edition.
Edited by Trevor J. Barnes and Eric Sheppard.
© 2019 John Wiley & Sons Ltd. Published 2019 by John Wiley & Sons Ltd.

linguistically and conceptually so it has meaning for contemporary geographers in France? Despite attempts during the 1970s to politicize French geography, it remains deeply indebted to the Vidalian intellectual legacy (Buttimer 1971). It is true that Yves Lacoste promoted a Marxist approach to geography in *Hérodote*, the journal he founded in 1976. But Lévy argued that there was a significant difference between the Marxism of Anglophone radical geography and the Marxism of Lacoste's *géopolitique*. Lacoste was always deeply committed to international political action rather than the construction of theory as such (Lévy 1984). That said, since the 1990s with the globalization of the academy, contemporary French geography has become more connected than ever with Anglophone geography. So, are there now signs of an emerging Anglophone-inflected community of French radical geographers?

Certainly, there seem to have been changes recently. For example, sessions on critical and radical geography are now held during the *Festival International de Géographie*, a forum to connect academics to the outside public. Critical and radical geography are also now taught to students within the university. In 2015, candidates who sat for the *agrégation de géographie* were asked about the links between geography, geographers, and power.[1]

This visibility of radical geography is the result of a process that started three decades ago. During the 1980s, the very popular handbook, *Les Concepts de la Géographie Humaine*, provided a comprehensive review of radical geography (De Koninck 1984). In 2003, Jean-Bernard Racine dedicated an essay to radical geography in the dictionary edited by Jacques Lévy and Michel Lussault (Racine 2003). Radical geography was thus admitted to mainstream French geography. It produced two kinds of issues: first, its reception in France; and second, the specific form it took, which became *géographie (critique et) radicale*.[2] My chapter examines how, why, and when Anglophone radical geography crossed the Atlantic; and how these epistemological, theoretical, critical, and methodological proposals were then received in France and assimilated within its geographical tradition.

In this short chapter, I make use of two main sources of information. The first are various databases used to document academic activities, such as *Calenda*[3] (a calendar for Arts, Humanities, and Social Sciences that has promoted academic events for more than 15 years). I also use three bibliographical databases to track references and citations to radical geography, or to its main authors: *Persée*,[4] *OpenEdition Journals*,[5] and *Cairn*.[6] The second source are various theoretical and programmatic papers, journals, and books recently published in France (translated from the English or written by French academics) that are concerned with radical geography. The close reading I give them enables me to analyze their circulation and reception, allowing the documentation of French radical geography in-the-making.[7]

The first part of the chapter is about the increasing impact of radical geography in France, providing an explanation of the growing interest in radical and critical thought within French social science, especially Geography. The second part focuses on the links between geography and Marxism in the French academic context. The final section discusses the specificities of the *géographie critique et radicale "à la française"* (Morange and Calbérac 2012). I suggest that approach takes a hybrid form, combining a strong French tradition of an emphasis on social and political issues, and an Anglophone emphasis on critical and radical research.

## The Contemporary Emergence of Radical Geography in France

According to Cécile Gintrac, the height of radical geography in France was around 2010 to 2012 (Gintrac 2017). During these three years, radical geography gained its greatest prominence and its broadest audience. In documenting this period, I focus on: editorial activities; academic events; and the geographical sites of radical geography along with their main actors.

### A thriving publishing market

According to Matthieu Giroud in his preface to the French translation of David Harvey's *Paris, Capital of Modernity* (Harvey 2012a), and Serge Weber (Weber 2012) who interviewed the pool of translators who worked on that book, contemporary interest in radical geography in France is because of the recent availability of French translations of Anglophone writings by radical geographers.

Radical geography was known in France long before 2010. Paul Claval (1977) wrote about it in the 1970s, and De Konnick (1984) and Racine (2003) provided reviews in respectively a handbook and a dictionary. However, the authors of the latter two publication, did not live and teach in France, but in respectively Canada and Switzerland. Even though they spoke and taught in French, their intellectual communities were Anglophone. They recognized the lack of interest in radical geography in France and tried to fill that gap using their own familiarity with the English-language literature.

There was further interest following the publication of Jean-François Staszak's (2001) edited collection *Géographies Anglo-Saxonnes*. While French academics are most used to publishing handbooks that offer a comprehensive view of a field, Staszak preferred the unusual format of a reader to introduce a French audience to the diversity and complexity of contemporary Anglophone geography. The book consisted of an introduction, several translated founding texts, and a set of individual chapters each about a major field in Anglophone geography. Topics included feminism, postcolonialism, and of course radical geography. Béatrice Collignon's introduction explained the historical and ideological backgrounds of radical geography, and there were three newly translated papers: Harvey's (1992), Merrifield's (1995), and Peet's (1997). The book played a crucial role in allowing a French audience to read some Anglophone major texts that previously were not easily accessible, as well as introducing brand new perspectives such as postmodernity,[8] and critical and radical geography.

Things accelerated after 2007 thanks to the fact that David Harvey's papers and books were now normally translated into French. Indeed, Harvey dominated French discussions of radical geography, becoming effectively its sole representative. His papers were translated in journals such as the Marxist *Actuel Marx* (Harvey 2007a, 2007b, 2007c), and in compendiums. His *Géographie et Capital: Vers un Matérialisme Historico-Géographique* (Harvey 2014c) was specially designed for French readers[9]: It is a collection of papers selected and organized to illustrate Harvey's spatial and historical materialism.

Harvey's books are also regularly translated: *The New Imperialism* (Harvey 2010), *Paris, Capital of Modernity* (Harvey 2012a), *A Companion to Marx's Capital* (Harvey 2012b), *A Brief History of Neoliberalism* (Harvey 2014a), and *Rebel Cities: From the Right to the City to the Urban Revolution* (Harvey 2015). As the increasing pace of translation suggests, Harvey has become a major author in France. His primacy occludes other radical geographers and the specifics of their radical approaches, however. Further, the timings of the translations are often linked to events and processes in France. For example, *Paris, Capital of Modernity* was translated only when it was clear that gentrification was occurring in Paris, and associated with a debate around the right to the city (Clerval 2008). Or again, the non-translation into French of *The Condition of Postmodernity* (Harvey 1990) was a result of disinterest by French geographers in the cultural turn (Claval and Staszak 2008). Other of his books, especially from the 1980s and 1990s, have also not been translated. Nonetheless, Harvey has become *the* author to read. His books are regularly reviewed (for instance Clerval 2012) and he is interviewed (Mangeot *et al.* 2012). Ironically, David

Harvey is now better known than when in 1995 he was awarded in France, in Saint-Dié-des-Vosges, the IGU Vautrin-Lud Prize, the "Nobel Prize for Geography."

## A radical calendar

Alongside these publications, various academic events have been organized in French universities to promote radical approaches. Some have an epistemological purpose, to clarify historical, ideological and theoretical backgrounds. That was the goal of the first issue of the 2012 volume of the journal, *Les Carnets de Géographes* (Morange and Calbérac 2012). It was the first collective reflection in French geography aimed at clarifying a radical approach "à la française." On the one hand, it elucidated the meaning of critical and radical approaches (Gintrac 2012) including the specific practices of radical academics in the U.S. (Vergnaud 2012). On the other hand, the journal grappled with how to perform radical approaches within a French context. Meanwhile, a workshop, "*Épistémologie des Savoirs Géographiques,*"[10] was organized on May 24, 2012 in Lyon by Cécile Gintrac and Martine Drozdz.[11] It pointed to the two different roots of radicality within French geography: the legacy of Elisée Reclus from the end of the nineteenth century; and American radical geography from the 1960s and 1970s.

Two major symposia have also punctuated the French critical and radical debate. The first organized in March 2008 at Nanterre University was dedicated to spatial justice[12] – "*Justice et Injustice Spatiale*" – a field opened by radical geography during the 1970s that paid tribute to Henri Lefebvre's legacy at the University in which he taught (Lefebvre 1968). This event gave an opportunity to evaluate Lefebvre's influence on both the social sciences (Revol 2012) and radical geography (Brennetot 2011). That symposia led to a special issue of *Annales de Géographie* (Gervais-Lambony and Dufaux 2009), and an edited book (Bret *et al.* 2010). Moreover, a journal dedicated to spatial justice – *JS/SJ*[13] – was founded following the symposium. This stimulated great interest in radical geography, opening the way for numerous translations. The second symposium – "*Espace et Rapports Sociaux de Domination: Chantiers de Recherche*" – was on September 2012[14] in Marne-la-Vallée University. It can be considered the birth place of French radical geography. It was designed as a dialogue between prestigious U.S. radical geographers (such as Neil Smith, Don Mitchell, Pierpaolo Mudu, and Salvatore Engel-Di Mauro) and young Francophone researchers who presented their Marxist-inspired geographical work. Its proceedings are now an indispensable reference (Clerval *et al.* 2015).

*Mapping, embodying, and historicizing french
radical geography*

Mapping these events and locating the main actors demonstrates
that French radical geography has its own geography. Key have been the
universities of Nanterre and Marne-la-Vallée. Both are in Paris's suburbs. Nanterre was created at the beginning of the 1960s in the West of
Paris, famously known as the starting point of the May 1968 crisis.
Marne-la-Vallée was inaugurated in the 1990s in an Eastern suburb of
Paris. Academics in both universities (geographers and, more broadly,
social scientists) work on common political and social issues. There is
also an important geography outside the university. For example, there
are the sites of the publishing houses eager to promote Marxists theories
such as *Les Prairies Ordinaires*, and Harvey's first French publisher. Its
founder, Nicolas Vieillescazes (an English-to-French translator), played a
crucial role in introducing Harvey to Francophone academics.

Within this wider French geography, specific actors promoted radical
approaches. On the one hand, there were Marxists primarily interested
in theory. While they praised Harvey, it was because he was a Marxist
theorist not because he was a geographer.[15] On the other hand, there
were geographers who were interested in Marxism because they wanted
new approaches to the discipline. They were composed of the generation
born during the 1960s (Jean-François Staszak, Béatrice Collignon,
Philippe Gervais-Lambony, or Frédéric Dufaux) who trained in foreign
(especially Anglophone) universities; and the generation born at the end
of the 1970s (Anne Clerval, Matthieu Giroud, Serge Weber Julien
Rebotier) or at the beginning of the 1980s (Cécile Gintrac or Martine
Drozdz). This latter group were fluent in English, curious about the
Anglo-American discipline, and conscious of the linguistic domination
of Anglophone geography (Houssay-Holzschuch and Milhaud 2013).

Those academic events dedicated to radical geography and especially
to its leaders during the 2000s and 2010s in France prompt us to focus
on the potential links between French geography and Marxism.

## A New Space for Marxism in French Geography

The various elements discussed in the first part of this chapter put the stress
on two phenomena, deeply linked: First, the contemporary interest of French
for English-language geography (including a radical one); and second, the
growing interest by French geographers in Marxism, even though interest in
Marxism has been declining within the human sciences since the end of the
"Golden age" during the 1960s and 1970s (Cusset 2003).

## Has French geography become radical?

More than a craze for radical approaches,[16] the contemporary context reveals a keen interest for English-language geography that started during the 1990s and the 2000s. Geographers born during the 1960s and the 1970s, who read English and completed their education abroad, strongly urged French academics not to remain isolated in the periphery but instead to be at the center. Translations – such as Staszak's (2001) – empowered French geographers to participate in a worldwide debate. Meanwhile, because of the Internet, the English-language literature was increasingly opened to French readers. French libraries typically did not systematically collect foreign books and journals. Consequently, it was not until the 2000s that more and more French geographers were finally able to catch up with what had occurred in the English-language geography during the prior few decades. Both the radical geography of the 1960s and 1970s and the later critical and cultural geography of 1980s and 1990s thus arrived at the same time. That's why this English-language geography is called *géographie critique et radicale*, paying no attention to the differences between the two approaches: *critique* because it is about undermining foundations, continually renewing the academic tradition; and *radical*, because of its connection to Marxism, which continues to assert both some intellectual foundation and a broad academic tradition.[17] Consequently, radical geography appears diluted in its French language version.

French geography, therefore, has not massively adopted the radical turn. Because French geography has been defined by an empirical approach rather than by theory and political commitment; when radical geography has been adopted, it prompted geographers to focus on substantive topics such as domination or inequality rather than abstract conceptual innovation.

Paradoxically, the radicalness of radical geography has not been demonstrated by geographers, but by non-geographers. It is French Marxist sociologists, philosophers, and historians (especially those who first translated Harvey's texts in journals such as *Actuel Marx* or *Vacarme*, or in books published by Les Prairies ordinaire), who paid most attention to David Harvey as an exegete of Marx. Only then, after these translations, did those French Marxists invite geographers to consider and explore Harvey's thought, and more broadly, the project of radical geography (Harvey 2012a).[18]

## A brand new marxism?

The increased interest in *géographie radicale* reflects a larger shift that has occurred within French Marxism since the 1960s and 1970s (Keucheyan 2010). Then the dominant form of French Marxism was structuralism (Dosse 1991, 1992). After the fall of the U.S.S.R. and the rise of neoliberalism during the 1980 and 1990 (Cusset and Meyran 2016), French Marxism was reinvented. Whereas French Marxism of the 1960 and the 1970 was linked to the very powerful *Parti Communiste Français* and its structuralist interpretation, Marxists of the 1990s and the 2000s were eager to turn to original texts and interpretations. That partly explains why David Harvey became a major reference: His *Companion to Marx's Capital* (Harvey 2012b) echoes and replaces Althusser's *Lire "le Capital"* (Althusser *et al.* 1965).

Because of David Harvey and others, reference to the spatial is now *de rigueur* (Soja 1989; Warf and Arias 2009). French geographers consequently began promoting critical and radical approaches. Forty years after it first appeared, French geographers have now assimilated radical geography. But what kind of radical geography is this French *géographie radicale*?

## Radical Geography "à la Française"

To understand the French form of radical geography, we need to scrutinize prefaces, introductions to translations, proceedings of conference, book reviews, and programmatic manifestos. In them, we find judgments made of Anglophone radical geography, as well as strategies for bringing that approach into the French geographical tradition.

### Encountering anglophone and francophone geographies

Reflections about radical geography in France inevitably start with the history of French geography. The interest in social power, domination, and inequalities began in French geography long before American radical geography existed. It started with Reclus at the end of the nineteenth century, then Renée Rochefort during the 1960s (Claval 1967), Lacoste during the 1970s (Hepple 2000), and finally the French school of social geography from the 1980s (Frémont *et al.* 1984; Di Méo and Buléon 2005). All these approaches made social and political issues central. This inventory of French geography both upsets the usual history that radical geography was born in North America during the 1960s, and acts as a prompt to discover more about a seemingly lost French intellectual tradition. While French geography may now appear to be in debt to Anglophone radical geography, in the past it exported its own radical approaches.

Furthermore, it is important to remember that radical geography was not the only new approach to enter French geography from Anglo-America. There were other critical approaches assimilated during the same period such as humanistic geography and the "cultural turn." In this sense, radical geography was only one element in a broader *géographie critique et radicale* (Morange and Calbérac 2012) that reshaped French geography.

## Fieldwork against theory?

The inclusion of radical geography within the general category of *géographie critique et radicale* circumvents a debate long bedevilling French geography that goes to the role of theory within the discipline. The reluctance to discuss theory in French geography goes back at least to Vidal de La Blache who promoted fieldwork as the main (the only?) method for geographers (Volvey *et al.* 2012). In their introduction to the proceedings of Marne-la-Vallée symposium, the editors focused not on the novelty of radical geographical theory, but rather on how radical geography could usefully illuminate concrete studies of social inequality or domination *in combination with field work*.

Separating Marxist theory and its empirical study serves the purpose of blocking broad discussions of the legacy of Marxism. The place of Marxism in French social sciences (Barbe 2014), and geography in particular (Lévy 1985; Pailhé 2003), has always remained a sore spot. The move to justify the use of Marxism in geography as an empirical rather than a theoretical approach can therefore be seen as a strategy by French radical geographers to avoid contentious, fraught, even potentially embarrassing debate about Marxism. *Géographie critique et radicale*'s use is practical, to study concrete cases of social discrimination and domination.

## How to be a critical and radical academic

How to be a critical and radical academic is addressed by Camille Vergnaud (2012), Cécile Gintrac (2015b), Neil Smith (2015), Don Mitchell (2015), and Pierpaolo Mudu (2015). All these authors invited readers to break with the principle of academic neutrality, which is still axiological in French universities. French researchers should practice instead self-engagement, adopting ethical and reflexive approaches, and which is what they are increasingly doing. Reflexivity is more and more becoming a necessary procedure to validate methodological procedures (Calbérac and Volvey 2014).

More broadly, *géographies critiques et radicales* is a hybrid between French social geography and Anglophone critical and radical geographies. The importation of Anglophone radical geography has profoundly renewed the French geographical tradition. *Géographies critiques et radicales* is now one of the most exciting fields in the discipline, led by young and talented researchers. To answer Jacques Lévy's question, the words *critical* and *radical* perfectly fit the French context. In the *Commentary* he wrote to celebrate its thirtieth anniversary, Richard Peet said that the future of radical geography, "should not mainly consist in finding still more French authors to quote" (Peet 2000: 952–953). He is wrong. The vitality of *géographies critiques et radicales* shows that French authors are exactly those who should be quoted!

## Notes

1   *Agrégation* is a high-level competitive exam to recruit teachers to secondary schools, but it is also to distinguish academics who work in a university. *Agrégation* symbolises the French conservative and elite academic system.
2   *Géographie radicale* is the French geographical designation of *radical geography*. It suggests that French geographers perceive radical geography as originating from the U.S., and it also connotes the form taken by radical geography.
3   www.calenda.org.
4   www.persee.fr.
5   journals.openedition.org.
6   www.cairn.info.
7   This chapter deals with only the French context, and it is written from France. Hence, bibliographical materials used are mostly in French. The chapter also aims to decolonialize the dominant Anglophone geography by scrutinizing what occurs on its Gallic margins.
8   Thanks to that book and its authors, postmodernity became an issue that was debated in French geography from the beginning of the 2000s (Collignon and Staszak 2004).
9   Its title, given by the editors, sounds so French!
10   https://calenda.org/208599.
11   Cécile Gintrac both witnessed the rise of radical geography (Gintrac 2015a), and was a participant (Gintrac and Giroud 2014). Martine Drozdz is a critical and radical geographer (Drozdz *et al.* 2012). Especially, she is a critic of the neo-liberal city.
12   Call for papers: http://calenda.org/192718; program: http://calenda.org/194494.
13   *Justice Spatiale/Spatial Justice* is a bilingual online journal: https://www.jssj.org/.
14   Call for papers: http://calenda.org/204854; program: http://calenda.org/209020.

15  This point of view is still quite uncommon in France insofar as the university is structured by disciplines (geography, sociology…) rather than by field studies (gender studies, Marxist studies…) (Monteil and Romerio 2017).
16  On the history of Marxism in French geography, see Pailhé (2003).
17  In French, *radical* designates the moderate left-wing of the political chessboard.
18  Only after the pioneer translation by Béatrice Collignon in Staszak (2001), did geographers seriously consider David Harvey's works for the first time.

# References

Althusser, L., Balibar, E., and Establet, R. (1965). *Lire "le Capital"*. Paris: Maspéro.
Barbe, N. (2014). Présentation. *Le Portique. Revue de Philosophie et de Sciences Humaines*, 32. https://leportique.revues.org/2714.
Brennetot, A. (2011). Les géographes et la justice spatiale: généalogie d'une relation compliquée. *Annales de Géographie* 678: 115–134.
Bret, B., Gervais-Lambony, P., Hancock, C., and Landy, F. eds (2010). *Justice et Injustices Spatiales*. Nanterre: Presses Universitaires de Paris Ouest.
Buttimer, A. (1971). *Society and Milieu in the French Geographic Tradition*. Chicago: AAG.
Calbérac, Y. and Volvey, A. (2014). Introduction. J'égo-géographie… *Géographie et Cultures*, 89–90: 5–32.
Claval, P. (1967). Géographie et profondeur sociale. *Annales. Économies, Sociétés, Civilisations* 22 (5): 1005–1046.
Claval, P. (1977). Le Marxisme et l'espace. *Espace Géographique* 6 (3): 145–164.
Claval, P, and Staszak, J.-F. (2008). Où en est la géographie culturelle? Introduction. *Annales de Géographie* 660–661: 3–7.
Clerval, A. (2008). *La gentrification à Paris intra-muros: dynamiques spatiales, rapports sociaux et politiques publiques*. Ph.D., Université Panthéon-Sorbonne.
Clerval, A. (2012). David Harvey et le matérialisme historico-géographique. *Espaces et Sociétés*, 147: 173–85.
Clerval, A, Fleury, A., Rebotier, J., Weber, S., and Université de Marne-la-Vallée, eds (2015). *Espace et Rapports de Domination*. Rennes: Presses universitaires de Rennes.
Collignon, B. and Staszak, J.-F. (2004). Que faire de la géographie postmoderniste? *L'espace Géographique* 2004/1: 38–42.
Cusset, F. (2003). *French Theory. Foucault, Derrida, Deleuze & Cie et les Mutations de la Vie Intellectuelle aux Etats-Unis*. Paris: La Découverte.
Cusset, F. and Meyran, R. (2016). *La Droitisation Du Monde: Conversation Avec Régis Meyran*. Paris: Textuel.
De Koninck, R. (1984). La géographie critique. In A. Bailly, ed., *Les Concepts de la Géographie Humaine*, pp. 121–131. Paris: Masson.
Di Méo, G. and Buléon, P. eds (2005). *L'espace Social. Une Lecture Géographique des Sociétés*. Paris: Armand Colin.

Dosse, F. (1991). Histoire du Structuralisme 1. *Le Champ du Signe*. Paris: La Découverte.

Dosse, F. (1992). *Histoire du Structuralisme 2. Le Chant du Cygne*. Paris: La Découverte.

Drozdz, M., Gintrac, C., and Mekdjian, S. (2012). Actualités de la géographie critique. *Carnets de Géographes*, 4. https://cdg.revues.org/1014.

Frémont, A., Chevalier, J., Hérin, R., and Renard, J. (1984). *Géographie Sociale*. Paris: Masson.

Gervais-Lambony, P. and Dufaux, F. (2009). Justice... Spatiale! *Annales de Géographie* 2009/1 (665–666): 3–15.

Gintrac, C. (2012). Géographie critique, géographie radicale: Comment nommer la géographie engagée? *Les Carnets de Géographes* 4. http://cdg.revues.org/972

Gintrac, C. (2015a). *Au seuil critique de la ville: trois groupes de géographie engagée*. Ph.D. thesis, Université Paris Ouest Nanterre La Défense.

Gintrac, C. (2015b). Quels positionnements pour quelle(s) géographie(s) critique(s)? In A. Clerval, A. Fleury, J. Rebotier, and S. Weber, eds, *Espace et Rapports de Domination*. Rennes: Presses Universitaires de Rennes.

Gintrac, C. (2017). La fabrique de la géographie urbaine critique et radicale. *EchoGéo*, 39. http://echogeo.revues.org/14901.

Gintrac, C. and Giroud, M. eds (2014). *Villes Contestées: Pour une Géographie critique de L'urbain*. Collection Penser/Croiser. Paris: les Prairies ordinaires.

Harvey, D. (1990). *The Condition of Postmodernity*. Cambridge and Oxford: Blackwell.

Harvey, D. (1992). Social Justice, Postmodernism and the City. *International Journal of Urban and Regional Research* 16 (4): 588–601.

Harvey, D. (2007a). Le "Nouvel Impérialisme": accumulation par expropriation. *Actuel Marx* 35: 71–90.

Harvey, D. (2007b). L'urbanisation du capital, The urbanization of capital. *Actuel Marx* 35: 41–70.

Harvey, D. (2007c). Réinventer la géographie, Reinventing Geography. *Actuel Marx* 35: 15–39.

Harvey, D. (2010). *Le Nouvel Impérialisme*. Paris: Les Prairies ordinaires.

Harvey, D. (2012a). *Paris, Capitale de la Modernité*. Paris: Les Prairies ordinaires.

Harvey, D. (2012b). *Pour Lire "Le Capital"*. Montreuil: La Ville brûle.

Harvey, D. (2014a). *Brève Histoire du Néolibéralisme*. Paris: Les Prairies ordinaires.

Harvey, D. (2014b). *Géographie et Capital: Vers un Matérialisme Historico-Géographique*. Paris: Syllepse.

Harvey, D. (2015). *Villes Rebelles: Du Droit à la Ville à la Révolution Urbaine*. Paris: Buchet/Chastel.

Hepple, L. (2000). Géopolitiques de Gauche: Yves Lacoste, Hérodote and French radical geopolitics. In K. Dodds and D. Atkinson, eds, *Geopolitical Traditions. A Century of Geopolitical Thought*, pp. 268–302. London: Routledge.

Houssay-Holzschuch, M. and Milhaud, O. (2013). Geography after Babel. A view from the French Province. *Geographica Helvetica* 1: 1–5.

Keucheyan, R. (2010). *Hémisphère Gauche. Une Cartographie des Nouvelles Pensées Critiques*. Paris: La Découverte.

Lefebvre, H. (1968). *Le Droit à la Ville...* Paris: Editions Anthropos.

Lévy, J. (1984). Les lieux des hommes: Un nouveau départ pour la géographie. *La pensée* 1984/3 (239): 30–45.

Lévy, J. (1985). French geographies of today. *Antipode* 17 (2–3): 9–12.

Mangeot, P., Dupart, D., Gintrac, C. and Vieillescazes, N. (2012). Marx and the City. entretien avec David Harvey. *Vacarme* 2012/2 (59): 218–249.

Merrifield, A. (1995). Situated knowledge through exploration: Reflections on Bunge's geographical expeditions. *Antipode* 27 (1): 49–70.

Mitchell, D. (2015). Devenir et rester un géographe radical aux Etats-Unis. In A. Clerval, A. Fleury, J. Rebotier, and S. Weber, eds, *Espace et Rapports de Domination*, pp. 45–56. Rennes: Presses Universitaires de Rennes.

Monteil, L. and Romerio, A. (2017). Des disciplines aux "studies". *Revue D'anthropologie des Connaissances 11*, 3 (3): 231–244.

Morange, M. and Calbérac, Y. (2012). Géographies critiques "à la française"? *Carnets de Géographes* 4. http://cdg.revues.org/976.

Mudu, P. (2015). Devenir et rester un géographe radical en Italie et Ailleurs. In A. Clerval, A. Fleury, J. Rebotier, and S. Weber, eds, *Espace et Rapports de Domination*. Rennes: Presses Universitaires de Rennes.

Pailhé, J. (2003). Références marxistes, empreintes marxiennes, géographie française. *Géocarrefour* 78 (1): 55–60.

Peet, R. (1997). The cultural production of economic forms. In R. Lee and J. Wills, eds, *Geographies of Economies*, pp. 37–46. London: Arnold.

Peet, R. (2000). Celebrating thirty years of radical geography. *Environment and Planning A* 32 (6): 951–953.

Racine, J.-B. (2003). Radical geography. In J. Lévy and M. Lussault, eds, *Dictionnaire de La Géographie et de l'éspace Des Sociétés*. Paris: Armand Colin.

Revol, C. (2012). Le succès de lefebvre dans les urban studies Anglo-Saxonnes et les conditions de sa redécouverte en France. *L'Homme et La Société* 2012/3 (185–186): 105–118.

Smith, N. (2015). The future is radically open. *ACME. An International E-Journal for Critical Geography* 14 (3): 954–964.

Soja, E. W. (1989). *Postmodern Geographies. The Reassertion of Space in Critical Social Theory*. London, New York: Verso.

Staszak, J.-F. ed. (2001). *Géographies Anglo-Saxonnes. Tendances Contemporaines*. Paris: Belin.

Vergnaud, C. (2012). Qu'est-ce que cela signifie être enseignant-chercheur "critique"? *Carnets de Géographes* 4. http://cdg.revues.org/992.

Volvey, A., Calbérac, Y., and Houssay-Holzschuch, M. (2012). Terrains de je. (du) sujet (au) géographique. *Annales de Géographie* 687–688: 441–461.

Warf, B. and Arias, S. eds (2009). *The Spatial Turn. Interdisciplinary Perspectives*. London, New York: Routledge.

Weber, S. (2012). Traduire et lire Harvey: la pensée géographique par la brèche. *EchoGéo*, 22. http://echogeo.revues.org/13284.

# Conclusion

## Eric Sheppard and Trevor J. Barnes

Through the various essays making up this book, we seek to retell the
early years of the emergence of radical geography within North America:
From its beginnings to the early 1980s. To date, the default origin narrative for this history was one place: Clark University, Worcester, MA. As
the nine chapters making up the first half of this book demonstrate,
there was much more to it than Clark, important as it was. Radical
geography emerged in variegated ways across multiple places. The
spatial history in Part I also teases out how uneven networked connectivities linking places, along with a disciplinary politics of scale, shaped
what came to be taken as radical geography by the time David Harvey
(1982) published his seminal *The Limits to Capital*. To the extent
possible, we told these stories through the recollections of those active at
the time: A central motivation for the volume was to record these voices
while they still are available. Notwithstanding the considerable research
found within these chapters, it is important to acknowledge some
caveats, however.

First, this history of radical geography is temporally limited, to the
early years. For reasons elaborated next, it was a period during which
radical geography was dominated and was even defined by geographical
Marxist political economy, and constructed by largely white, male
geographers. While contemporary radical/critical geography feels very
different from that period, that past nevertheless shaped the present.

Second, the account is spatially limited. We acknowledge that all the
spaces of early radical geography were not captured, and at this point
will likely never be fully captured. We look forward to others working to

*Spatial Histories of Radical Geography: North America and Beyond*, First Edition.
Edited by Trevor J. Barnes and Eric Sheppard.

uncover different sites, however, supplementing the necessarily incomplete account in this volume.

Third, we stress that each chapter's narrative itself is a situated account, shaped by its authors' predilections and conceptions of radical geography, their interpretations, and by those interviewed and their particular understandings. We acknowledge that accounts by other authors of the same people and places would likely be different. Indeed, the chapter by Warren *et al.* (this volume) compares two contrasting historical accounts (HIStory and HERstory) of one of our sites, the Detroit Geographical Expedition and Institute. Again, we look forward to seeing these, both to supplement and to contest what is here. In our view, no proper geographical account of knowledge production can claim to be definitive, and we certainly do not make that claim here.

Fourth, we acknowledge that North America should not be taken as a privileged region from which to understand global radical geography. This is just one of many possible staring points, and the purpose of Part II of this volume is to place the North American story in conversation with the trajectories of radical geography in selected other countries. Those chapters uncover ways in which North American radical geography shaped its trajectory elsewhere, but also include examples of different, earlier and distinct origins that had little to do with North America.[1]

In this concluding chapter, we seek to draw the book's threads together to summarize what did (and did not) happen during these early decades, to trace the links between this past and contemporary North American radical geography, and to reflect on the implications of the success of Anglophone radical/critical geography within the discipline and the quandaries this poses for knowledge production.

## The Emergence of a Radical Geography Canon

As noted, the early years can be summarized as a time period during which North American radical geography became canonical. By 1982, it was not only the case that the education of radical geographers started by reading Karl Marx's *Capital*, but that being a Marxist was a precondition for admission into the radical geography camp. This certainly was not pre-ordained. As Chris Philo documented (see also the Introduction and Huber *et al.*, this volume), radical geography's hallmark organ – *Antipode* – was quite eclectic in its first six volumes. Philo (1998: 3) writes: "Marxist geography was very much a minority offering in the journal, and in fact Marx received virtually no name-checks at all until Harvey's 1972 paper." Marxist radical geography also was contested. Klein (this volume) notes that Marxist versions of radical geography did not receive widespread support in Francophone Québec, except for the group he discusses, *Groupe de recherche sur l'espace, la dépendance et les inégalités* (GREDIN). Kobayashi (this volume) also offers a radical counterpoint, tracing the emergence of a black radical tradition in U.S. geography that sat alongside the Marxist canon, albeit largely ignored by it (but see Wilson 2000a, 2000b). Blomley and McCann, and Huber *et al.* (both this volume) further document conflicts between Marxists and both feminist and community-oriented visions of radical geography (which then shaped the nature of radical geography respectively in Vancouver and at Clark). Finally, Peake and Lauria *et al.* (both this volume) show how eclecticism persisted through connectivities radiating from the Union of Socialist Geographers, particularly in the U.S. Midwest. Yet networks are not flat; their connectivities are uneven. As discussed by Sheppard and Barnes (this volume) and Huber *et al.* (this volume), these continent-wide networks came to be dominated by the axis connecting Baltimore (Johns Hopkins University) with relatively proximate Worcester (Clark University). In turn, these nodes were dominated by, respectively, David Harvey and Richard Peet, whose influence, initiative, intellect, and parallel turns to Marx steadily narrowed the remit of *Antipode* and the definition of radical geography.[2]

The emergence of this canon also reflected other factors. On the one hand as discussed in the introduction, radical geography emerged within the context of the politics of 1968 and the emergence of a "new left" deeply influenced by the specter of Marx (Derrida 1994; Dosse 1998), particularly its influential Anglophone *New Left Review*. It was also deeply shaped by the demographics of North American geography, at that time largely a masculine field whose protagonists (then, but also now) were predominantly white. Radical geography, premised on the notion of tearing up the existing discipline by its roots, should have been an inclusive space where all kinds of geographers would flourish characterized by differently constructed genders, races, and sexualities, and including geographers from global peripheries as well as the core. Yet the precondition of identifying oneself singularly as a Marxist made that space overly narrow and exclusive

Thinking back to his early Hopkins days, Harvey acknowledges that the white radical politics emerging from that university in the early 1970s, in which he was deeply engaged, failed to connect with black radicalism and Baltimore's black panthers (Sheppard and Barnes, this volume). A few select white radical geographers concerned themselves with race as constructed in the U.S. (Kobayashi, this volume), yet their perceptions of what this meant could depart radically from the lived experiences of black urban residents with whom they expressed kinship (see the discussions of William Bunge in Warren *et al.* and Kobayashi, both this volume).

With respect to gender, the intellectual struggles between male Marxists and feminists within radical geography's early nodes triggered feminist alternatives. Huber *et al.* (this volume) document the sexism experienced by women involved in the production of *Antipode*, and Blomley and McCann (this volume) trace the marginalization of women's voices in Vancouver. Nevertheless, Suzanne Mackenzie and others in Vancouver (Blomley and McCann, this volume) took this as a starting point to craft what was initially a socialist feminist account of urban life, triggering a feminist geography that became increasingly influential and eclectic – coevolving with second and third wave feminism. A similar process occurred at Clark, where Julie Graham and Kathy Gibson began as classical Marxist economic geographers but by the 1990s had shifted to a feminist and queer theoretical post-Marxism (Huber *et al.*, this volume. See also Mackenzie and Rose 1983; Gibson, Graham and Shakow 1989; Gibson-Graham 1996; Nagar 2018). With respect to sexuality, the revolutionary politics of several white, straight males did not extend to an openness to LBGTQ identities. Bunge was particularly blunt and insensitive in upbraiding radical geography students and Detroit Expedition practitioners about the danger of homosexuality undermining the project of radical geography (Blomley and McCann; and Warren *et al.*, both this volume).

Radical geographies, and geographers, of the global periphery also gained little attention from a Marxist economic geography that focused largely on understanding uneven development in the North Atlantic realm. The Brazilian Milton Santos was a Marxist whose theorization revolved around the geographical concepts of "shared space" and the "double circuit" (Santos 1977, 1979). While his work was given initial attention, that quickly dissipated in North America notwithstanding his continuing influence in Latin America (Ferretti and Pedrosa 2018). In North America, Jim Blaut carried the flag of Marxist development theory in radical geography, attending to third world, but also to race (Kobayashi, this volume). His understanding of globalizing capitalism resonated with both world systems theory and certain subsequent postcolonial interventions (Blaut 1993, Sheppard 2005). Working in Africa, Ben Wisner and Phil O'Keefe pioneered what subsequently emerged as third world rural political ecology, subsequently shaped inter alia by Michael Watts (O'Keefe *et al.* 1976; Wisner *et al.* 1976, 1982, Watts 1983).[3] The most sustained attempt to put ideas of third world development in conversation with geographical Marxism was in Québec (Klein, this volume). Viewing Québec as located in the periphery of North American capitalism, the GREDIN group drew on Latin American dependency theory and the work of the Marxist development theorist Samir Amin to make sense of that province – an approach subsequently popularized in English under the label of internal colonialism (Hechter 1995).

Yet these exclusions also had the consequence that radical geography was increasingly recognized as having produced a separate, novel and internally cohesive set of geographical theories. By 1980 radical geography was widely acknowledged – alongside humanist geography – for its trenchant critiques of the quantitative revolution that during the 1960s and 1970s had defined geographical theory. Consequently, radical geography gained disciplinary status especially given the kudos that now attended theorization as a practice within human geography. That was also helped by a number of household names from the quantitative revolution, dissatisfied with the limits of positivist social engineering, swapping sides, to join radical geography. They included Harvey, Bunge, Gunnar Olsson, Bernard Marchand, Richard Morrill, Richard Peet, Allen Scott, and Michael Webber (Sheppard 1995, 2014).

Radical geography also more and more became an academic project. Community-oriented work, notably in the Detroit, Vancouver, and Toronto expeditions, faded as a result of external resistance and internal struggles (Warren *et al.*; Blomley and McCann, both this volume). Graduate students, now completing their radical geography theses, were looking for jobs.[4] The Union of Socialist Geographers, which had been central to tweaking mainstream geography's nose from the outside (Akatiff 2016), dissolved itself in favor of working to change the discipline from within, forming the AAG Socialist Geography Specialty Group (Peake, this volume). Some radical geographers with academic jobs were tenured – albeit in the face of considerable resistance (Peck and Barnes, this volume; Peake and Sheppard 2014). Radical geographers were also beginning to publish in the mainstream journals. In these ways, the Marxist canon was able to establish itself as a legitimate subfield of the discipline. A place had been cleared for radical geography at the discipline's academic table.

## Deconstructing the Canon

If radical geography's place was initially at the bottom of the table, it is now somewhere near, if not at, the top. Importantly, this elevation has been accompanied by a radical diversification of what counts as radical geography in North America. If the tension between theory and activism (see Barnes and Sheppard, this volume) was largely resolved by prioritizing the former, pluralism has flourished, as have the non-economic, feminism, and, more recently, critical race theory. "Radical" geography, meaning an emphasis on Marx, now coexists with the alternative framings of "critical" geography. This does not mean that Marxist geography has dissipated: Harvey's writing remains as widely discussed as ever, and is increasingly influential across the social sciences and humanities. Rather, it's that multifarious other radical/critical geographies have flourished concurrently. It is not either/or but both/and. For the purists, this has caused discomfort (footnote 2). For the tradition of radical thought in Geography, it has been a boon.

Assessing the state of affairs at the turn of the millennium, Noel Castree (2000: 956) argued that

> this expansion and pluralisation of dissident geographies cannot be attributed, at least in any simple or unmediated way, to the efforts of the first radical geographers. But it remains the case…that those early efforts were a condition of possibility for the subsequent emergence, in the 1980s and 1990s, of a serious and substantive corpus of Left geographical scholarship… Where, by the late 1970s, "radical geography" designated the then relatively small Left geographical community tout court [rather than any specific group or movement within it (though Marxist geography did loom very large)], so "critical geography" stands two decades on as an homologous umbrella term for that plethora of antiracist, disabled, feminist, green, Marxist, postmodern, poststructural, postcolonial, and queer geographies which now constitute the large, dynamic, and broad-based disciplinary Left.

Importantly, as a number of the chapters in this book trace, this decentering of the Marxist canon did not simply reflect subsequent shifts in intellectual fashion. Rather, the conditions of its possibility were built into radical geography from the get-go. The exclusions always were incomplete – internal contradictions and antitheses within the early radical geography project that would provide a nursery from which alternative radical visions could grow. As Katznelson (1997) argued, the very establishment of an alternative to orthodoxy, radical geography, provided the conditions for further alternatives. In this case, pre-existing alternatives such as radical feminism and eco-socialism and third worldist perspectives had not disappeared, even if they were not very visible within radical geography. Yet by the mid-1980s they were impossible to ignore. In 1983, Pion Press expanded its stable of *Environment and Planning* journals to include an alternative radical geography outlet: *Environment and Planning D: Society and Space*. Its first editor Michael Dear, a former quantitative geographer turned postmodernist, introduced its mission in this way (Dear 1983: 1–2):

> Currently unresolved questions between competing perspectives will be directly confronted (for example, between Marxian and Weberian

approaches). We welcome papers which examine any of these aspects of the philosophy and practice of social theory, as well as papers emphasizing such substantive topics as class and class formation; production and reproduction relations in everyday life; gender and patriarchy; theory of the state; spatial division of labour and the reproduction of labour power; theories of under-development; 'world-system' approaches; industrial restructuring and uneven development; multinational corporations, capital movement and the urban hierarchy; language, culture and ideology; and urban social movements.

Within a few years, *Society and Space* emerged as a popular, interdisci-plinary venue for an eclectic radical body of scholarship on socio-spatial theory, also hosting high profile debates in which Weberians and feminists took on Harvey as their stalking horse for the Marxist canon (Saunders and Williams 1986; Harvey 1987; Deutsche 1991; Massey 1991; Harvey and Haraway 1995; Harvey 1998).

When Joe Doherty and Eric Sheppard were asked by Peet to take over as co-editors of *Antipode* (also tasked with shepherding its transition to Basil Blackwell – which in those days was less than supportive), their opening editorial similarly sought an explicitly open door for various forms of radical scholarship:

> As co-editors we intend that *Antipode* will continue to be a forum for the publication of significant contributions to a radical (Marxist/socialist/feminist/anarchist) geography. With the assistance of an active interna-tional board of editors, we will seek to maintain the traditional areas of strength in environmental questions, urban political economy and development issues. We intend to improve the journal's coverage of femi-nist approaches, and to keep it at the forefront of debate in all aspects of geography. (Sheppard and Doherty 1987: 2)

Yet other journals emerged as spaces for exploring alternative radical visions – themselves a sign of the increasing influence of radical/critical geography. *Gender, Place and Culture: A Journal of Feminist Geography* began publication in 1994, and *Ethics, Place and Environment: A Journal of Philosophy and Geography* followed in 1998.

When the journal *ACME: An International Journal of Critical Geographies* opened its virtual doors in 2002, its editors positioned *ACME* as not only challenging the political economy of publishing through its open access format (a format subsequently colonized by mainstream publishers), but also further opening up the canon (Moss, Berg and Desbiens 2002: 3):

> Other journals publishing critical geography, often organized along the lines of a theoretical or political approach, tend to focus on a particular

set of theories or politics... Pulling together an array of these varying approaches to theory and praxis is the critical imaginary of the journal... hoping to evoke notions of an amalgam of critical thinking, radical analysis, and politicized activities. We are interested in work done from a variety of critical and radical perspectives, as for example, anarchist, antiracist, environmentalist, feminist, Marxist, postcolonial, poststructuralist, queer, situationist, and socialist.

This diversification has become a hallmark of radical/critical geography – and *Antipode: A Radical Journal of Geography*. Its 2017 editorial collective of *Antipode*, announced itself as follows

[W]hile we look different, we continue to push Geography's radical and critical edge in a number of ways, many of which will be familiar, inspired as they are by Marxist, socialist, anarchist, anti-racist, anticolonial, feminist, queer, trans, green, and postcolonial thought. ... [W]e are also committed to the new, the innovative, the creative, and the heretofore unthought radical edges of spatial theorisation and analysis. (Sharad Chari, Tariq Jazeel, Andy Kent, Katherine McKittrick, Jenny Pickerill and Nik Theodore, https://antipodefoundation.org/about-the-journal-and-foundation/a-radical-journal-of-geography/, accessed June 30, 2018)

A decentered radical geography has not escaped its own vigorous internal masculinist disagreements (e.g., Amin and Thrift 2005, 2007; Smith 2006; Harvey 2006) – a disciplinary variant on the old adage that left politics is consistently undermined by its own internecine fighting. Yet there remains a productive tension between disagreement and engaged pluralism surrounding the strange attractor that is radical geography.

This is not the place for a comprehensive analysis of the last 30 years of radical/critical geography but suffice to say that it has flourished. On the one hand, it can be characterized by a series of twists and turns that, often setting out from and/or differentiating themselves with respect to Marxist geographies, have created a highly diverse body of scholarship that shares a critical perspective on the status quo. Those perspectives include political ecology, the cultural and relational turns, post-prefixed, critical race and queer geographies, discourse and representation, performativity and non-representational theory, and most recently settler colonial and decolonization approaches. Beyond this have been forays to engage with and transform areas of the discipline that seemed potentially resistant to critical geography, subsequently generating active research in such areas as critical physical geography, and critical and qualitative GIS (Cope and Elwood 2009; Lave 2014; Dempsey 2016; Wilson 2017).

Radical geography also has developed an increasingly global scope. British and Australian geographers produced issues of the *USG Newsletter*. Norcup (this volume) describes how, during the 1980s, North American Marxist initiatives were joined with British anti-racist, post-colonial, and feminist radical geographic pedagogy in the journal *Contemporary Issues in Geography and Education*. Internationalization involved the diffusion of aspects of North American radical geography, for example into South Africa and Mexico in the 1990s (Maharaj and Crossa, both this volume). They cross-fertilized and existed in tension with already existing indigenous traditions of Marxist and anarchist geography in Japan and France (Mizuoka and Calbérac, both this volume). By the late 1990s, supra-national organizations of radical geographers emerged to shape this field across national borders, persisting to this day. The International Critical Geography Group has organized triennial conferences, beginning in Vancouver in 1997, followed by Daegu, Békéscsaba, Mexico City, Mumbai, Frankfurt, and Ramallah (https://internationalcriticalgeography.org/conferences/, accessed July 18, 2018). EARCAG, the East Asian Regional Conference on Alternative Geography, stemming from the 1997 ICCG conference, has met regularly across east Asia since 1999 (http://lab.geog.ntu.edu.tw/lab/r401/earcag/, accessed July 18, 2018).

Indeed, radical scholarship has become more influential within Anglophone Geography than within any other Anglophone North American social science, attracting considerable notice from beyond the discipline. In part, this is because of Geography's relatively marginal position within the U.S. and Canadian academic power structure (see Bell Hooks, 1989, on the politics of the margin). But it is also because of the success of radical/critical geographers in scaling the heights of the publication- and citation-oriented "key performance indicators" of a neoliberalizing academy, as exemplified by David Harvey (2017: 4–5, Huber *et al.*, and Blomley and McCann, both this volume) who consistently cracks the top 20 list of social science citation leaders. Now, a full half century after *Antipode* first made an appearance, there exists an extensive corpus of radical geographical research. It can be found in mainstream as well as radical journals; in its own new outlets; in books published by both the most reputable university presses and radical publishing houses; in highly ranked ISI citation metrics; and in large research grant awards. In short, as Castree (2000) remarked, there has been a professionalization of radical/critical geography. This poses its own quandaries as radical/critical geographers ponder their achievement of unprecedented disciplinary power.

## Radical Geography's Contemporary Quandaries[5]

Radical geographers have written passionately about how the neoliber-alization of universities exerts pressure on academic life (Mitchell 2000; Mullings *et al.* 2016): Increased pace of work; short-termism; increased competitive individualism; minimum publishable units; the expansion of the space-time of work to envelop more and more of everyday life; and persistent power hierarchies (such as white supremacy, masculinism, and heterosexism). As we have seen, radical geography has been academi-cally highly successful. But it raises the issue of whether there is a tension between that success, which some would argue has only been achieved by conforming to norms of the neoliberal university including competi-tiveness, the production of commodified knowledge, and techniques of self-governance, and radical geography's claim to pursue politically progressive (non-capitalist) forms of academic research and activism.

No doubt, some of the recently intensified internecine struggles over the identity and direction of radical geography – waged at both the scale of personal politics (who gets to speak on behalf of critical geography), and that of epistemologies (what counts as radical knowledge) – can be traced to these pressures.

Radical geographers should reflect long and hard on the implications of playing the neoliberal game. It leads to trend-following rather than trend-setting scholarship, because the latter is slower to generate cita-tions and publications, and discourages imaginative and out-of-the-box thinking. Playing the neoliberal game means: opting for quick publica-tions rather than ambitious and careful scholarship; aligning with the agendas and priorities of publishers that pursue higher profit-producing companions, dictionaries, and encyclopedias than original scholarly monographs; valuing influence, productivity, and salaries over autonomy, slow scholarship, and the pursuit of knowledge for knowledge's sake; pursuing quantity over quality, and big-grant team projects over smaller grant money individual projects.

Paralleling other areas of practice associated with neoliberal globaliza-tion, the nation-state scale is by-passed in favor of global networking. Renowned critical geographers fly to meetings around the globe and are read in English worldwide. What is presented as decentered and collabo-rative knowledge production is in fact the interaction of select centers of calculation within a small set of elite spaces of Europe and white settler societies (Paasi 2005). Such a practice only reinforces an individualized culture of scholarly writing, with very little true collaboration.

This Anglicization of radical geography may seem unremarkable and inevitable, a straightforward reflection of trajectories of homogeniza-tion, including that of languages (Nettle and Romaine 2002), associated with neoliberal globalization. Some welcome English as an effective lingua franca for preserving the geographical diversity of knowledge (Rodríguez-Pose 2004). Many argue the opposite, however; that Anglicization marginalizes and impoverishes thinking that emerges from other linguistic traditions (Gutiérrez and López-Nieva 2001; Raju 2002; Gregson et al. 2003; Viaou 2003; García-Ramón 2004; Hones 2004; Ramírez 2004; Simonsen 2004; Timár 2004; Belina 2005). The pressure to publish in English-language journals and gain respect in major Anglophone nodes of knowledge exchange pressures authors to align with the theoretical debates and methodological trends currently fashionable therein. To act otherwise is to risk rejection and disinterest.

There is also the danger of defining what counts as knowledge as only that which originates from the U.S./U.K. America and Britain become the sole source of general principles and theoretical labor, with places outside these two countries simply providing case studies and manual labor. These unequal and asymmetric connectivities, shaping the uneven development of radical geography, are not reducible to a simple hierarchy of global core and periphery. There are also international gradients separating the U.S./U.K. from other Anglophone nations such as Canada, Australia, New Zealand, and South Africa, and those in turn from other European languages. This triggers shifting neocolonial knowledge production hierarchies located further east or south (Timár 2004).

Further, voices originally from the global south but now located in the U.S. and U.K. (non-white and non-elite), and from indigenous populations in white settler societies, remain overlooked within largely white radical geography. The migration of scholars from elsewhere into the Anglo-American realm suggests the possibility of hybridity, often celebrated as the flip side of the homogenization and diffusion associated with globalizing capital. Yet, this hybridity is demographic rather than intellectual. Theoretical and philosophical turns within Anglo-American critical geography (occurring at speeds that even locals find dizzying), combined with control over knowledge dissemination and truth-making networks, mean that immigrant scholars find themselves playing intellectual catch-up rather than being in a position to bring new ideas that enrich radical geography. Many of these scholars remain located and embedded within Anglophone debates, practicing intellectual assimilation in order to gain acceptance.

It is thus hard for subaltern academic voices to create space from which to add their distinctive positionalities to the production of knowledge within radical geography (Raju 2002). Gayatri Spivak has herself turned away from what is often interpreted as skepticism about this possibility, to embrace the significance of "learning from below"

(Sharpe and Spivak 2002). As critical geographers, we should be particularly aware of the importance of diversifying voices that matter, incorporating within the conversation those located socially and geographically in the global south (Mohanty 2003), no matter how difficult (Sangtin Writers' Collective and Nagar 2006; Nagar and Geiger 2007).

The focus on professional success also marginalizes radical geographic practices focused outside the academy. From the beginning, radical geography was riven by tensions between the academic and the activist path – tensions that the bulk of graduate students embarking on this path continue to struggle to negotiate. Participatory action research, community engagement, and the coproduction of geographical knowledge remain important and academically fashionable. Yet they are harder to practice than preach, running up against both power hierarchies separating the ivory tower and low-income communities, and incentive structures within the university that marginalize such activities. Further, even this is a long way from street-wise political activism – a lodestone for early radical geography.

Finally, with academic influence has come academic power. Radical geographers are now Deans, Provosts, and University Presidents and Vice Chancellors across the Anglophone academy. It has proven largely impossible, however, to use those positions of power to radically change the academy itself. Those occupying such positions are often handstrung, their potential actions already aligned with and structured by neoliberalizing national academic agendas. Further, in holding those administrative positions (now former?) radical geographers appear to see the academy differently. They prosecute a restricted notion of meritorious academic performance: numbers of publications, peer-reviewed journals, citations, H-scores, and impact factors; only this is now excellence (Readings 1996). Such policies are maintained despite these administrators having their offices occupied by students, or being picketed by faculty and staff unions – actions they may well have participated in earlier in their careers – also suffering public censure for the glaring contradictions between their past radical scholarship and current behavior.

These are the quandaries of radical geography's academic success, as it has emerged from the foundations created in and beyond North America between the mid-1950s and the early 1980s. But where we should go from here? We believe that radical geography's long-term vitality as an intellectual and political project must be based on a relentless desire to interrogate critically our own presuppositions. Success such as we have described for radical geography can produce self-confidence, even arrogance, which of course we must continually guard against. In part to do that, to avoid smug complacency, we must search for alternatives to current trajectories. In the name of engaged pluralism we must work to avoid internecine struggles for control over the direction of radical geography (Barnes 2007; Barnes and Sheppard 2010). We should proactively work to include a broader range of radical voices. We should collaborate with like-minded scholars across the academy to effect political change in the institutions of post-secondary education. We should approach teaching, as well as scholarship, as an emancipatory activity to be pursued within and outside the academy. Finally, we should unbound critical geography by working to produce knowledge that takes on board the desires, voices, and experiences of those marginalized positionalities we profess to prioritize.

The role of universities as sites of critique of the status quo is under unprecedented peril. As radical geographers, we should challenge any threat to the university's capacity to engage in critical thought; we should determine where critical inquiry is best produced and participate in these venues, wherever they may be. The future trajectory of radical geography is unknowable, but its future is secure if we keep this goal at the center of what we do.

## Notes

1   Since the beginning of this decade, Lawrence Berg and Ulrich Best have been attempting to collect a more even-handed set of accounts of the emergence of radical geography in different countries into a book provisionally titled *Placing Critical Geographies*.

2   Indeed, alienated by the turn of radical/critical geographers away from Marx particularly from the 1990s, in 2008 Peet founded a new, avowedly Marxist journal, again "produced by academic labor" at Clark University: *Human Geography: A New Radical Journal* (https://hugeog.com/about-us, accessed June 24, 2018)

3   The more-than-human world received remarkably little attention among Marxist geographers at this time, for a discipline that defines itself as having core expertise in nature-society relations. Neil Smith was the main figure seeking to interrogate the role of nature in capitalism: Smith and O'Keefe (1980) and Smith (1984). There was a parallel strand of geographical eco-socialism propagated by Larry Wolfe (the University of Cincinnati) and Wilbur Zelinsky (Penn State, an original USG member). In 1971 they founded a parallel radical geography group, the Socially and Ecologically Responsible Geographers (SERGE), publishing the mimeographed journal *Transition* until 1986. This was largely tolerantly dismissed by the Marxists.

4   Others, disillusioned with the academic turn, abandoned the academy or the discipline, although community-oriented work remained the concern of some radical academic geographers, particularly in the U.S. Midwest and in Québec (Blomley and McCann; Lauria *et al.*, and Klein, all this volume).

5   The section draws on the 2006 *Antipode* lecture by Eric Sheppard: "Quandaries of Critical Geography".

# References

Akatiff, C. (2016). Roots of radical geography. *Yearbook of the Association of Pacific Coast Geographers* 78: 258–278.

Amin, A. and Thrift, N. (2005). What's left? Just the future. *Antipode* 37: 220–238.

Amin, A. and Thrift, N. (2007). On being political. *Transactions of the Institute of British Geographers NS* 32: 112–115.

Barnes, T. J. (2007). "Not only...but also": Quantitative and critical geography. *The Professional Geographer* 61: 1–9.

Barnes, T. and Sheppard, E. (2010). "Nothing includes everything". Towards engaged pluralism in Anglophone economic geography. *Progress in Human Geography* 34: 193–214.

Belina, B. (2005). Anglophones: If you want us to understand you, you will have to speak understandably! *Antipode* 37: 853–855.

Blaut, J. (1993). *The Colonizer's Model of the World*. New York: Guilford Press.

Castree, N. (2000). Professionalisation, activism, and the university: Whither 'critical geography'? *Environment and Planning A* 32: 955–970.

Cope, M. and Elwood, S. (2009). *Qualitative GIS: A Mixed Methods Approach*. Thousand Oaks, CA: Sage.

Dear, M. J. (1983). Society and Space: An introduction. *Environment and Planning D: Society and Space* 1: 1–2.

Dempsey, J. (2016). *Enterprising Nature: Economics, Markets, and Finance in Global Biodiversity Politics*. Chichester: John Wiley & Sons, Ltd.

Derrida, J. (1994). *Specters of Marx: The State of the Debt, the Work of Mourning, & the New International*. London: Routledge.

Deutsche, R. (1991). Boys Town. *Environment and Planning D: Society and Space* 9: 5–30.

Dosse, F. (1998). *A History of Structuralism*. Minneapolis, MN: University of Minnesota Press.

Ferretti, F. and Pedrosa, B. V. (2018). Inventing critical development: A Brazilian geographer and his Northern networks. *Transactions of the Insititue of British Geographers* 43 (4): 703–717

García-Ramón, M.-D. (2004). The spaces of critical geography: an introduction. *Geoforum* 35: 523–524.

Gibson, K., Graham, J. and Shakow, D. (1989). Calculating economic indicators in value terms: the Australian economy and industrial sectors, 1974/75 and 1978/79. *The Journal of Australian Political Economy* 25: 17–43.

Gibson-Graham, J. K. (1996). *The End of Capitalism (As We Knew It)*. Oxford, UK: Blackwell.

Gregson, N., Simonsen, K. and Vaiou, D. (2003). Writing (across) Europe: On writing spaces and writing practices. *European Urban and Regional Studies* 10: 5–22.

Gutiérrez, J. and López-Nieva, P. (2001). Are international journals of human geography really international? *Progress in Human Geography* 25: 53–69.

Harvey, D. (1982). *The Limits to Capital*. Oxford: Basil Blackwell.

Harvey, D. (1987). Three myths in search of a reality in urban studies. *Environment and Planning D: Society and Space* 5: 367–376.

Harvey, D. (1998). The body as an accumulation strategy. *Environment and Planning D: Society and Space* 16: 401–421.

Harvey, D. (2006). Editorial: The geographies of critical geography. *Transactions of the Institute of British Geographers NS* 31: 409–412.

Harvey, D. (2017). "Listen, Anarchist!" A personal response to Simon Springer's "Why a radical geography must be anarchist". *Dialogues in Human Geography* 7: 233–250.

Harvey, D. and Haraway, D. (1995). Nature, politics, and possibilities: A debate and discussion with David Harvey and Donna Haraway. *Environment and Planning D: Society and Space* 13: 507–527.

Hechter, M. (1995). *Internal Colonialism: The Celtic Fringe in Bristish National Development*. Transaction Publishers.

Hones, S. (2004). Sharing academic space. *Geoforum* 35: 549–552.

Hooks, B. (1989). Choosing the margin as a space of radical openness. *Framework: The Journal of Cinema and Media* 36: 15–23.

Katznelson, I. (1997). From the street to the lecture hall: The 1960s. *Daedalus* 126: 311–332.

Lave, R. (2014). Engaging within the academy: a call for critical physical geography. *ACME: An International E-Journal for Critical Geographies* 13: 508–515.

Mackenzie, S. and Rose, D. (1983). Industrial change, the domestic economy and home life. In J. Anderson, S. Duncan and R. Hudson, eds, *Redundant Spaces in Cities and Regions? Studies in Industrial Decline and Social Change*, pp. 155–199. London: Academic Press.

Massey, D. (1991). Flexible sexism. *Environment and Planning D: Society and Space* 9: 31–57.

Mitchell, K. (2000). The value of academic labor. *Environment and Planning A* 32: 1713–1718.

Mohanty, C. T. (2003). "Under western eyes" revisited: Solidarity through anti-capitalist struggles. In *Feminism without Borders: Decolonizing Theory, Practicing Solidarity*, pp. 221–251. Durham, NC: Duke University Press.

Moss, P., Berg, L., and Desbiens, C. (2002). The political economy of publishing in geography. *ACME: An International Journal for Critical Geographies* 1: 1–7.

Mullings, B., Peake, L., and Parizeau, K. (2016). Cultivating an ethic of wellness in Geography. *The Canadian Geographer* 60: 161–167.

Nagar, R. (2018). Hungry translations: The world through radical vulnerability: The 2017 Antipode RGS – IBG Lecture. *Antipode* 51: 3–24.

Nagar, R. and Geiger, S. (2007). Reflexivity, positionality and identity in feminist fieldwork: Beyond the impasse. In A. Tickell, T. Barnes, J. Peck and E. Sheppard, eds, *Politics and Practice in Economic Geography*, pp. 267–278. Thousand Oaks, CA: Sage.

Nettle, D. and Romaine, S. (2002). *Vanishing Voices: The Extinction of the World's Languages*. Oxford: Oxford University Press.

O'Keefe, P., Westgate, K. and Wisner, B. (1976). Taking the naturalness out of natural disasters. *Nature* 260: 566–567.

Paasi, A. (2005). Globalization, academic capitalism, and the uneven geographies of international journal publishing spaces. *Environment and Planning A* 37: 769–789.

Peake, L. and Sheppard, E. (2014). The emergence of radical/critical geography within North America. *ACME: An International Journal for Critical Geographies* 13: 305–327.

Philo, C. (1998). *Eclectic Radical Geographies: Revisiting the early Antipodes*. Unpublished manuscript: The University of Glasgow. Available at http://eprints.gla.ac.uk/116527/, accessed November 27, 2018.

Raju, S. (2002). We are different, but can we talk? *Gender, Place and Culture* 9: 173–177.

Ramírez, B. (2004). The non spaces of critical geography in Mexico. *Geoforum* 35: 545–548.

Readings, B. (1996). *The University in Ruins*. Cambridge, MA: Harvard University Press.

Rodríguez-Pose, A. (2004). On English as a vehicle to preserve geographical diversity. *Progress in Human Geography* 20: 1–4.

Sangtin Writers' Collective and Nagar, R. (2006). *Playing with Fire: Feminist Thought and Activism through Seven Lives in India*. Minneapolis, MN: University of Minnesota Press.

Santos, M. (1977). Spatial dialectics: The two circuits of urban economy in underdeveloped countries. *Antipode* 9: 49–60.

Santos, M. (1979). *The Shared Space: The Two Circuits of the Urban Economy in Underdeveloped Countries*. London: Methuen.

Saunders, P. and Williams, P. (1986). The new conservatism: some thoughts on recent and future developments in urban studies. *Environment and Planning D: Society and Space* 4: 393–399.

Sharpe, J. and Spivak, G. (2002). Conversation with Gayatri Chakravorty Spivak: Politics and the imagination. *Signs: Journal of Women in Culture and Society* 28: 609–624.

Sheppard, E. (1995). Dissenting from Spatial Analysis. *Urban Geography* 16: 283–303.

Sheppard, E. (2005). Jim Blaut's model of the world. *Antipode* 37: 956–962.

Sheppard, E. (2014). We have never been positivist. *Urban Geography* 35: 636–644.

Sheppard, E. and Doherty, J. (1987). Editorial. *Antipode* 19: 93–94.

Simonsen, K. (2004). Differential spaces of critical geography. *Geoforum* 35: 525–528.

Smith, N. (1984). *Uneven Development: Nature, Capital and the Production of Space*. Oxford: Basil Blackwell.

Smith, N. (2006). What's left? Neo-critical geography, or the flat pluralist world. *Antipode* 37: 887–899.

Smith, N. and O'Keefe, P. (1980). Geography, Marx and the concept of nature. *Antipode* 12: 30–39.

Timár, J. (2004). More than "Anglo-American", it is "Western": hegemony on geography from a Hungarian perspective. *Geoforum* 35: 533–538.

Viaou, D. (2003). Radical debate between "local" and "international": A view from the periphery. *Environment and Planning D: Society and Space* 21: 133–137.

Watts, M. (1983). *Silent Violence: Food, Famine and Peasantry in Northern Nigeria*. Berkeley: University of California Press.

Wilson, B. M. (2000a). *America's Johannesburg: Industrialization and Racial Transformation in Birmingham*. Lanham, MD: Rowman and Littlefield.

Wilson, B. M. (2000b). *Race and Place in Birmingham: The Civil Rights and Neighborhood Movements*. Lanham, MD: Roman and Littlefield.

Wilson, M. W. (2017). *New Lines: Critical GIS and the Trouble of the Map*. University of Minnesota Press.

Wisner, B., Weiner, D., and O'Keefe, P. (1982). Hunger: a polemical review. *Antipode* 14: 1–16.

Wisner, B., Westgate, K., and O'Keefe, P. (1976). Poverty and disaster. *New Society* 37: 546–548.

# Index

*Spatial Histories of Radical Geography: North America and Beyond*, First Edition.
Edited by Trevor J. Barnes and Eric Sheppard.
© 2019 John Wiley & Sons Ltd. Published 2019 by John Wiley & Sons Ltd.

Yada Faction, The, 27
  advent, 305–307
  North American radical geography, contrast, 303
Yamakawa, Mitsuo, 306
Yamamoto, Kenji, 306

Young, Coleman, 65
  meeting, 72
Your Place, running, 95

Zelinsky, Wilbur, 11, 350
Zinn, Harold, 200